AF748961

Lignin Biorefineries for Green Industries

This book focuses on lignin biorefinery, with special emphasis on lignin extraction, its chemical structure, depolymerization and degradation techniques, and its applications in the production of green industrial products. It provides an overview of lignin chemistry and biorefinery processes, as well as the potential of lignin in the synthesis of industrial chemicals. This book covers pretreatment methods for lignocellulosic biomass using various extraction techniques and includes the chemical analysis of lignin structures obtained from different methods. It also explains the catalytic valorization of lignin and its biological degradation processes.

Key features:

- Covers the latest lignin extraction and valorization technologies.
- Provides updated information on pretreatment technologies for lignin synthesis from biomass.
- Provides insights into the conversion of depolymerized lignin into biopolymers, resins, and other high-value products.
- Discusses the exploration of lignin-based modern medicines and smart nano-biomaterials.
- Provides a discussion of the environmental and market impacts of lignin-based biorefinery.

This book is intended for researchers and graduate students in chemical engineering, lignin chemistry, and bioresource engineering.

Lignin Biorefineries for Green Industries

Edited by

Dolly Kumari and Ravi Kant Bhatia

CRC Press
Taylor & Francis Group
Boca Raton London New York

CRC Press is an imprint of the
Taylor & Francis Group, an **informa** business

Designed cover image: Shutterstock

First edition published 2026
by CRC Press
2385 NW Executive Center Drive, Suite 320, Boca Raton FL 33431

and by CRC Press
4 Park Square, Milton Park, Abingdon, Oxon, OX14 4RN

CRC Press is an imprint of Taylor & Francis Group, LLC

© 2026 selection and editorial matter, Dolly Kumari and Ravi Kant Bhatia; individual chapters, the contributors

Reasonable efforts have been made to publish reliable data and information, but the author and publisher cannot assume responsibility for the validity of all materials or the consequences of their use. The authors and publishers have attempted to trace the copyright holders of all material reproduced in this publication and apologize to copyright holders if permission to publish in this form has not been obtained. If any copyright material has not been acknowledged please write and let us know so we may rectify in any future reprint.

Except as permitted under U.S. Copyright Law, no part of this book may be reprinted, reproduced, transmitted, or utilized in any form by any electronic, mechanical, or other means, now known or hereafter invented, including photocopying, microfilming, and recording, or in any information storage or retrieval system, without written permission from the publishers.

For permission to photocopy or use material electronically from this work, access www.copyright.com or contact the Copyright Clearance Center, Inc. (CCC), 222 Rosewood Drive, Danvers, MA 01923, 978-750-8400. For works that are not available on CCC please contact mpkbookspermissions@tandf.co.uk

For Product Safety Concerns and Information please contact our EU representative GPSR@taylorandfrancis.com. Taylor & Francis Verlag GmbH, Kaufingerstraße 24, 80331 München, Germany.

Trademark notice: Product or corporate names may be trademarks or registered trademarks and are used only for identification and explanation without intent to infringe.

ISBN: 9781041099161 (hbk)
ISBN: 9781041100676 (pbk)
ISBN: 9781003653189 (ebk)

DOI: 10.1201/9781003653189

Typeset in Times
by codeMantra

Contents

About the editors

Dolly Kumari is an Assistant Professor of Chemistry in the Faculty of Chemical Sciences, Institute of Natural Sciences and Humanities, Shri Ramswaroop Memorial University, Lucknow-Deva Road, Barabanki, U.P., India, since 2022. She earned her Ph.D. in Chemistry from Dayalbagh Educational Institute, Dayalbagh, Agra, India. With a strong background in chemistry, her research spans renewable energy generation, pretreatment of lignocellulosic biomass, agricultural waste management, liquid and gaseous biofuel production, bioremediation, and wastewater treatment. Her primary focus is on understanding pretreatment of lignocellulosic biomass using harmless chemicals or substances for biofuel production, reflected in over 20 publications in esteemed national and international journals having a cumulative impact factor of 59.5 with more than 1100 citations, and she has filed one patent on a pretreatment method. She has also started work on green nanomaterial synthesis for catalytic applications. She has supervised the projects of postgraduate students. She also plays a vital role in academic and non-academic activities at Shri Ramswaroop Memorial University.

Dr. Ravi Kant Bhatia is a distinguished Indian scientist and academician renowned for his significant contributions to environmental biotechnology, bioenergy, and microbial enzyme technology. He currently serves as an Assistant Professor in the Department of Biotechnology at Himachal Pradesh University (HPU), Shimla. Dr. Bhatia has established himself as a leading figure in sustainable bioresource management and microbial innovation. He possesses extensive experience in both academia and industry and is actively engaged in teaching and cutting-edge research. In recognition of his scientific contributions, he has been listed among the World's Top 2% Scientists in 2025 in the field of Biotechnology. Dr. Bhatia has qualified prestigious national examinations including UGC-NET, ICAR-NET, and GATE, and was awarded the UGC Postdoctoral Fellowship (UGC-PDF) to pursue his postdoctoral research. He has also received the SERB International Travel Support (ITS) fellowship to present his research findings at an international conference in Scotland. He has authored over 100+ research publications with a cumulative impact factor exceeding 200 and has received more than 5,200+ citations. His research interests include microbial enzyme technology, biofuels, nano-biotechnology, environmental microbiology, and waste valorization. Dr. Bhatia has successfully led and contributed to several research projects funded by national agencies, securing grants worth approximately ₹15 crore. He has completed 6 projects as Co-Principal Investigator and 4 as Principal Investigator. He has also represented India at various international conferences. His collaborations extend globally, working with renowned scientists and institutions worldwide. In addition, he serves as a reviewer for several prestigious national and international journals and has edited more than 10 books with reputed publishers. Beyond research, Dr. Bhatia is deeply committed to mentoring students and promoting innovation through incubation and entrepreneurial initiatives. His interdisciplinary approach, integrating fundamental research with practical applications, continues to drive advancements in environmental biotechnology and sustainable development. Dr. Ravi Kant Bhatia's dedication and impactful contributions have significantly advanced the field and continue to inspire the next generation of scientists in India and beyond.

Contributors

Sa-ad Abubakari
King Fahd University of Petroleum and Minerals (KFUPM)
Dhahran, Saudi Arabia

Irshad Ahmad
King Fahd University of Petroleum and Minerals (KFUPM)
Dhahran, Saudi Arabia

Hussein Alhassan
King Fahd University of Petroleum and Minerals (KFUPM)
Dhahran, Saudi Arabia

Parul Bali
Department of Microbiology
Himachal Pradesh University
Shimla, India

Nikita Basant
Centre of Excellence (AI & ML), Amity Institute of Biotechnology
Amity University
Lucknow, India

Mateo Bello Villarino
Department of Chemistry
Umeå University
Umeå, Sweden

Anne Bhambri
Department of Biotechnology
Shri Guru Ram Rai University
Uttarakhand, India

Pragati Bhardwaj
Department of Biotechnology
Guru Nanak Dev University
Amritsar, India

Ravi Kant Bhatia
Department of Biotechnology
Himachal Pradesh University
Shimla, India

Tanima Bhattacharya
Sustainable Nanomaterials & Biopolymer Laboratory, Department of Paediatrics
Center for Global Health Research
Saveetha Medical College, Saveetha Institute of Medical and Technical Sciences
Chennai, India

Aubrey Dickson Chigwada
Department of Environmental Sciences
College of Agriculture and Environmental Sciences
University of South Africa (UNISA)
Roodepoort, South Africa

Pushpak Dahiya
Department of Biotechnology
Himachal Pradesh University
Shimla, India

Mark Dakurah
King Fahd University of Petroleum and Minerals (KFUPM)
Dhahran, Saudi Arabia

Alok Prasad Das
Department of Life Sciences
Rama Devi Women's University
Bhubaneswar, India

Fatemeh Dodangeh
Department of Chemical and Biochemical Engineering
Faculty of Engineering, Western University
London, Canada

Glen Cletus Dsouza
Faculty of Agriculture
Dalhousie University
Nova Scotia, Canada

Farheen Firdaus
Department of Biotechnology
Patna Women's College (Autonomous), Patna University
Patna, India

Payam Ghiaci
High-Throughput Centre
RISE Research Institutes of Sweden
Örnsköldsvik, Sweden

Meenu Hans
Department of Biotechnology
Guru Nanak Dev University
Amritsar, India

Mattias Hedenström
Department of Chemistry
Umeå University
Umeå, Sweden

Kopal Jain
Department of Biotechnology
Manav Rachna International Institute of Research & Studies
Faridabad, India

Leif J. Jönsson
Department of Chemistry
Umeå University
Umeå, Sweden

Santosh Kumar Karn
Department of Biochemistry & Biotechnology
Sardar Bhagwan Singh University
Dehradun, India

Sachin Kumar
Department of Biochemical Engineering
Harcourt Butler Technical University
Kanpur, India

Suraj Kumar
Department of Biotechnology
Shoolini Institute of Life Sciences and Business Management
Solan, India

Rohit Kumar Singh
Department of Biotechnology
Himachal Pradesh University
Shimla, India

Dolly Kumari
Faculty of Chemical Sciences
Institute of Natural Sciences and Humanities
Shri Ramswaroop Memorial University
Barabanki, India

Moni Kumari
Department of Biochemistry & Biotechnology
Sardar Bhagwan Singh University
Dehradun, India

Unnati Kushavah
Biochemistry and Structural Biology Division
CSIR-Central Drug Research Institute
Lucknow, India

Anas Abdulwahab Lawal
King Fahd University of Petroleum and Minerals (KFUPM)
Dhahran, Saudi Arabia

Gulab Singh Maurya
Faculty of Physical Sciences
Institute of Natural Sciences and Humanities
Shri Ramswaroop Memorial University
Barabanki, India

Vasudha Maurya
Department of Biotechnology
DAV University
Jalandhar, India

Henry Joseph Oduor Ogola
Department of Environmental Sciences
College of Agriculture and Environmental Sciences, University of South Africa (UNISA)
Roodepoort, South Africa

Alok Patel
Department of Civil, Environmental, and Natural Resources Engineering
Luleå University of Technology
Luleå, Sweden

Amisha Patel
Department of Chemistry
Umeå University
Umeå, Sweden

Aman Prakash
Department of Marine Biology
Leon H. Charney School of Marine Sciences
University of Haifa
Haifa, Israel

Rohit Rai
Faculty of Applied Medical Sciences
Lovely Professional University
Phagwara, India

Sachchida Nand Rai
Centre of Experimental Medicine & Surgery
Institute of Medical Sciences
Banaras Hindu University
Varanasi, India

Ambika Raita
Department of Biotechnology
Shoolini Institute of Life Sciences and Business Management
Solan, India

Vivek Raj
Department of Biotechnology
Patna Women's College (Autonomous)
Patna University
Patna, India

Vishnu D. Rajput
Department of Nanobiotechnology
Southern Federal University
Rostov, Russia

Ranju Kumari Rathour
Department of Biotechnology
Chandigarh College of Technology, CGC
Mohali, India

Suman Salhotra
Department of Biotechnology
Guru Nanak Dev University
Amritsar, India

Aditya Sharma
Department of Biotechnology
Himachal Pradesh University
Shimla, India

Chhavi Sharma
University Centre for Research and Development
University Institute of Biotechnology, Chandigarh University
Mohali, India

Shruti Shukla
Faculty of Biosciences
Institute of Biosciences and Technology, Shri Ramswaroop Memorial University
Barabanki, India

Amandeep Singh
Department of Chemistry
Indian Institute of Technology (IIT) Ropar
Punjab, India

Narayan Singh
Department of Biotechnology
DAV University
Sarmastpur, India

Roshini Singh
Amity Institute of Biotechnology
Amity University
Lucknow, India

Memory Tekere
Department of Environmental Sciences
College of Agriculture and Environmental Sciences, University of South Africa (UNISA)
Roodepoort, South Africa

Kirti Thakur
Department of Biotechnology
Shoolini Institute of Life Sciences and Business Management
Solan, India

Meenu Thakur
Department of Biotechnology
Shoolini Institute of Life Sciences and Business Management
Solan, India

Neeraj Tiwari
Faculty of Biosciences
Institute of Biosciences and Technology
Shri Ramswaroop Memorial University
Barabanki, India

Saleem Ullah
VTT Technical Research Centre of Finland Ltd.
Tekniikantie 21, Finland

Deepika Umrao
Department of Biotechnology
Dr. BR Ambedkar National Institute of Technology
Jalandhar, India

Virat Vaibhav
Department of Biosciences and Bioengineering
Indian Institute of Technology Guwahati
Guwahati, India

Gayathri Balaji Venkata
Department of Chemical and Biochemical Engineering
Western University
London, Canada

Preface

The transition toward a sustainable, low-carbon, and circular bioeconomy has become a defining challenge of the 21st century. Lignocellulosic biomass is one of the most abundant and renewable biological resources available on the earth. Among its various complex matrices, lignin is the most prevalent aromatic biopolymer and is frequently considered as one of the underappreciated components. However, lignin is now a vital feedstock for green industries due to recent scientific and technological advancements. The purpose of this book, *Lignin Biorefineries for Green Industries*, is to change this perception and showcase lignin as a key component of upcoming biorefineries rather than just as a leftover by-product. Because of its exceptional abundance and aromatic structures, lignin is a priceless sustainable substitute for aromatics obtained from fossil fuels, which are utilized in various materials, chemicals, fuels, and polymers. With its emphasis on resource efficiency, carbon neutrality, and waste valorization, the idea of a lignin-centric biorefinery is in line with global sustainability goals. Lignin biorefineries provide methods to generate high-value products such as bio-based adhesives, resins, carbon fibers, phenolic chemicals, antioxidants, fuels, and functional biomaterials by combining biological, chemical, and thermochemical conversion mechanisms.

This book aims to provide a comprehensive and multidisciplinary analysis of these new opportunities. Various chapters of this book examine lignin's basic chemistry, structure, and technological developments in lignin isolation and fractionation, and creative approaches to lignin depolymerization and upgradation into valuable products. Biological and enzymatic methods are given particular attention, which reflects the increased interest in mildly operating, highly selective processes that are safe for the environment. One intriguing approach to effective and scalable lignin valorization is the coupling of microbial and enzymatic systems with chemical catalysis. Furthermore, life-cycle assessment, sustainability indicators, and techno-economic analysis are emphasized to close the gap between laboratory research and industrial applications in various sectors, such as bioenergy, bioplastics, building materials, agriculture, and environmental remediation, in light of the global trend toward green industries. Recent advancements and case studies show how lignin-derived products can lower greenhouse gas emissions, support rural and industrial bioeconomies, and replace their petroleum-based counterparts.

Researchers in biotechnology, chemical engineering, materials science, and sustainability studies, as well as graduate students, business executives, and policymakers, are the target audience for *Lignin Biorefineries for Green Industries*. This book aims to stimulate new research avenues and expedite the conversion of lignin valorization technologies into workable green industrial solutions by combining basic science, technological advancements, and industrial viewpoints. We hope that this volume will help redefine lignin's place in the bioeconomy and further the world's shift to resilient and sustainable industry.

Dr. Dolly Kumari
Faculty of Chemical Sciences
Institute of Natural Sciences and Humanities
Shri Ramswaroop Memorial University
Lucknow-Deva Road, Barabanki, U.P., India

Dr. Ravi Kant Bhatia
Assistant Professor,
Department of Biotechnology, Himachal Pradesh University
Summer Hill, Shimla-171005, H.P., INDIA

Acknowledgments

The editors wish to express their sincere gratitude to all those who have contributed, directly and indirectly, to the successful completion of the book *Lignin Biorefineries for Green Industries*. This volume is the result of a collective effort by researchers, academicians, industry professionals, and institutions who share a common vision of advancing sustainable and circular bio-based technologies. First and foremost, we extend our heartfelt thanks to the contributing authors for their scholarly dedication, timely submissions, and willingness to share their expertise. Their diverse perspectives and in-depth analyses have enriched this book, making it a comprehensive reference on lignin valorization, biorefinery concepts, and green industrial applications. We are equally grateful to the reviewers for their critical insights and constructive suggestions, which significantly improved the scientific rigor, clarity, and overall quality of the chapters.

We extend our sincere appreciation to the publishing team for their professional support, patience, and meticulous editorial assistance. Their efforts in coordinating reviews, formatting manuscripts, and ensuring high production standards were instrumental in bringing this book to fruition. We also acknowledge the support of our respective institutions (Shri Ramswaroop Memorial University, Barabanki, U.P., INDIA, and Himachal Pradesh University, Shimla, H.P., INDIA) and funding agencies (National Mission on Himalayan Studies Project Management Unit (NMHS-PMU), MoEF & CC, Govt. of India, New Delhi, [Grant number: NMHS/2022-23/SG 81/01/280]), whose research ecosystems and infrastructural facilities enabled the generation of much of the knowledge presented in this volume. We gratefully acknowledge the encouragement received from colleagues, research scholars, and students during discussions and deliberations on lignin biorefinery concepts.

We express our deep personal gratitude to our families for their unwavering support, understanding, and encouragement during the preparation of this book. We hope that *Lignin Biorefineries for Green Industries* will serve as a valuable resource for researchers, industry stakeholders, and policymakers and will contribute meaningfully to the global pursuit of sustainable and green industrial development.

Finally, we would like to express our gratitude to everyone who has supported us over the years but whose names we failed to include.

Dr. Dolly Kumari
SRMU, Barabanki, U.P., INDIA

Dr. Ravi Kant Bhatia
HPU, Shimla, INDIA
(Editors)

Foreword

The transition from a fossil-based economy to a sustainable, circular bioeconomy is one of the defining scientific and industrial challenges of the present time. In this context, lignin, nature's most underutilized component of lignocellulosic biomass, stands at the crossroads of opportunity and necessity. *Lignin Biorefineries for Green Industries* arrives at a critical moment, offering a comprehensive and timely perspective on how lignin can be transformed from an industrial by-product into a cornerstone of green manufacturing. For decades, lignin has primarily been viewed as a low-value residue of the pulp, paper, and bioethanol industries, often relegated to combustion for process heat. Yet, lignin's complex and highly functional aromatic structure holds immense potential as a renewable feedstock for fuels, chemicals, materials, and advanced functional products. Unlocking this potential requires an integrated biorefinery approach, one that combines advances in biomass fractionation, catalytic conversion, biotechnology, and systems engineering. This book thoughtfully addresses these dimensions, presenting lignin not as a waste stream but as a strategic resource for sustainable industrial development.

The chapters of this volume bridge fundamental understanding of lignin structure and biosynthesis with cutting-edge technologies for depolymerization, valorization, and product development. Chemical, thermochemical, and biological routes are discussed with clarity, highlighting both their promise and practical constraints. By addressing analytical tools, process integration, and techno-economic considerations, this book moves beyond laboratory-scale curiosity and engages with the realities of industrial translation. Equally important is this book's strong alignment with the principles of green chemistry and sustainability. The emphasis on low-energy processes, benign catalysts, waste minimization, and life-cycle assessment reflects a mature understanding that technological innovation must be coupled with environmental and societal responsibility. In an era where industries are under increasing pressure to decarbonize and reduce environmental footprints, lignin-based solutions described here offer credible pathways toward greener value chains in sectors ranging from chemicals and materials to energy and agriculture.

This volume will serve a diverse audience. Researchers and graduate students will find it a valuable reference for understanding the state of the art and identifying future research directions. Industry professionals and policy planners will appreciate the insights into scalable technologies, market opportunities, and strategic frameworks for integrating lignin valorization into existing biorefinery and manufacturing infrastructures. By fostering dialogue across disciplines, this book contributes to the collaborative ethos required to advance sustainable technologies. *Lignin Biorefineries for Green Industries* is more than a compilation of scientific knowledge; it is a forward-looking statement on how renewable carbon can underpin the industries of the future. By illuminating the pathways through which lignin can be efficiently and responsibly converted into high-value products, this book makes a significant contribution to the global pursuit of sustainable development. It is my hope that the ideas and innovations presented herein will inspire continued research, industrial adoption, and policy support, ultimately helping to realize a resilient and greener bio-based economy.

Dr. Narender Sharma
Director (Scientist F)
Central Pollution Control Board (CPCB),
MoEFCC, Government of India

1 Introduction to lignin and its potential

Dolly Kumari and Gulab Singh Maurya

1.1 INTRODUCTION

There is a significant push to increase our reliance on sustainable methods due to the population's exponential expansion and energy demands (Kumari et al., 2025; Delgado et al., 2019). The development of techniques that rely on renewable or sustainable resources is the main focus of current research. Agricultural and forestry wastes contain large amounts of lignin, a non-food polymer with a carbon-neutral notion (Melro et al., 2021; Rathour et al., 2023). Researchers are always trying to create high-quality technological methods for lignin isolation in order to convert it into a variety of value-added compounds, given the low use of these lignin resources and their potential in many chemical sectors (Melro et al., 2021). The most plentiful renewable resource on Earth is lignocellulosic biomass, which is made up of cellulose, hemicellulose, and lignin that have been entangled (Chauhan, 2020; Kumari & Singh, 2020). In the tissues of vascular plants and algae, lignin, a type of intricate and stiff organic polymer, provides crucial structural and reinforcing support. After cellulose, lignin is the second most prevalent terrestrial polymer on Earth. Lignin has long been considered an industrial leftover in pulp and paper waste, agricultural wastes, and other hydrolytic industries due to its stability and high recalcitrance (Bhatt et al., 2017; Karthäuser et al., 2021). One of the most important tactics for creating a sustainable and green bioeconomy is the effective use of lignin, an intricate polymer found in plant cell walls. However, because lignin is resistant, bioconversion presents a major obstacle. Through a variety of enzymatic pathways, microorganisms—especially bacteria and fungi—play a critical role in the biodegradation of lignin. *Pseudomonas putida* is regarded as a suitable host for the breakdown and valorization of lignin among bacteria because of its strong and adaptable metabolism and resistance to a wide range of harmful substances (Rathour et al., 2024; Zhou et al., 2025).

Millions of tons of lignin are now available each year as a byproduct of industrial lignocellulose hydrolysis and use, thanks to the development of biorefineries recently (Brienza et al., 2024). About 100 million tons of lignin is becoming available as a desirable but limited-use feedstock in the pulp and paper sector alone (Zhang et al., 2020a). For the biotechnological production of different biofuels and biochemicals (Kumari & Jain, 2020; Kumari et al., 2018), cellulose and hemicellulose fractions are easily utilized as substrates in subsequent biorefinery and fermentation scale-ups (Kumari et al., 2021a, 2021b). The majority of lignin, however, cannot be effectively

DOI: 10.1201/9781003653189-1

used. Heat and energy are primarily created through combustion, and raw lignin is mainly used in manufacturing glues, resins, and asphalt, which together account for the majority of lignin's use (Demuner et al., 2021). From 2024 to 2030, the global lignin market is projected to expand at a compound annual growth rate (CAGR) of 4.5%, from its estimated USD 1.08 billion in 2023 (Lignin Market, 2024). As a result, interest in using lignin in a more beneficial and efficient way has grown recently. Degradation of lignin is essential for the recycling and value-adding of plant biomass and is a key component of ecosystems' carbon and nutrient cycles. The effective breakdown and depolymerization of the complex and resistant polymer is the first and most crucial stage in the value-adding process of lignin (van Erven et al., 2024). Although the chemical breakdown of lignin has been documented, the result is a poisonous, complicated mixture whose composition is hard to pinpoint. Chemical hydrolysates of this type are not very helpful for later added-value applications (Linger et al., 2014).

There has been a widespread adage in industry that "One can make everything from lignin, except money" (Ragauskas et al., 2014). This uncomfortable scenario has prompted a large amount of work in the industrialization of lignin. In 2007, a paper from DOE analyzed the function of lignin as a regenerative material source and analyzed the technical impediments to lignin utilization (PNNL, 2007). Long-term (more than 10 years and requiring significant new essential knowledge and technology development), medium-term (5 to perhaps 20 years), and near-term (current uses and those that seem possible within 3–10 years) commercialization opportunities for lignin products were identified in this report. Figure 1.1 shows the steps for the valorization of lignin for value-added products. The present chapter summarizes the

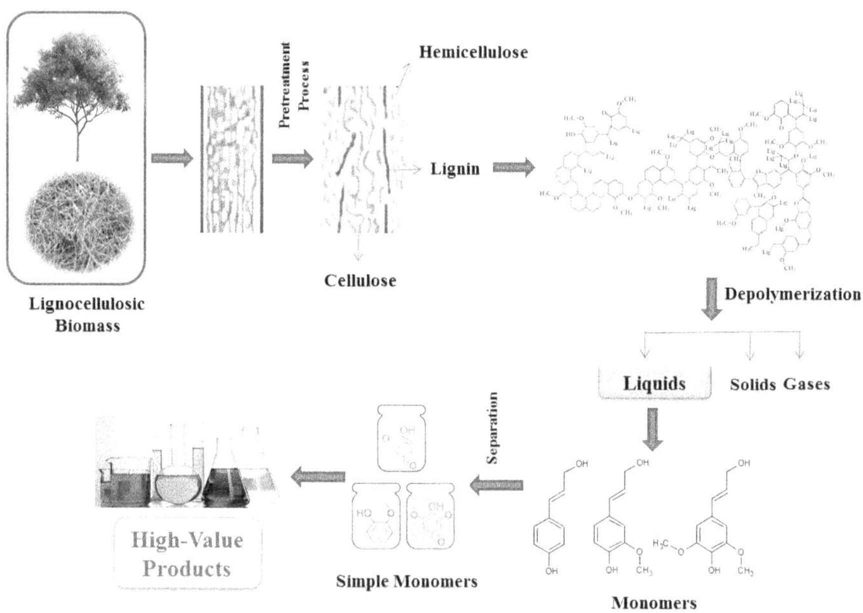

FIGURE 1.1 Lignin valorization process for the production of value-added products

basic structure of lignin, hindrances in lignin valorization, and the industrial potential of lignin valorization.

1.2 WHAT IS LIGNIN?

The most prevalent non-carbohydrate component in plant cell walls, lignin gives cells their stiffness, hydrophobicity, and defense against microbial invasion (Smith, 2019). These characteristics are thought to be the main cause of lignocellulose recalcitrance because of these characteristics (Zoghlami & Paës, 2019). It is an amorphous polymer composed of oxygenated p-propylphenol units with a networked three-dimensional structure. The three main components of lignin are p-coumaryl alcohol, which yields p-hydroxyphenyl units (H); coniferyl alcohol, which yields guaiacyl units (G); and sinapyl alcohol, which yields syringyl units (S). These three types of units are distinguished from each other by the quantity of methoxy groups at the ortho-positions of aromatic rings: H has no methoxy groups, G has one, and S has two. Within the lignocellulose matrix, lignin and hemicellulose are connected by ether and ester linkages to form hybrid structural motifs known as lignin-carbohydrate complexes (LCCs) (Zhao et al., 2020; Giummarella et al., 2019; Carvalho et al., 2020).

The main linkage types that are frequently observed in LCCs are benzyl ester, benzyl ether, phenyl glycosidic, ferulate ester, and diferulate ester. While sugar uronic acids in hemicellulose are connected to hydroxyl groups in lignin structures by glycosyl and benzyl ester links, mannosyl units of carbohydrates are connected to phenolic or hydroxyl groups in lignin by phenyl glycosidic and benzyl ether linkages. The amount of the different LCC patterns in lignocellulose varies depending on the feedstock. For example, hardwood LCCs have a high content of phenyl glycosidic linkages, while softwood LCCs have a significant number of benzyl ether structures (Zhao et al., 2017). The molecular weight (MW) of lignin polymers is challenging to determine because of the strong connections that bind lignin to hemicellulose, making it more difficult to isolate pure native lignin. Nonetheless, it has been found that lignin fractions that are near to native have MWs between 2,500 and 10,000 g mol^{-1} (Schutyser et al., 2018).

1.3 STRUCTURAL CHARACTERISTICS OF LIGNIN

Plant cell walls include lignin, which has been known for millennia to be the primary component responsible for giving cell walls their stiffness. Cellulose, hemicellulose, and lignin make up the majority of the cell wall. In literature, cellulose and hemicellulose are used in a variety of ways, although lignin is rarely used because of its intricate polymeric structure (Jiang et al., 2022; Kumari & Singh, 2018). As previously stated, lignin is made up of different aromatic units that function as methoxy and aldehyde groups. Depending on factors, including biomass type, growth, and environmental conditions, various plant species have varying typical compositions of these monomer units (Sawatdeenarunat et al., 2015). Plants primarily use three fundamental building blocks to produce the complex aromatic polymer: p-coumaryl alcohol, coniferyl alcohol, and sinapyl alcohol. As a result, the three distinguishable basic units that make up the complex heterogenic polymeric lignin network are H, G, and S moieties (Faix, 1991).

Guaiacyl-based lignin (G) is primarily found in gymnosperms (softwood, such as conifers, pines, cedars, and spruces), while guaiacyl-syringyl-based lilac lignin (G-S) is primarily found in dicotyledonous plants (legumes, beans, sunflowers, tomatoes, etc.), and guaiacyl-syringyl-based lilac lignin (G-S-H) is primarily found in monocotyledonous plants (maize, wheat, rice, cane, etc.) (Mansfield et al., 2012). The production of oxidative radicals from these structural units, followed by sequential radical coupling, is how lignin polymerization takes place in plants (Vanholme et al., 2010). It makes up roughly 15% to 25% of herbaceous plants, 25%–35% of hardwood, and 15%–25% of softwood. Methoxy (CH_3O-), carboxyl (-COOH), carbonyl (C=O), phenolic, and aliphatic hydroxyl groups are among the functional groups that make up lignins. The increased chemical reactivity of lignins due to these functional groups increases their potential uses. It can be utilized to create adhesives, coatings, bioplastics, fuel, and other products based on its structure, which is closely related to the lignin isolation process (Tribot et al., 2019). Although their composition varies, it is generally accepted that each cell wall has an average of 15%–25% lignin, 30%–40% cellulose, and 15%–30% hemicellulose, with a small percentage of other components such as starch, pectin, protein, carbohydrates, and minerals. (Isikgor et al., 2015). In contrast to the majority of natural polymers, which are constituted of a single monomer, lignin is made up of 3D copolymers joined by ether linkages of phenylpropanoid units (Frigon & Guiot, 2010).

1.4 TYPES OF LIGNIN AND THEIR EXTRACTION

Plants' lignin composition and structure are strongly influenced by their plant origins and the method of extraction. Lignin was first extracted as a byproduct of the paper industry or by hydrolyzing carbohydrates via strong acids like hydrochloric acid and sulfuric acid (Mazar et al., 2021). The lignin structure underwent significant chemical changes as a result of these techniques. Later, more sophisticated procedures were created that made it possible to separate and extract lignin from other byproducts. The lignin structure is better preserved by these more recent, accurate extraction techniques, including organosolv, enzymatic hydrolysis, or ionic liquid extraction, therefore opening the door for its use in a variety of applications. Sulfur-driven and sulfur-free extraction processes coming from the paper and pulp sector, which initially generated lignin as a side product (kraft, sulfite, and soda processes), and biorefinery operations intended to transform biomass into various materials and chemicals are the two types of commercial procedures that yield so-called technical lignins. They make it possible to isolate very pure and reactive lignins (organosolv, ionic liquid, and hydrolysis) (Ruwoldt et al., 2023).

The kraft process is the most widely used industrial pulping method in the pulp and paper sector; it produces more than 90% of chemical pulp and around two-thirds of virgin pulp worldwide (Tran & Vakkilainnen, 2012). A significant amount of lignin is produced as a byproduct of the kraft process; the majority of this lignin is not separated for commercial use. The most generally accessible and economically produced form of lignin is lignosulfonates, which are created by sulfite pulping operations. The market for lignosulfonates is expected to expand at a CAGR of 3.6% from USD 12 billion in 2022 to USD 1.4 billion by 2027 (Mehra, 2025). The growing need for lignosulfonates in various industries, such as oil well additives, livestock feed binders, and construction additives, is what is causing this expansion. Lignosulfonates possess a sulfur concentration of 2.1%–9.4% and can be recovered through filtration (Ruwoldt et al., 2023). Soda pulping, on the other hand, produces lignin containing a low hemicellulose percentage and is sulfur-free. By using sodium hydroxide (NaOH) at temperatures between 140 and 170°C under high pressure (6 to 10 bar), it produces lignin that is different from kraft and sulfite lignins (Ruwoldt et al., 2023).

1.5 INDUSTRIAL SOURCES OF LIGNIN

Lignin is a significant byproduct of lignocellulose pretreatments. Because of their quick spread, lignin production from lignocellulose biorefineries is predicted to reach almost 225 million tons annually by 2030, compared to the 50–70 million tons produced annually by pulp and paper factories (Bajwa et al., 2019). Therefore, increasing the commercial viability of second-generation biorefineries requires the development of techniques for the efficient valorization of lignin. Various lignin-derived products, such as materials (e.g., thermosets, thermoplastics, and elastomers), fine chemicals (e.g., vanillin and eugenol), and bulk chemicals (e.g., benzene, toluene, xylene, and phenol), can be produced, thanks to the development of alternative conversion routes for lignin streams isolated from biomass pretreatment, even though the lignin fraction is frequently used as an energy source in pulping processes (Gillet et al., 2017; Wong et al., 2020).

Various lignins (kraft, soda, lignosulfonates, and organosolv) that are now being investigated rely on novel biorefinery concepts that target the entire biomass composition within an "optimized systems" approach, also referred to as "integrated biorefinery" (Mazar et al., 2020). Additionally, so-called "lignin-first" isolation approaches aim largely to extract lignin and valorize it as a major useful component of biomass, generally in the form of hydrocarbons with lower MWs that could be used in fuel blends (Tardy et al., 2023). Lignin-first and other highly conserved lignins, which may nevertheless be different from the natural lignins, are still exploratory and cannot be generated at industrial scales. In general, only ~0.075 Mt/year of hydrolysis lignin from biorefineries and soda lignin through small-scale businesses pulping of annual plants is produced and separated. Estimates further show that approximately 730 Mt/year of lignin might be collected from shoots and straw as leftovers from agricultural streams, from which less than one half has to be reintroduced to the soil (Tardy et al., 2023). Table 1.1 summarizes the various lignin feedstocks, their products, and their applications.

Additionally, organosolv techniques were first created for the production of paper, but they eventually changed their focus to the extraction of lignin. This technique makes use of several organic solvent blends, like ethanol/water (Alcell process) (Benhamou et al., 2025). Although the lignins precipitate from the spent liquor, this procedure enables the segregation of lignin and hemicelluloses from cellulose that stays solid. Organosolv lignin lacks sulfur and has a structure more similar to native lignin. Depending on how well the extraction procedure worked, several organosolv lignins include less than 1% carbs, while others have significant carbohydrate content (Ruwoldt et al., 2023). Enzymatic hydrolysis has at last

TABLE 1.1
Summarizing lignin feedstocks, their products, and applications

Lignin Feedstock	Lignin Content	Product	Applications	References
Softwood kraft lignin	50%	Nanoparticles	Deliver sorafenib and benzazulene	Figueiredo et al. (2017)
Kraft lignin	Upto 80%	Hydrogels	Drug delivery system	Farhat et al. (2017)
Kraft lignin	50%	Carbon materials	Production of biochar, carbon fibers, carbon black, bio-oil, syngas, and activated carbon	Puziy et al. (2018)
Organosolv hardwood lignin	40–60%	Polymer	Production of nylon-based 3D printing materials	Nguyen et al. (2018)
Alkali lignin	Upto 60% (volume)	Polymer	Production of polyamide 12-based 3D printing materials with high porosity	Ajdary et al. (2021)

become a practical technique for extracting lignin, providing the benefit of digesting down cellulose and hemicellulose in a targeted manner while preserving the structural integrity of lignin (Benhamou et al., 2025). This technique uses certain enzymes, like cellulases or lignin-degrading enzymes, such as peroxidases or laccases, to specifically break down lignin-carbohydrate disruption and biomass polysaccharide complexes.

1.6 CHALLENGES IN LIGNIN UTILIZATION

1.6.1 Structural heterogeneity and recalcitrance

Since lignin is the greatest sustainable source of aromatic building blocks, it has emerged as one of the top contenders to replace fossil-based resources. However, the chemical industry needs to depolymerize polymeric lignin's complicated structure into simpler building units (Beckham et al., 2016). One of the most effective depolymerization methods is reductive depolymerization. Currently, under investigation in pilot-scale operations for upscaling to industrial scale are two process configurations: (i) reductive catalytic fractionation (RCF) and (ii) reductive catalytic depolymerization (RCD) (Daelemans et al., 2025). Phenoxy radicals are produced by extracellular oxidases and reassemble the monomeric H, S, and G units to form various C–C and C–O bonding arrangements. Aryl ether bonds, which are characterized by the β-O-4 motif, are the most prevalent of these native linking motifs (Lahive et al., 2020). Effective lignin valorization is hindered by the significant extent of chemical structure and molecular mass distribution diversity (Key & Bozell, 2016).

Lignin's heterogeneity results in a variety of characteristics that present significant challenges for chemical interaction and process development when used in the production of materials and chemicals (Pang et al., 2021). Nonetheless, it has been demonstrated that controlled catalytic depolymerization is a useful technique for disassembling the complex molecules that make up lignin into more manageable components. In addition to providing basic chemicals (aromatics) that might be further valued, this lessens the heterogeneity of lignin. Many technologies, such as photocatalysis, electrocatalysis, reductive depolymerization (Jiang et al., 2015), oxidative depolymerization, and acidolysis (Deuss et al., 2015), have been developed in recent years to enhance lignin depolymerization. One of these methods is reductive depolymerization, which produces large yields of a specific set of phenolic monomers by selectively catalyzing aryl ether C–O bond breakage under relatively mild circumstances.

1.6.2 Difficulty in product selectivity and depolymerization

Alkyl phenolic monomers with a range of functional groups, including p-sidechains and o-methoxy groups, are produced when lignin is reductively depolymerized. When altered (for instance, by amination), these compounds exhibit prospective as building blocks for chemical and pharmaceutical applications. In this way, bio-oil is better suited for use in platform chemicals due to the higher monomer yields produced by RCF processes. However, even with its potential, obtaining sufficient selectivity for a single type of monomer in the bio-oil emulsion remains challenging. The final composition from lignin or biomass decomposition can be tailored based on specific process parameters, catalyst selection, and biomass type used (Daelemans et al., 2025).

For example, using Ru/C as a catalyst in the RCF of biomass results in 75% selectivity for 4-propyl phenolics; Pd/C, however, encourages the production of propanol-substituted 91% selective phenolics (birch wood, 250°C, methanol) (Liu et al., 2020). More recently, Ouyang et al. (2019) used a methanol/water (1:2 v/v) mixture under hydrogen-free circumstances (birch wood, 230°C, Pt/Al_2O_3, methanol/water (1:2)) to obtain a 49% yield of monomers of phenol with 82% selectivity for 4-propyl syringol (v/v). Additionally, they showed that modifying the MeOH/H_2O ratio and the reaction temperature may cause selectivity to change in the direction of monomers replaced with propyl or propenyl.

1.6.3 Low-value applications despite high-volume availability

As technology for reductive depolymerization develops effectively, it is now crucial to use the phenolic products for the commercial viability and eventual success of these strategies. The results of lignin depolymerization usually include a mixture of monomers, dimers, and bigger oligomers, which reflects the complex structure that comprises lignin itself. The efficiency of the catalytic depolymerization procedure determines the ratios of these groups, which in turn affects the mixture's complexity and ease of separation. Applications in fine chemicals are probably better suited for mixtures with fewer but more chemically and physically comparable components (Daelemans et al., 2025). On the other hand, highly variable product combinations that are difficult to separate might be more appropriate for uses like solvents, printing inks (Liao et al., 2020), or polymer formulations like polyurethane (Vendamme et al., 2020) and epoxy resins (Feghali et al., 2020), where particular physical or chemical characteristics are more significant than purity.

1.7 POTENTIAL OF LIGNIN AS A RENEWABLE RESOURCE

1.7.1 Rich aromatic structure unique among biopolymers

The abundance of the biochemical sites in lignin's chemical structure suggested that lignin could be a crucial precursor in the generation of aromatic compounds. A few more unconventional monolignol constituents are also detected in small proportions (Gellerstedt & Henriksson, 2008). Additionally, these polyphenolic rings give aqueous systems hydrophobicity, which eventually gives plants resistance to biological and chemical degradation. According to research, lignin is a powerful source of aromatic chemical compounds, making it a useful source for producing phenolics (Alonso et al., 2010). According to earlier studies (Graglia et al., 2015), lignin can hydrolyze into phenolic resins and alkylphenols, which can function as extremely effective fuels and energy sources. The lignin-derived compounds can also be employed in the production of polyurethane resins, polyols, foams, antifungal agents, and numerous products of industrial application (Mahmood et al., 2016; Rößiger et al., 2018). The creation of char as a byproduct and the difficulty of product separation and purification are the main obstacles to the conversion of lignin.

1.7.2 Potential feedstock for value-added chemicals and materials

Being a polymer, lignin has promise for industrialization in materials applications. Currently, roughly three-fourths of the commercial derivatives from lignin take advantage of this feature (PNNL, 2007). Because lignin is stubborn, the depolymerization process requires a lot of energy and is expensive. Therefore, it is promising to employ lignin polymer directly as a precursor to produce value-added products without further depolymerization (Upton & Kasko, 2016). Some of the extensively investigated lignin-based macromolecular compounds are carbon fibers and polyurethane. Section 1.8 describes in detail various aromatic chemicals, polymers, nanomaterials, and biofuels derived from lignin.

1.7.3 Importance in circular bioeconomy and sustainable development

The integration of technological products into the biological cycle is made possible by the lignin-based bioeconomy, according to the cradle-to-cradle concept (Braungart et al., 2007), as extracellular oxidative enzymes depolymerize lignin to produce lower molecular mass aromatic compounds (Riyadi et al., 2020). Wastes are transformed into nutrients in this way, avoiding their buildup in riverine, marine, or land settings. In order to incorporate lignin into a circular economy, it is important to take into account three factors: (1) new ideas for reusing, repurposing, upcycling, and downcycling; (2) extraction and manufacturing processes toward a resource-efficient and sustainable industry; and (3) end-of-life, such as biological processing (such as biodegrading or use as feed) and thermal transformation. In the upcoming years, there are significant prospects for the resource-efficient use of lignin, making money from leftovers and waste, and merging upcycling and cascading principles. We predict a change from the direct energetic usage (energy from waste, waste into fuel) (Bhatt et al., 2017) toward value-added substances (bioplastics, lignin nanoparticles, carbon materials, fine chemicals), permitting items with prolonged durability (Tardy et al., 2023).

1.8 EMERGING APPLICATIONS OF LIGNIN

Lignin obtained from lignocellulosic biomass through various extraction processes can be utilized for the production of a variety of valuable products (Figure 1.2).

1.8.1 Bio-based chemicals: phenols, vanillin, and benzene derivatives

As a substitute, defunctionalization techniques are becoming increasingly popular to increase the purity and selectivity of monomers from lignin depolymerization, offering a more environmentally friendly method of producing bulk platform chemicals. Predominantly, hydrodeoxygenation (HDO) is used to defunctionalize the phenolic core, producing simpler compounds. Various end products, including (n-alkyl) phenols, (n-alkyl) cyclohexanols, aromatics, or aliphatics, are produced depending on the degree of deoxygenation and defunctionalization (Schutyser et al., 2018). Usually made from benzene obtained from fossil fuels, phenol is an essential platform chemical used mostly for polyamides, phenolic resins, and bisphenol A (BPA). However, it can also be made from monomers derived from lignin (Weber et al., 2020). With uses in agrochemicals, polymers, tastes, perfumes, medicines, and corrosion inhibitors,

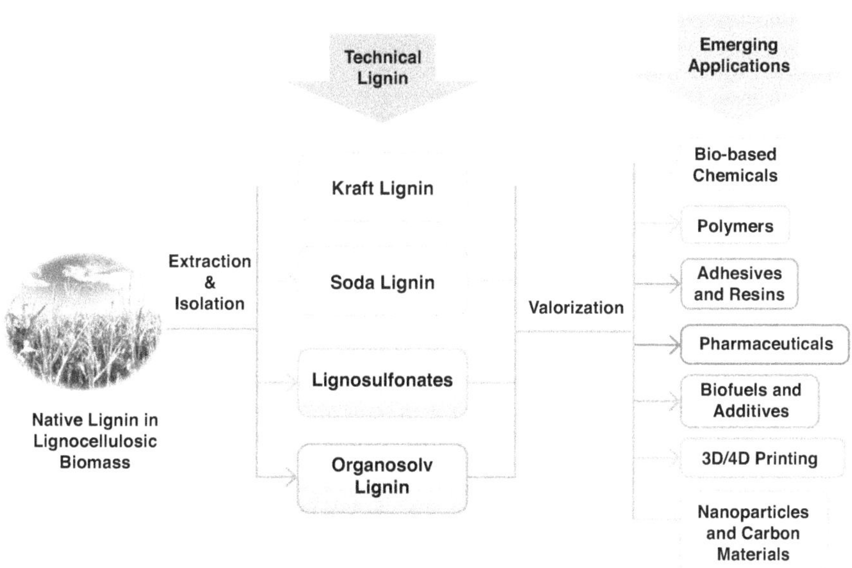

FIGURE 1.2 Valorization processes of common lignin types for emerging applications

catechol is a useful chemical (Park et al., 2021). Its industrial manufacture from phenol generated from fossil benzene frequently suffers from severe circumstances and low selectivity.

1.8.2 Polymers and additives

Lignin is a very intriguing renewable resource for a variety of material applications because of the availability of aliphatic and aromatic hydroxyl groups. In addition to adding antioxidative, antibacterial, and antistatic qualities as well as enhanced UV stability to materials, lignin's distinct molecular structure may enhance the mechanical and rheological characteristics of polymeric networks and blends (Diment et al., 2024; Thys et al., 2023). However, in order to overcome problems associated with a highly heterogeneous molecular structure, replacing fossil polymer building blocks and additives with lignin ideally requires a feedstock that has a lower polydispersity and molecular mass (compared to technical lignin side streams) as well as management of functional groups (Vishtal & Kraslawski, 2011). Oligomeric fractions of depolymerized lignin have been shown to be adequate—or perhaps even better—for making polymers or resins of desired properties (Han et al., 2023).

Lignin can be utilized as an additive and functional filler ingredient instead of being covalently coupled into a thermoset polymer network. To increase the mechanical and thermal resilience of thermoplastics, lignin particles were mixed with them in numerous research studies (Fazeli et al., 2024; Bertella & Luterbacher, 2020). Because it has a significant impact on the lignin's dispersion in the matrix material, the particle size of lignin is of greater importance for mechanical performance than the molecular weight. However, improved antioxidant qualities of lignin/polymer blends have been linked to a decreased molecular mass (Sadeghifar & Argyropoulos, 2015). This finding appears to be directly related to a molecular structure with a high concentration of phenolic hydroxy groups (Diment et al., 2024).

1.8.3 Carbon materials

1.8.3.1 Carbon fibers

Petroleum-derived polyacrylonitrile (PAN) is used in more than 90% of the world's carbon fiber production. In comparison to PAN, lignin has over 60% carbon content and is a sustainable substance, which is available in vast amounts (Norberg, 2012). Using polyvinyl alcohol as a softening agent and lignosulphonate as a precursor made from lignin, Nippon Kayaku Co. in Japan has commercialized the production of carbon fiber. Carbon fiber derived from lignin offers significant potential for adding value to biomass. It might increase the biorefinery's overall earnings by 30%–300% and reduce CO_2 emissions (Agrawal et al., 2014). Because carbon fiber composites have a high modulus, high strength, and reduced weight, they are a good alternative to titanium and aluminum alloys in aircraft applications. Carbon fiber composites are used in the professional sports and recreational market sectors that include tennis rackets, hockey sticks, golf clubs, fishing rods, archery arrows, and bows (Agrawal et al., 2014).

The global market for carbon fibers (CFs) was estimated at 46,000 tons in 2011 and is projected to reach 140,000 tons by 2020, according to a number of materials research publications. A crucial element in the actual manufacturing of carbon nanofibers is polyacrylonitrile (PAN). Due to the high cost of PAN, CFs are rarely used in numerous industrial uses (Gill et al., 2016). In reality, CFs range in price from $5 to $44 per kg (high-quality CF price per kg for commercial uses, 2025). This cost might be considerably reduced by lignin, making CFs more widely available for a range of industrial applications.

1.8.3.2 Activated carbon

Activated carbons are adsorbents that are used in industry to deodorize and purify process streams, as well as to disinfect liquid and gaseous effluents (Jain et al., 2024). Activation of lignin can be done chemically or physically to produce activated carbon. Chemical activation at high temperatures leads to higher product yields (Mussattoa et al., 2010). When it comes to metal finishing and electroplating solution purification, activated carbon performs exceptionally well. The additional applications comprise gas and water purification, caffeine reduction, gold purification, metal recovery, medicine, treatment of sewage, air filters in gas masks, breathing apparatus in compressed air, etc. The pharmaceutical sector could profit from activated CFs, which may absorb microorganisms (Agrawal et al., 2014).

1.8.4 Biofuels: lignin-derived oil and syngas

Lignin's large-scale availability and aromatic chemical structure make it a promising source for sustainable biofuels (Kumari et al., 2026). Bioethanol, which is made from the conversion of corn, and biodiesel, which is made via the hydrogenation of palm oil and used vegetable oil, currently dominate the biofuel industry (Kumari et al., 2024; Rathour et al., 2023; Kumari & Singh, 2022). Biofuels made from lignin can assist in decarbonizing difficult-to-electrify industries like sustainable aviation fuels by providing a reduced carbon footprint alternative to bioethanol (Obydenkova et al., 2017; Webber et al., 2024). Conventional hydrogenolysis produces phenolic mixtures (such as cresols, guaiacols, syringols, and catechols) that are unsuitable for direct application as drop-in biofuels. To improve the hydrogenolysis oil, a further catalytic hydrotreatment step is required. Although there are several possibilities for entry for lignin-derived biofuels due to the wide range of fuels in commercial use, it is challenging to establish a single goal benchmark price (Bhatia et al., 2021).

1.8.5 Pharmaceuticals

Gel networks have gained substantial attention due to their superior water absorption and solubility. This is explained by the creation of hydrogen connections between lignin's hydroxyl groups and water molecules, which makes it easier to absorb and hold onto water. Specifically, gels made from natural products have achieved importance in the pharmaceutical industry due to their superior biocompatibility and degradability. Their unique structural characteristics allow their applications as frameworks and transporters in both in vivo and in vitro situations. Moreover, the natural, harmless, and disintegrating characteristics of lignin make it potentially useful in medicinal applications, particularly in medicine delivery and food packaging. Such films must be biodegradable, including no immunological reaction, no inflammatory responses, no toxicity, and no additional negative consequences (Zhao et al., 2025).

The emergence of drug-resistant bacterial diseases necessitates alternatives to antibiotics. Lignin, accumulated in plant cell barriers in anticipation of pathogen invasion, serves as a mechanical barrier and has antibacterial properties (Kim et al., 2020). Lignin-derived biomaterials exhibit potential as multifunctional substances that can replicate the extracellular matrix, encouraging wound healing and tissue regeneration restoration. They can serve as crosslinks, fillers, and accelerators with other polymers to form strong networks and encapsulate cytokines (Legrand & Martino, 2022). Furthermore, they act as transporters for DNA (Dai et al., 2021), proteins (Zhu et al., 2023), medicinal compounds (Liu et al., 2024), antioxidants (Maqsoudlou et al., 2022), and disinfectants (Fadaka et al., 2021), enabling tailored drug delivery (Yu et al., 2020).

1.8.6 Adhesives and resins

Particleboard, plywood, and animal feed pellets are just a few of the products that use lignin as a natural adhesive or binder in phenolic resins. Additionally, it is used as a filler or addition in rubber and plastics to improve thermal stability and UV resistance while lowering dependency on components produced from fossil fuels (Yan et al., 2025). There are three primary types of lignin-based polymers: vitrimers, thermoplastics, and thermosetting resins (polyurethanes, phenolic resins, and epoxy resins) (Wang et al., 2016). These groups cover a spectrum of operational characteristics, with vitrimers exhibiting traits between thermosetting and thermoplastic elements. Epoxy resins, a representative thermosetting polymer substance, are a famous example where lignin has been applied as an alternative for bisphenol A and employed in combination with ethylene oxide and polyether amine (Jin et al., 2019).

1.8.7 Lignin for 3D to 4D printing

Because lignin has a large number of phenolic and aliphatic hydroxyl groups, it can be functionalized to produce extremely reactive polyol derivatives that are helpful in various applications (Duigou et al., 2020). Hydroxyl group alteration can be done through alkylation, esterification, etherification, and phenolation (Wu et al., 2025). Jiang et al. (2022) examined alkylation using 1, 6-dibromohexane, boosting lignin's molecular mass by 1643% and dimming its color substantially. Zhang et al. (2020b) highlighted the potential use of lignin in 3D printing by generating alkylated derivatives of dealkaline lignin (DAL) using an efficient single-step esterification involving undecanoyl and dodecanoyl chlorides, yielding the modified forms DAL-11ene and DAL-12ane. These derivatives displayed dramatically increased photoinitiation effectiveness relative to unmodified DAL, with DAL-11ene's versatile group facilitating direct integration into the polymer network. This change encourages the safe and efficient usage of 3D printing.

A new kind of 3D printing called 4D printing gives 3D-printed things the ability to modify their forms, characteristics, and functions in response to external inputs (Yan et al., 2023). Smart lignin polymers show potential for 4D printing. The particular chemical modification of lignin materials permits 3D-printed things to be susceptible to external stimuli, such as heat, light, and humidity. A predictable and predetermined time-dependent modification in the functionality of 3D printable goods could be accomplished by 4D printing.

1.8.8 Nanoparticles

Synthesis of nanomaterials from lignin includes various techniques, like enzyme-mediated synthesis (Jiang et al., 2020), interfacial polymerization/crosslinking (Huang et al., 2019), spray drying (Spicher et al., 2024), self-assembly (Jergens et al., 2023), and solvent exchange (Li et al., 2017), which are utilized in various fields, especially in pharmaceuticals. Multifunctional lignin nanoparticles containing curcumin have been produced utilizing several chemical solvents, like methanol, tetrahydrofuran, ethylene glycol, dimethyl sulfoxide (DMSO), and 70% ethanol (Singh et al., 2024). Remarkably, the nanoparticles' form, size, and physicochemical properties were similar across multiple solvents and displayed extraordinary thermal stability. These nanoparticles' small size enables efficient cellular penetration, which lowers intracellular pH and eventually diminishes bacterial ATP supply and causes bacterial cell death (Meng et al., 2023). Furthermore, nanosilver has been added to carboxymethyl-modified lignin using ultrasonics, resulting in nanosilver-doped lignin matrices with photothermal antibacterial qualities (Jiang et al., 2024).

1.9 CONCLUSION

Historically a waste output of paper, pulp, and biorefining, lignin is currently receiving more attention than before. The subject of lignin valorization has witnessed substantial growth and breakthroughs in the past decade. Lignin's aromatic characteristics have considerable market potential. The use of lignin is advantageous for the environment and sustainability. However, the use of lignin is currently mostly restricted to very cheap fuel and power sources using present technologies. The industrialization of lignin as an antecedent for highly efficient macromolecular compounds and aromatic compounds requires further study and development. There are still impediments remaining between the investigation of the intrinsic lignin framework, the processing of lignin, the breakdown of polymers, and the final product use. Although each of these fields is being thoroughly researched, there is a severe lack of cooperation and synergy. Because of their natural origin, low toxicity, and biodegradability, lignin-based biomaterials have become a revolutionary platform in materials science and biomedicine. These materials demonstrate tremendous potential in tissue rejuvenation, wound healing, hemostasis, antibacterial therapy, and drug delivery systems. Additionally, lignin has the potential to be effectively used in the chemical industry for the production of polymers, additives, adhesives, printing materials, and nanomaterials.

REFERENCES

Agrawal, A., N. Kaushik, S. Biswas. 2014. Derivatives and applications of lignin – an insight. *Scitech J.* 1(7): 30–36.

Ajdary, R., N. Kretzschmar, H. Baniasadi et al. 2021. Selective laser sintering of lignin-based composites. *ACS Sustainable Chem. Eng.* 9: 2727–35. https://doi.org/10.1021/acssuschemeng.0c07996

Alonso, D. M., J. Q. Bond, J. A. Dumesic. 2010. Catalytic conversion of biomass to biofuels. *Green Chem.* 12(9): 1493–13. https://doi.org/10.1039/C004654J

Bajwa, D. S., G. Pourhashem, A. H. Ullah, S. G. Bajwa. 2019. A concise review of current lignin production, applications, products and their environmental impact. *Ind. Crops Prod.* 139: 111526. https://doi.org/10.1016/j.indcrop.2019.111526

Beckham, G. T., C. W. Johnson, E. M. Karp, D. Salvachúa, D. R. Vardon. 2016. Opportunities and challenges in biological lignin valorization. *Curr. Opin. Biotechnol.* 42: 40–53. https://doi.org/10.1016/j.copbio.2016.02.030

Benhamou, A. A., L. Abid, I. Calvez et al. 2025. Advances in lignin chemistry, bonding performance, and formaldehyde emission reduction in lignin-based urea-formaldehyde adhesives: a aeview. *ChemSusChem.* 18: e202500491. https://doi.org/10.1002/cssc.202500491

Bertella, S., J. S. Luterbacher. 2020. Lignin functionalization for the production of novel materials. *Trends Chem.* 2: 440–53. https://doi.org/10.1016/j.trechm.2020.03.001

Bhatia, S. K., S. S. Jagtap, A. A. Bedekar et al. 2021. Renewable biohydrogen production from lignocellulosic biomass using fermentation and integration of systems with other energy generation technologies. *Sci. Total Enviro.* 765: 144429. https://doi.org/10.1016/j.scitotenv.2020.144429

Bhatt, A. K., R. K. Bhatia, S. Thakur, N. Rana, V. Sharma, R. K. Rathour. 2017. Fuel from waste: a review on scientific solution for waste management and environment conservation. *Prospect Alterna. Transp. Fuels.* 205–33. https://doi.org/10.1007/978-981-10-751 8-6_10

Braungart, M., W. McDonough, A. Bollinger. 2007. Cradle-to-cradle design: creating healthy emissions – a strategy for eco-effective product and system design. *J. Clea. Prod.* 15 (13–14): 1337–48. https://doi.org/10.1016/j.jclepro.2006.08.003

Brienza, F., D. Cannella, D. Montesdeoca, I. Cybulska, D. P. Debecker. 2024. A guide to lignin valorization in biorefineries: traditional, recent, and forthcoming approaches to convert raw lignocellulose into valuable materials and chemicals. *RSC Sustain.* 2: 37. https://doi.org/10.1039/D3SU00140G

Carvalho, D. M. D., M. H. Lahtinen, M. Lawoko, K. S. Mikkonen. 2020. Enrichment and identification of lignin–carbohydrate complexes in softwood extract. *ACS Sustainable Chem. Eng.* 8(31): 11795–04. https://doi.org/10.1021/acssuschemeng.0c03988

Chauhan, P. S. 2020. Role of various bacterial enzymes in complete depolymerization of lignin: a review. *Biocatal. Agric. Biotechol.* 23: 101498. https://doi.org/10.1016/j.bcab.2020.101498

Daelemans, B., B. Sridharan, P. Jusner et al. 2025. From lignin to market: a technical and economic perspective of reductive depolymerization approaches. *Green Chem.* 27: 13160–78. https://doi.org/10.1039/D5GC02316E

Dai, Z. W., J. Li, Y. F. Lin, Z. G. Wang, Y. Huang. 2021. Facile construction of a solely-DNAbased system for targeted delivery of nucleic acids. *Nanomaterials* 11(8): 1967. https://doi.org/10.3390/nano11081967

Delgado, N., F. Ysambertt, G. Chavez, B. Bravo, D. E. Garcıa, J. Santos. 2019. Valorization of kraft lignin of different molecular weights as surfactant agent for the oil industry. *Waste Biomass Valor.* 10: 3383–95. https://doi.org/10.1007/s12649-018-0352-4

Demuner, I. F., F. J. B. Gomes, J. S. Gomes et al. 2021. Improving kraft pulp mill sustainability by lignosulfonates production from processes residues. *J. Clean. Prod.* 317: 128286. https://doi.org/10.1016/j.jclepro.2021.128286

Deuss, P. J., M. Scott, F. Tran, N. J. Westwood, J. G. de Vries, K. Barta. 2015. Aromatic monomers by in situ conversion of reactive intermediates in the acid-catalyzed depolymerization of lignin. *J. Am. Chem. Soc.* 137: 7456–67. https://doi.org/10.1021/jacs.5b03693

Diment, D., O. Tkachenko, P. Schlee et al. 2024. Study toward a more reliable approach to elucidate the lignin structure–property–performance correlation. *Biomacromolecules* 25: 200–12. https://doi.org/10.1021/acs.biomac.3c00906

Duigou, A. L., D. Correa, M. Ueda, R. Matsuzaki, M. Castro. 2020. A review of 3D and 4D printing of natural fibre biocomposites. *Mater. Des.* 194: 108911. https://doi.org/10.1016/j.matdes.2020.108911

Fadaka, A. O., N. R. S. Sibuyi, A. M. Madiehe, M. Meyer. 2021. Nanotechnology-based delivery systems for antimicrobial peptides. *Pharmaceutics* 13(11): 1795. https://doi.org/10.3390/pharmaceutics13111795

Faix, O. 1991. Classification of lignins from different botanical origins by FT-IR spectroscopy. *Holzforschung* 45: 21–8. https://doi.org/10.1515/hfsg.1991.45.s1.21

Farhat, W., R. Venditti, N. Mignard, M. Taha, F. Becquart, A. Ayoub. 2017. Polysaccharides and lignin based hydrogels with potential pharmaceutical use as a drug delivery system produced by a reactive extrusion process. *Int. J. Biol. Macromol.* 104: 564–75. https://doi.org/10.1016/j.ijbiomac.2017.06.037

Fazeli, M., S. Mukherjee, H. Baniasadi et al. 2024. Lignin beyond the *status quo*: recent and emerging composite applications. *Green Chem.* 26: 593–30. https://doi.org/10.1039/D3GC03154C

Feghali, E., D. J. van de Pas, A. J. Parrott, K. M. Torr. 2020. Biobased epoxy thermoset polymers from depolymerized native hardwood lignin. *ACS Macro Letters* 9(8): 1155–60. https://doi.org/10.1021/acsmacrolett.0c00424

Figueiredo, P., K. Lintinen, A. Kiriazis et al. 2017. In vitro evaluation of biodegradable lignin-based nanoparticles for drug delivery and enhanced antiproliferation efect in cancer cells. *Biomaterials* 121: 97–108. https://doi.org/10.1016/j.biomaterials.2016.12.034

Frigon, J. C., S. R. Guiot. 2010. Biomethane production from starch and lignocellulosic crops: a comparative review. *Biofuels, Bioprod. Biorefin.* 4: 447–58. https://doi.org/10.1002/bbb.229

Gellerstedt, G., G. Henriksson. 2008. Lignins: major sources, structure and properties. In *Monomers, polymers and composites from renewable resources*, ed. M. N. Belgacem and A. Gandini, 201–24. Elsevier, Amsterdam. https://doi.org/10.1016/B978-0-08-045316-3.00009-0

Gill, A. S., D. Visotsky, L. Mears, J. D. Summers. Cost estimation model for PAN based carbon fiber manufacturing process. In *Proceedings of the International Manufacturing Science and Engineering Conference*, Novy Smokovec, Slovakia, 6–10 June 2016.

Gillet, S., M. Aguedo, L. Petitjean et al. 2017. Lignin transformations for high value applications: towards targeted modifications using green chemistry. *Green Chem.* 19: 4200–33. https://doi.org/10.1039/C7GC01479A

Giummarella, N., Y. Pu, A. J. Ragauskas, M. Lawoko. 2019. A critical review on the analysis of lignin carbohydrate bonds. *Green Chem.* 21: 1573–95. https://doi.org/10.1039/C8GC03606C

Graglia, M., N. Kanna, D. Esposito. 2015. Lignin refinery: towards the preparation of renewable aromatic building blocks. *ChemBioEng Rev.* 2(6):377–92. https://doi.org/10.1002/cben.201500019

Han, Y., B. A. Simmons, S. Singh. 2023. Perspective on oligomeric products from lignin depolymerization: their generation, identification, and further valorization. *Ind. Chem. Mater.* 1: 207–23. https://doi.org/10.1039/D2IM00059H

High-Strength Carbon Fiber Price per Kg for Professional Uses—Alibaba.Com. Available online: https://www.alibaba.com/showroom/carbon-fiber-price-per-kg.html (accessed on 26 November 2025).

Huang, J. L., K. Y. Leng, Y. Q. Chen et al. 2019. An interfacial crosslinking strategy to fabricate an ultrathin two-dimensional composite of silicon oxycarbide-enwrapped silicon nanoparticles for high performance lithium storage. *J. Mater. Chem. A.* 7: 22950–57. https://doi.org/10.1039/C9TA07738C

Isikgor, F. H., C. R. Becer. 2015. Lignocellulosic biomass: a sustainable platform for the production of bio-based chemicals and polymers. *Polym. Chem.* 6: 4497–59. https://doi.org/10.1039/C5PY00263J

Jain, Y., M. Dubey, Gunjan, D. Kumari. 2024. Production of activated carbon from banana. In *Recent Advances in Activated Carbon: Synthesis, Properties, and Applications*, ed. S. M. Ho, S. K. Heri, and Y. C. Sharma, 210–25. CRC Press, Taylor & Francis Group, Boca Raton. https://doi.org/10.1201/9781003488606

Jergens, E., S. de Araujo Fernandes-Junior, Y. Cui et al. 2023. DNA-caged nanoparticles via electrostatic self-assembly. *Nanoscale* 15: 9390–402. https://doi.org/10.1039/D3NR01424J

Jiang, W. Z., Y. C. Zhang, D. J. Yang, X. Q. Qiu, Z. X. Li. 2024. Ultrasonic-assisted synthesis of lignin-based ultrasmall silver nanoparticles for photothermal-mediated sterilization. *Int. J. Biol. Macromol.* 262: 129827. https://doi.org/10.1016/j.ijbiomac.2024.129827

Jiang, X., Z. Tian, X. Ji et al. 2022. Modification of cellulose with quaternary ammonium salts: a review. *Int. J. Biol. Macromol.* 201: 400–10.

Jiang, Y., Z. Li, X. Tang et al. 2015. Depolymerization of cellulolytic enzyme lignin for the production of monomeric phenols over Raney Ni and Acidic Zeolite Catalysts. *Energy Fuels* 29: 1662–68. https://doi.org/10.1021/ef5022297

Jiang, Y. H., Y. Y. Chen, D. Tian et al. 2020. Fabrication and characterization of lignin-xylan hybrid nanospheres as pesticide carriers with enzyme-mediated release property. *Soft Matter.* 16: 9083–93. https://doi.org/10.1039/D0SM01402H

Jin, F. L., M. Zhao, M. Park, S. J. Park. 2019. Recent trends of foaming in polymer processing: a review. *Polymers.* 11(6): 953. https://doi.org/10.3390/polym11060953

Karthäuser, J., V. Biziks, C. Mai, H. Militz. 2021. Lignin and lignin-derived compounds for wood applications—A review. *Molecules* 26: 2533. https://doi.org/10.3390/molecules26092533

Key, R. E., J. J. Bozell. 2016. Progress toward lignin valorization via selective catalytic technologies and the tailoring of biosynthetic pathways. *ACS Sustain. Chem. Eng.* 4: 5123–35. https://doi.org/10.1021/acssuschemeng.6b01319

Kim, S. H., P. Y. Lam, M. H. Lee, H. S. Jeon, Y. Tobimatsu, O. K. Park. 2020. The Arabidopsis R2R3 MYB transcription factor MYB15 is a key regulator of lignin biosynthesis in effector-triggered immunity. *Front Plant Sci.* 17(11): 583153. https://doi.org/10.3389/fpls.2020.583153

Kumari, D., R. Singh. 2018. Pretreatment of lignocellulosic wastes for biofuel production: a critical review. *Renew Sustain Energy Rev.* 90: 877–91. https://doi.org/10.1016/j.rser.2018.03.111

Kumari, D., R. Singh. 2020. Ultrasonic assisted petha waste water pretreatment of rice straw for optimum production of methane and ethanol using mixed microbial culture. *Renew Energy.* 145: 682–90. https://doi.org/10.1016/j.renene.2019.06.082

Kumari, D., R. Singh. 2022. Rice straw structure changes following green pretreatment with petha wastewater for economically viable bioethanol production. *Sci. Rep.* 12: 10443. https://doi.org/10.1038/s41598-022-14627-7

Kumari, D., G. S. Maurya, Y. Jain. 2026. Current status and challenges of biofuel sector. In *Biofuels: analytical insights and perspectives* (1st ed.), ed. R. K. Bhatia and M. Hans. CRC Press. https://doi.org/10.1201/9781003485636

Kumari, D., G. S. Maurya, R. Singh. 2024. Coupled pretreatment of rice straw with dairy wastewater and ultrasonication for biofuel production using mixed microbial source. *Bioresour. Technol. Reports.* 25: 101738. https://doi.org/10.1016/j.biteb.2023.101738

Kumari, D., G. S. Maurya, E. Tarif, Y. Jain. 2025. Sustainable biohythane production from lignocellulosic biomass and industrial by-products: a review of pretreatment technologies and economic feasibility. *J. Environ. Chem. Eng.* 13(6): 119936. https://doi.org/10.1016/j.jece.2025.119936

Kumari, D., P. Chahar, R. Singh. 2018. Effect of ultrasonication on biogas and ethanol production from rice straw pretreated with petha waste water and dairy waste water. *Int J Curr Eng Sci Res.* 5(1): 65–73.

Kumari, D., R. K. Bhatia, R. Singh. 2021. Technologies and challenges to recover energy products from carbohydrate rich effluents: a mini review. *Int J Plant Anim Environ Sci.* 11(3): 423–42. https://doi.org/10.26502/ijpaes.202113

Kumari, D., Y. Jain. 2020. Enhanced biofuel production from lignocellulosic biomass: an overview of advanced physico-chemical and biological technologies. In *Biorefineries: a step towards renewable and clean energy*, ed. P. Verma, 151–72. Springer, Singapore. https://doi.org/10.1007/978-981-15-9593-6_7

Kumari, D., Y. Jain, R. Singh. 2021. A study on green pretreatment of rice straw using Petha wastewater and Mausami waste assisted with microwave for production of ethanol and methane. *Energy Convers. and Manag.:* X. 10: 100067. https://doi.org/10.1016/j.ecmx.2020.100067

Lahive, C. W., P. C. J. Kamer, C. S. Lancefield, P. J. Deuss. 2020. An introduction to model compounds of lignin linking motifs; synthesis and selection considerations for reactivity studies. *Chem Sus Chem.* 13: 4238–65. https://doi.org/10.1002/cssc.202000989

Legrand, J. M. D., M. M. Martino. 2022. Growth factor and cytokine delivery systems for wound healing. *Cold Spring Harb Perspect Biol.* 14. https://doi.org/10.1101/cshperspect.a041234

Li, N. N., A. Z. Panagiotopoulos, A. Nikoubashman. 2017. Structured nanoparticles from the self-assembly of polymer blends through rapid solvent exchange. *Langmuir* 33: 6021–28. https://doi.org/10.1021/acs.langmuir.7b00291

Liao, Y., S. F. Koelewijn, G. Van den Bossche et al. 2020. A sustainable wood biorefinery for low-carbon footprint chemicals production. *Nicolaï, Science.* 367: 1385–90. https://doi.org/10.1126/science.aau1567

Lignin Market. 2024–2030. Market Analysis Report, 2024. Available online: https://www.grandviewresearch.com/industry-analysis/lignin-market (accessed on 14 October 2025).

Linger, J. G., D. R. Vardon, M. T. Guarnierietal et al. 2014. Lignin valorization through integrated biological funneling and chemical catalysis. *Proc. Natl. Acad. Sci. USA.* 111: 12013–18. https://doi.org/10.1073/pnas.1410657111

Liu, X., F. P. Bouxin, J. Fan, V. L. Budarin, C. Hu, J. H. Clark. 2020. Recent advances in the catalytic depolymerization of lignin towards phenolic chemicals: a review. *ChemSusChem.* 13: 4296–17. https://doi.org/10.1002/cssc.202001213

Liu, X., K. K. Huang, F. X. Zhang et al. 2024. Multifunctional nano-in-micro delivery systems for targeted therapy in fundus neovascularization diseases. *J Nanobiotechnol.* 20(22): 354. https://doi.org/10.1186/s12951-024-02614-1

Mahmood, N., Z. Yuan, J. Schmidt, M. Tymchyshyn, C. Xu. 2016. Hydrolytic liquefaction of hydrolysis lignin for the preparation of bio-based rigid polyurethane foam. *Green Chem.* 18(8): 2385–98. https://doi.org/10.1039/C5GC02876K

Mansfield, S. D., H. Kim, F. Lu, J. Ralph. 2012. Whole plant cell wall characterization using solution-state 2D NMR. *Nat. Protoc.* 7: 1579–89. https://doi.org/10.1038/nprot.2012.064

Maqsoudlou, A., E. Assadpour, H. Mohebodini, S. M. Jafari. 2022. The influence of nanodelivery systems on the antioxidant activity of natural bioactive compounds. *Crit Rev Food Sci Nutr.* 62: 3208–31. https://doi.org/10.1080/10408398.2020.1863907

Mazar, A., N. Jemaa, W. W. Al Dajani, M. Marinova, M. Perrier. 2021. Optimization of lignin recovery from the pre-hydrolysate of kraft-based dissolving pulp production processes. *Appl. Sci.* 11(1): 454. https://doi.org/10.3390/app11010454

Mazar, A., O. Ajao, M. Benali, N. Jemaa, W. W. Al-Dajani, M. Paleologou. 2020. Michael paleologou integrated multiproduct biorefinery for furfural production with acetic acid and lignin recovery: design, scale-up evaluation, and technoeconomic analysis. *ACS Sustainable Chem. Eng.* 8(47): 17345–58. https://doi.org/10.1021/acssuschemeng.0c04871

Mehra, A. 2025. Lignosulfonates Market – Global Forecast to 2025; *MarketsandMarkets*: https://www.marketsandmarkets.com/Research Insight/lignosulfonates-market.asp (accessed on 29 June, 2025).

Melro, E., A. Filipe, D. Sousa, B. Medronho, A. Romano. 2021. Revisiting lignin: a tour through its structural features, characterization methods and applications. *New. J. Chem.* 45: 6986–13. https://doi.org/10.1039/D0NJ06234K

Meng, T. T., Y. Ding, Y. Liu et al. 2023. In situ lignin adhesion for high-performance bamboo composites. *Nano Lett.* 23: 8411–18. https://doi.org/10.1021/acs.nanolett.3c01497

Mussattoa, S. I., M. Fernandes, J. M. R. George, J. M. O. José, A. T. José, I. C. Roberto. 2010. Production, characterization and application of activated carbon from brewer's spent grain lignin, *Bioresour. Technol.* 101: 2450–57. https://doi.org/10.1016/j.biortech.2009.11.025

Nguyen, N. A., S. H. Barnes, C. C. Bowland et al. 2018. A path for lignin valorization via additive manufacturing of high-performance sustainable composites with enhanced 3D printability. *Sci. Adv.* 4(12). https://doi.org/10.1126/sciadv.aat4967

Norberg, I. 2012. *Carbon fibers form kraft lignin, division of wood chemistry and pulp technology*. KTH Royal Institute of Technology, Stockholm.

Obydenkova, S. V., P. D. Kouris, E. J. M. Hensen, H. J. Heeres, M. D. Boot. 2017. Environmental economics of lignin derived transport fuels. *Bioresour. Technol.* 243: 589–99. https://doi.org/10.1016/j.biortech.2017.06.157

Ouyang, X., X. Huang, J. Zhu, M. D. Boot, E. J. M. Hensen. 2019. Catalytic conversion of lignin in woody biomass into phenolic monomers in methanol/water mixtures without external hydrogen. *ACS Sustain. Chem. Eng.* 7: 13764–73. https://doi.org/10.1021/acssuschemeng.9b01497

Pang, T., G. Wang, H. Sun, W. Sui, C. Si. 2021. Lignin fractionation: effective strategy to reduce molecule weight dependent heterogeneity for upgraded lignin valorization. *Ind. Crops Prod.* 165: 113442. https://doi.org/10.1016/j.indcrop.2021.113442

Park, J., M. A. Kelly, J. X. Kang et al. 2021. Production of active pharmaceutical ingredients (APIs) from lignin-derived phenol and catechol. *Green Chem.* 23: 7488–98. https://doi.org/10.1039/D1GC02158C

PNNL (Pacific Northern National Laboratory). 2007. *Top value added chemicals from biomass volume II: results of screening for potential candidates from biorefinery lignin*. PNNL, Richland, WA.

Puziy, A. M., O. I. Poddubnaya, O. Sevastyanova. 2018. Carbon materials from technical lignins: recent advances. *Top Curr. Chem.* 376: 95–128. https://doi.org/10.1007/s41061-018-0210-7

Ragauskas, A. J., G. T. Beckham, M. J. Biddy et al. 2014. Lignin valorization: improving lignin processing in the biorefinery. *Science.* 344. https://doi.org/10.1126/science.1246843

Rathour, R. K., M. Behl, K. Dhashmana et al. 2023. Non-food crops derived lignocellulose biorefinery for sustainable production of biomaterials, biochemicals and bioenergy: a review on trends and techniques. *Ind. CropsProd.* 204: 117220. https://doi.org/10.1016/j.indcrop.2023.117220

Rathour, R. K., M. Devi, P. Dahiya et al. 2023. Recent trends, opportunities and challenges in sustainable management of rice straw waste biomass for green biorefinery. *Energies* 16(3): 1429. https://doi.org/10.3390/en16031429

Rathour, R. K., N. Rana, V. Sharma, N. Sharma, A. K. Bhatt, R. K. Bhatia. 2024. Combatting synthetic dye toxicity through exploring the potential of lignin peroxidase from Pseudomonas fluorescence LiP RL5. *Environ. Sci. Pollut. Res.* 1–11. https://doi.org/10.1007/s11356-024-34400-9

Riyadi, F. A., A. A. Tahir, N. Yusof et al. 2020. Enzymatic and genetic characterization of lignin depolymerization by Streptomyces sp. S6 isolated from a tropical environment. *Sci. Rep.* 10(1): 7813. https://doi.org/10.1038/s41598-020-64817-4

Rößiger, B., G. Unkelbach, D. Pufky-Heinrich. 2018. Base-catalyzed depolymerization of lignin: history, challenges and perspectives. *Lignin Trends Appl.* 21: 99–120. https://doi.org/10.5772/intechopen.72964

Ruwoldt, J., F. H. Blindheim, G. Chinga-Carrasco. 2023. Functional surfaces, films, and coatings with lignin – a critical review. *RSC Adv.* 13: 12529–55. https://doi.org/10.1039/D2RA08179B

Sadeghifar, H., D. S. Argyropoulos. 2015. Correlations of the antioxidant properties of softwood kraft lignin fractions with the thermal stability of its blends with polyethylene. *ACS Sustain. Chem. Eng.* 3(2): 349–56. https://doi.org/10.1021/sc500756n

Sawatdeenarunat, C., K. C. Surendra, D. Takara, H. Oechsner, S. K. Khanal. 2015. Anaerobic digestion of lignocellulosic biomass: challenges and opportunities. *Bioresour. Technol.* 178: 178–86. https://doi.org/10.1039/D0NJ06234K

Schutyser, W., T. Renders, S. Van den Bosch, S. F. Koelewijn, G. T. Beckham, B. F. Sels. 2018. Chemicals from lignin: an interplay of lignocellulose fractionation, depolymerisation, and upgrading. *Chem. Soc. Rev.* 47: 852–08. https://doi.org/10.1039/C7CS00566K

Singh, G., K. Das, D. D. Mandal. 2024. Valorization of kraft lignin via its conversion into lignin nanoparticles to act both as DDS and biostimulator for plant. *Waste Biomass Valor.* 15: 4823–38. https://doi.org/10.1007/s12649-024-02479-3

Smith, M. D. 2019. An abbreviated historical and structural introduction to lignocellulose. In *Understanding lignocellulose*, ed. M. D. Smith, 1338: 1–15. ACS Symposium Series, American Chemical Society, Washington, DC. https://doi.org/10.1021/bk-2019-1338.ch001

Spicher, M. T., S. P. Schwaminger, D. von der Haar-Leistl, E. Fellenberg, S. Berensmeier. 2024. Process development for pilot-scale spray drying of ultra-small iron (oxyhydr) oxide nanoparticles. *Powder Technol.* 433: 119186. https://doi.org/10.1016/j.powtec.2023.119186

Tardy, B. L., E. Lizundia, C. Guizani, M. Hakkarainen, M. H. Sipponen. 2023. Prospects for the integration of lignin materials into the circular economy. *Mater. Today.* 65: 122–32. https://doi.org/10.1016/j.mattod.2023.04.001

Thys, M., G. E. Kaya, L. Soetemans et al. 2023. Bioaromatic-associated multifunctionality in lignin-containing reversible elastomers. *ACS Appl. Polym. Mater.* 5: 5846–56. https://doi.org/10.1021/acsapm.3c00491

Tran, H., E. K. Vakkilainnen. The kraft chemcial recovery process. *TAPPI Kraft Recover Course*, January 7–10, 2012, pp. 1–8.

Tribot, A., G. Amer, M. A. Alio et al. 2019. Wood-lignin: supply, extraction processes and use as bio-based material. *Eur. Polym. J.* 112: 228. https://doi.org/10.1016/j.eurpolymj.2019.01.007

Upton, B.M., A. M. Kasko. 2016. Strategies for the conversion of lignin to high-value polymeric materials: review and perspective. *Chem. Rev.* 116: 2275–06. https://doi.org/10.1021/acs.chemrev.5b00345

van Erven, G., V. J. P. Boerkamp, J. W. van Groenestijn, R. J. A. Gosselink. 2024. Choline and lactic acid covalently incorporate into the lignin structure during deep eutectic solvent pulping. *Green Chem.* 26: 7101–12. https://doi.org/10.1039/D4GC00909F

Vanholme, R., B. Demedts, K. Morreel, J. Ralph, W. Boerjan. 2010. Lignin biosynthesis and structure. *Plant Physiol.* 153: 895–05. https://doi.org/10.1104/pp.110.155119

Vendamme, R., J. Behaghel de Bueren, J. Gracia-Vitoria et al. 2020. Aldehyde-assisted lignocellulose fractionation provides unique lignin oligomers for the design of tunable polyurethane bioresins. *Biomacromolecules* 21: 4135–48. https://doi.org/10.1021/acs.biomac.0c00927

Vishtal, A., A. Kraslawski. 2011. Challenges in industrial applications of technical lignins. *BioResources* 6(3): 3547–68.

Wang, C., S. K. Stephen, A. V. Richard. 2016. Lignin-based thermoplastic materials. *ChemSusChem* 9(8): 770–83. https://doi.org/10.1002/cssc.201501531

Webber, M. S., J. Watson, J. Zhu et al. 2024. Lignin deoxygenation for the production of sustainable aviation fuel blendstocks. *Nat. Mater.* 23: 1622–38. https://doi.org/10.1038/s41563-024-02024-6

Weber, M., M. Weber, V. Weber. 2020. Phenol. In *Ullmann's encyclopedia of industrial chemistry*, 1–20. Wiley. https://doi.org/10.1002/14356007.a19_299.pub3

Wong, S. S., R. Shu, J. Zhang, H. Liu, N. Yan. 2020. Downstream processing of lignin derived feedstock into end products. *Chem. Soc. Rev.* 49: 5510–60. https://doi.org/10.1039/D0CS00134A

Wu, T., S. Sugiarto, R. Yang. 2025. From 3D to 4D printing of lignin towards green materials and sustainable manufacturing. *Mater. Horiz.* 12: 2789. https://doi.org/10.1039/d4mh01680g

Yan, L., A. J. H. Alonso, H. Liu, L. Dai, C. Si, M. H. Sipponen. 2025. Lignin polymerization: towards high-performance materials. *Chem. Soc. Rev.* 54: 6634. https://doi.org/10.1039/d4cs01044b

Yan, S., F. Zhang, L. Luo, L. Wang, Y. Liu, J. Leng. 2023. Shape memory polymer composites: 4D printing, smart structures, and applications. *Research* 6: 0234. https://doi.org/10.34133/research.0234

Yu, H. L., Z. H. Yang, F. Li, L. S. Xu, Y. Sun. 2020. Cell-mediated targeting drugs delivery systems. *Drug Deliv.* 27: 1425–37. https://doi.org/10.1080/10717544.2020.1831103

Zhang, S., J. Xiao, G. Wang, G. Chen. 2020. Enzymatic hydrolysis of lignin by ligninolytic enzymes and analysis of the hydrolyzed lignin products. *Bioresour. Technol.* 304: 122975. https://doi.org/10.1016/j.biortech.2020.122975

Zhang, X., S. Keck, Y. Qi, S. Baudis, Y. Zhao. 2020. Study on modified dealkaline lignin as visible light macromolecular photoinitiator for 3D printing. *ACS Sustainable Chem. Eng.* 8: 10959–70. https://doi.org/10.1021/acssuschemeng.0c03587

Zhao, B.C., B. Y. Chen, S. Yang, T. Q. Yuan, A. Charlton, R. C. Sun. 2017. Structural variation of lignin and lignin–carbohydrate complex in eucalyptus grandis E. urophylla during its growth process. *ACS Sustainable Chem. Eng.* 5: 1113–22. https://doi.org/10.1021/acssuschemeng.6b02396

Zhao, J., M. Zhu, W. Jin et al. 2025. A comprehensive review of unlocking the potential of lignin-derived biomaterials: from lignin structure to biomedical application. *J. Nanobiotechnol.* 23: 538. https://doi.org/10.1186/s12951-025-03604-7

Zhao, Y., U. Shakeel, M. Saif Ur Rehman, H. Li, X. Xu, J. Xu. 2020. Lignin-carbohydrate complexes (LCCs) and its role in biorefinery. *Cleaner Prod.* 253: 120076. https://doi.org/10.1016/j.jclepro.2020.120076

Zhou, Q., A. Fransen, H. de Winde. 2025. Lignin-degrading enzymes and the potential of Pseudomonas putida as a cell factory for lignin degradation and valorization. *Microorganisms* 13(4): 935. https://doi.org/10.3390/microorganisms13040935

Zhu, H. C., H. Luo, R. L. Chang et al. 2023. Proteinbased delivery systems for RNA delivery. *J Control Release.* 363: 253–74. https://doi.org/10.1016/j.jconrel.2023.09.032

Zoghlami, A., G. Paës. 2019. Lignocellulosic biomass: understanding recalcitrance and predicting hydrolysis. *Front Chem.* 7: 874. https://doi.org/10.3389/fchem.2019.00874

2 Strategies for lignin extraction from lignocellulosic biomass

Deepika Umrao, Suman Salhotra, Pragati Bhardwaj, Sachin Kumar, Rohit Rai, and Meenu Hans

2.1 INTRODUCTION

Lignin is one of the three major structural components of lignocellulosic biomass (LCB), along with cellulose and hemicellulose. Lignin accounts for ~15–30% of LCB and plays the role of providing stiffness, support, and resistance from microbial deterioration to the plant (Akpan, 2019; Rathour et al., 2025; Kumari et al., 2025). Lignin, a highly complex natural polymer, is composed of phenolic subunits: p-coumaryl, coniferyl, and sinapyl alcohols, interconnected by ether and carbon–carbon (C–C) bonds Kumari & Singh, 2018). Its unique structural diversity makes it a promising renewable resource for the synthesis of bio-based chemicals, fuels, and other high-value products. However, due to its heterogeneity, recalcitrance, and covalent attachment to other polysaccharides (cellulose, hemicellulose) in the plant cell wall, lignin is extremely difficult to remove and use completely (Chakar & Ragauskas, 2004).

The expensive and energy-demanding extraction of lignin from biomass is one of the key challenges for the development of a sustainable biorefinery. Nevertheless, with the emerging global focus on biomass valorization, biorefinery construction, and green chemistry, more attention should be paid to the extraction and utilization of lignin. Lignin valorization, through the production of high-value-added products, holds the potential to significantly enhance the economic viability of biomass biorefineries (Ali et al., 2024). The development of reductive catalytic fractionation (RCF) results in lignin extraction by enhancing the monomeric structure and its proper function (Jindal et al., 2023). Brienza et al. (2024) have shown that polymers derived from lignin could replace naphtha-derived polymers. Abu-Omar et al. (2021) observed that lignin depolymerization can produce monomers with about 70% efficiency. Lignin derived from biomass has potential for biodegradable plastics, adhesives, and coatings, providing an eco-friendly alternative to petrochemical-based materials. Figure 2.1 represents the multidisciplinary applications of lignin toward sustainable technologies.

DOI: 10.1201/9781003653189-2

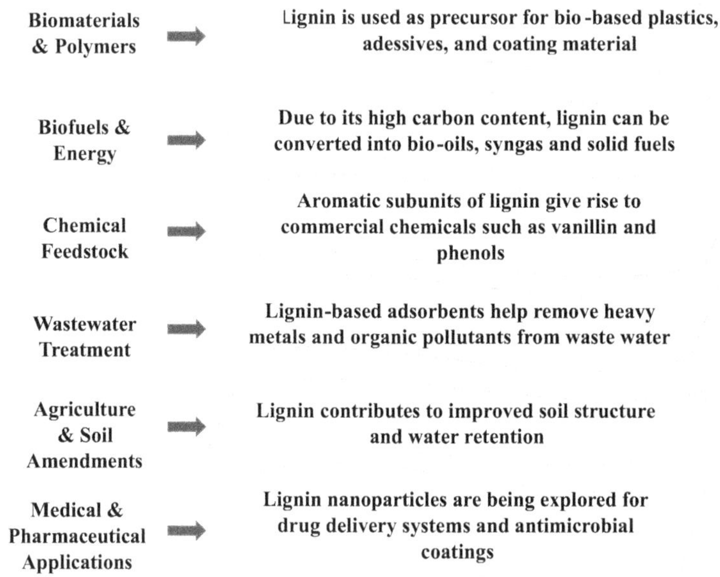

FIGURE 2.1 Applications of lignocellulosic biomass-derived lignin

Selection of the extraction method significantly influences lignin yield and purity, thereby shaping its structural attributes and potential applications. To effectively valorize lignin, a comprehensive understanding of how these methods alter its structure and properties is essential. Such insight is crucial not only for advancing lignin-based technologies but also for driving sustainable innovation within the bioeconomy (Lehto & Alén, 2015). This chapter critically discusses these traditional extraction methods, outlining their chemical foundation, guiding principles, advantages, limitations, and implications for future use.

2.2 CONVENTIONAL METHODS OF LIGNIN EXTRACTION

Lignin can be extracted using a variety of techniques, such as conventional pulping methods as well as recent catalytic and solvent-based strategies. An extensive comparison among various lignin extraction techniques is represented in Table 2.1. Pulp and paper industry demand has traditionally played a role in the development of strategies to recover lignin from LCB (Chakar & Ragauskas, 2004). Kraft, sulfite, soda, and organosolv processes have been traditionally used to maximize cellulose and minimize lignin as a byproduct (Haykir & Viell, 2024; Carvajal et al., 2016; Lehto & Alén, 2015). The following sections outline the major conventional methods employed for lignin extraction from biomass, highlighting their principles, advantages, and limitations.

2.2.1 Kraft process

The most widely used method in the industrial production of lignin is the kraft process, which accounts for approximately 90% of the world's pulp production

(Carvajal et al., 2016). In alkaline pulping, wood chips are treated with sodium hydroxide and sodium sulfide under elevated temperature (150–170°C) and pressure (7–10 atm), where sulfide ions catalyze the cleavage of β-O-4 ether linkages in lignin, enabling its dissolution into black liquor while preserving the cellulose matrix. While the kraft process is highly effective in removing lignin, the resulting kraft lignin is highly condensed, contains sulfur, and is structurally derivatized under strong basic conditions (Norgren & Edlund, 2014). This structural deterioration restricts its direct application as a high value-added product without further post-treatment. The main advantage of the kraft process is its lignin removal efficiency, as it separates lignin from LCB with comparatively low ash content and an integrated nature. The robust nature of lignin is a significant barrier to the production of high value-added bioproducts. Current research focuses on enhancing the efficiency of the kraft process so that a higher quality of lignin can be isolated at earlier stages (Saadan et al., 2024).

2.2.2 Sulfite process

Sulfurous acid (H_2SO_3) and a base such as calcium, magnesium, sodium, or ammonium were used in the 19th century to solubilize the lignin as lignosulfonates. The reaction is primarily aimed at the carbon–carbon and ether inter-unit linkages in lignin under relatively acidic conditions (pH 1.5-5) at 120–180°C, depending on the desired products (Saadan et al., 2024). Because lignosulfonates are water-soluble, they can be used as dispersants, adhesives, and additives in a range of industrial applications. Nevertheless, the introduction of sulfonic groups to lignin significantly alters the chemical properties and reduces its quality for further valorization to highly pure aromatic chemicals. Although the sulfite process yields lignin with greater chemical uniformity than the kraft process, lignosulfonate continues to occupy a marginal niche due to environmental concerns such as sulfur emissions and wastewater treatment challenges, and its limited valorization beyond a few low-value products.

2.2.3 Soda process

The soda pulping process is an alkaline, sulfur-free process that dissolves lignin using sodium hydroxide at 140–170°C. It is suitable for non-wood biomass such as wheat straw and bagasse due to lower lignin content than wood (Saadan et al., 2024). A key advantage of this process is that sulfur is not incorporated into soda lignin, resulting in higher-quality materials and chemical derivatives. However, under high-temperature alkaline conditions, lignin may undergo extensive condensation, diminishing its reactivity and functional potential. Nonetheless, soda pulping presents a more environmentally friendly alternative to sulfur-based methods and aligns well with the principles of green chemistry.

2.2.4 Organosolv process

The organosolv method dissolves lignin at low temperature (120–200°C) and pressure using organic solvents, such as ethanol, methanol, and acetone, and acids, such as sulfuric acid. This methodology efficiently cleaves lignin-polysaccharide linkages and facilitates the dissolution of lignin into the organic phase, preserving its native structure more effectively than alkaline or sulfite-based techniques (Capolupo & Faraco, 2016). Organosolv lignin is a highly attractive bio-product for manufacturing biopolymers, adhesives, and aromatic monomer production due to its high purity, low molecular weight, and reduced sulfur impurities. One of the economic limitations of the organosolv process is the cost of solvent recovery and corrosion-resistant reactors. The use of recyclable green solvents such as γ-valerolactone (GVL) is a recent breakthrough that has the potential to improve process sustainability (Hu et al., 2024).

2.3 NEW AND DEVELOPING METHODS FOR THE EXTRACTION AND RECOVERY OF LIGNIN

There are some advanced methods for increasing selectivity as well as maintaining the structural integrity of lignin, thus overcoming the limitations of conventional extraction methods. Some novel approaches for extracting lignin without structural modification and/or solubilization with the goal of directly or indirectly converting it into chemicals. Recent progress in this field is critically addressed in this section under the perspectives regarding the principle, characteristics, challenges, and industrial prospects.

2.3.1 Deep eutectic solvents (DES) and ionic liquids (ILs)

Deep eutectic solvents (DES) and ionic liquids (IL) have drawn interest as green alternative solvents for lignin recovery. DESs are generally a combination of a hydrogen-bond donor and an acceptor, e.g., choline chloride with urea or organic acids that combine to create an eutectic with a melting point less than that of its constituents. On the other hand, ILs are low-temperature liquid salts with low vapor pressure and excellent thermal stability. Studies have demonstrated that DESs can solubilize the lignin while maintaining cellulose integrity under moderate temperatures (80–130°C) and mild acidic or basic conditions (Verdía Barbará et al., 2025). The lignin-carbohydrate bond can also be successfully broken by ILs, particularly those based on imidazolium, which makes it easier to remove lignin and even fractionate it into low molecular weight phenolics. DESs offer a desirable and affordable alternative to ILs that are typically less expensive and toxic. However, DES-based techniques for degradation and scalability are not yet mature. Further studies have now aimed to create customized, task-specific DESs for biomass fractionation and lignin valorization (Singh et al., 2024).

2.3.2 Catalytic fractionation: reductive catalytic fractionation (RCF)

Oxidative and reductive catalytic fractionation (OCF and RCF) are two primary types of catalytic fractionation. The RCF process is revolutionary because it integrates lignin extraction and depolymerization in a single-step process. In RCF, biomass is reacted with hydrogen gas in the presence of a metal catalyst (e.g., Ru/C, Pd/C, or Ni/C) and a solvent (frequently methanol, ethanol, or water) at mild temperatures

(150–250°C) (Jindal et al., 2023). Cellulose and hemicellulose remain mostly intact, while lignin is depolymerized into stable aromatic monomers. High yields of valuable phenolic compounds with less condensation of RCF open new possibilities to produce bioacrylate polymers, resins, and biofuels. RCF holds significant promise for integrated biorefineries and comprehensive lignin valorization. However, key challenges, including catalyst deactivation through fouling and poisoning, solvent management, and process intensification, remain major bottlenecks to its broader implementation. Recent developments investigate heterogeneous catalyst patterns and flow-based continuous RCF systems to overcome these challenges (Renders et al., 2019). However, the OCF technique uses aqueous alkaline solvents or alkaline solvents that mimic RCF but produce distinct products, such as phenols with carboxylic acids, aldehydes, and related oxygenated functional groups instead of alkyl side chains. In comparison to RCF-derived products, aromatics obtained through OCF exhibit greater water solubility, an advantage for downstream processes such as microbial funneling, where mixed aromatic compounds are converted into distinct value-added products (Tarabanko et al., 2017).

Luo et al. (2021) reported an oxidative fractionation method in which O_2 in acetone as the solvent and a heterogeneous non-precious metal Co-N-C catalyst are used; the process yields 15 wt% phenolic compounds containing aldehydes (vanillin, syringaldehyde) and carboxylic acids (p-hydroxybenzoic acid, vanillic acid, syringic acid), correlating with the alkylated phenols produced by the current RCF technique. This process facilitates the preservation of cellulose, which remains insoluble under reaction conditions and is recovered as a solid. The resulting oxygenated aromatics possess favorable properties for applications in polymer synthesis and biological funneling into value-added products.

2.3.3 Steam explosion and hydrothermal pretreatment

Steam explosion is a pyrolytic pretreatment process in which biomass is exposed to high-pressure steam (160–240°C) and sudden release, causing breakdown of biomass constituents and partial lignin removal. Furthermore, hydrothermal treatments can be conducted without applying any chemical reagents, causing lignin and hemicellulose to solubilize by using sub-critical water at 170–230°C (Hoang et al., 2023). These operations enable the selective removal of lignin fragments and enhance the enzymatic digestibility of cellulose for its subsequent fermentation. Hydrothermal treatment and steam explosion offer the benefits of low chemical input and operational simplicity. However, the resulting distillation lignin is often highly heterogeneous and structurally condensed, limiting its suitability for direct use in high-value-added applications (Wang et al., 2024). Hybrid approaches combining steam explosion with mild chemical pretreatments, such as acid- or base-catalyzed methods, are increasingly being investigated to improve lignin quality.

2.3.4 Enzymatic and microbial extraction methods

Biological treatment strategies utilizing ligninolytic microorganisms or lignin-degrading enzymes (laccases, peroxidases) provide environmentally feasible and selective alternatives for lignin depolymerization. Catalytic oxidation of lignin bonds by enzymes is carried out at mild conditions, i.e., at room temperature and neutral pH, leaving the carbohydrate fractions undisturbed. Novel synthetic biology tools are being developed to design microbial strains that can break down lignin and directly convert it to the desired chemicals (Kumar et al., 2024). Such enzymatic and microbial processes are characterized by unmatched selectivity and extremely low environmental load. However, their slow kinetics, enzyme instability, and scale-up challenges make them unsuitable for industrial purposes at present. Research is increasingly focused on enhancing enzyme robustness, engineering microbial consortia, and integrating biological treatments into hybrid chemical-biological processing frameworks (Zhuo et al., 2021).

2.4 PARAMETERS INVOLVED IN LIGNIN EXTRACTION FROM LIGNOCELLULOSIC BIOMASS

The extraction of lignin from LCB is a multivariate process with several factors affecting the performance and quality of the extract. Many factors affect the success of lignin extraction from LCB, which needs to be adopted based on the application and operational scale. The nature of the biomass is one of the most important factors, which varies with composition and structure (wood, straw, agricultural residues) and ultimately influences the extraction process and yield (Lobato-Peralta et al., 2021). The purity and structural integrity of the extracted lignin are greatly influenced by the technique (chemical, enzymatic, organosolv, or other extraction methods). The type of solvent used directly affects lignin solubility as well as its ability to get separated and recovered after extraction. Other factors that affect the kinetics of the process are temperature, pH, and reaction time, and they favor lignin solubilization.

Further, pressure and humidity influence the stability and dissolution of lignin. The purity and recovery of extracted lignin are important, as they determine whether the lignin can be processed for biofuels, biopolymers, or high-value chemicals (Adebayo & Olanrewaju, 2025). The cost of raw materials, solvents, and energy consumption influence the economic viability of lignin extraction. The scale of operation further dictates the choice of process, equipment, and operational complexity. Additionally, lignin's molecular weight and functional group composition play a critical role in determining its performance in downstream applications. Finally, key considerations from both financial and environmental perspectives include the toxicity of by-products, which requires appropriate mitigation strategies, and the potential for solvent recycling. Figure 2.2 summarizes the important parameters that should be considered while extracting lignin from LCB.

Jindal et al. (2023) showed that well-defined metal catalysts reduce the recondensation by stabilizing the intermediates in addition to aiming at the bond cleavage in RCF. Further, precise control of residence time and pH is essential to optimize lignin removal while minimizing undesirable condensation reactions (Gui et al., 2025). Mass transfer is influenced by particle size; the finer the particles, the faster the extraction rates, but reducing particle size also requires higher energy (Adewuyi et al., 2020). Ultimately, an optimum moisture level is important, as low moisture content might exacerbate the reaction's intensity, while it may also impede the

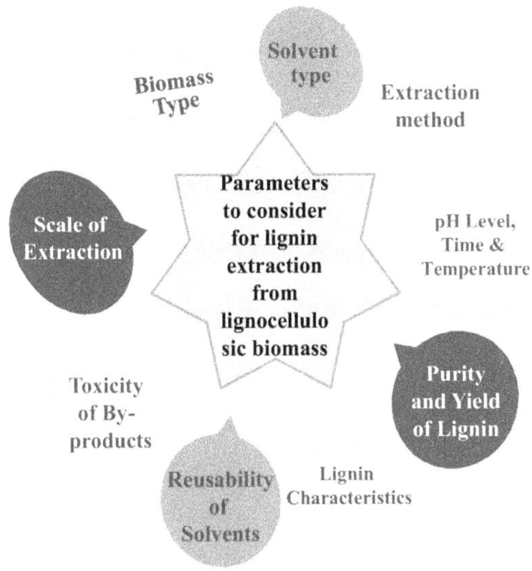

FIGURE 2.2 Critical factors affecting lignin extraction from lignocellulosic biomass

efficiency of the process, whereas a high concentration of water can decrease the reaction rates (Cousins, 1976).

2.5 CHALLENGES OF LIGNIN EXTRACTED FROM BIOMASS AND ITS SOLUTION

Lignin separation and utilization are crucial steps toward advancing economic development and sustainable innovation. While lignin holds immense potential across diverse applications, its widespread adoption remains constrained by several challenges. Developing thoughtful strategies with targeted solutions can significantly contribute to both the economic and environmental goals of lignin valorization. Key hurdles include its structural heterogeneity, low purity, resistance to degradation, and chemical inertness. Additionally, environmental concerns such as toxic waste from conventional kraft and sulfite pulp alongside high energy demands, limited economic competitiveness, scalability constraints, and variability in feedstock composition continue to impede progress, as outlined in Table 2.2.

2.6 ROADMAP FOR LIGNIN EXTRACTION FROM BIOMASS

A shift from lignin recovery to the lignin biorefinery is gaining attention due to sustainability targets as well as cost-effective solutions in technology (Chauhan et al., 2025). In the short-term perspective, green solvent chemistry based on ILs and DESs is breaking new ground in lignin recovery by providing high selectivity and recyclability under low-temperature operating conditions. Machine-learning-driven solvent design and molecular dynamics simulations are already allowing the prediction of lignin solubility and preservation of structure. Pilot-scale demonstrations with an

TABLE 2.1
Overview of methods for the extraction of lignin from biomass

Parameter	Industrial Methods (Kraft, Sulfite, Soda)	Chemical Methods (Acid, Alkaline, Organosolv)	Biological Methods (Fungi, Enzymes)	Other Methods (Ionic Liquids, DES, Supercritical CO_2)	Reference
Brief description	Uses harsh chemicals ($NaOH$, Na_2S, etc.) at high temperature and pressure to break down lignocellulose; common in the paper industry.	Employs acids or alkalis to solubilize lignin. Organosolv uses organic solvents under mild conditions for cleaner lignin.	Relies on lignin-degrading fungi or enzymes like laccases and peroxidases to selectively degrade lignin.	Uses green solvents like ionic liquids (ILs), deep eutectic solvents (DES), or supercritical fluids to extract lignin with minimal degradation.	Ragauskas et al. (2014)
Lignin purity	Moderate (sulfur contamination common in kraft)	Moderate to high (organosolv gives high purity)	High purity, low yield	High purity, near-native structure	Morais et al. (2020)
Extraction efficiency	High	Medium to high	Low	High	Beckham et al. (2016)
Environmental impact	High (toxic waste, sulfur emissions)	Moderate to high (chemical residues, neutralization required)	Low (biodegradable, non-toxic)	Low to moderate (depends on solvent recyclability and toxicity)	Figueiredo et al. (2018)
Process cost	Low (well-established infrastructure)	Medium (reagent and energy costs)	High (enzyme cost, slow process)	High (expensive solvents, recovery systems needed)	Zijlstra et al. (2020)
Scalability	High (used globally in pulp and paper)	Medium to high (organosolv scaling ongoing)	Low (still at lab/pilot scale)	Medium (some pilot studies but limited industrial adoption)	Zakzeski et al. (2010)
Energy requirement	High (e.g., kraft needs high T and P)	Moderate to high	Low	Moderate (pressurization or heating for some solvents)	Saadan et al. (2024)

(Continued)

TABLE 2.1
(Continued)

Parameter	Industrial Methods (Kraft, Sulfite, Soda)	Chemical Methods (Acid, Alkaline, Organosolv)	Biological Methods (Fungi, Enzymes)	Other Methods (Ionic Liquids, DES, Supercritical CO_2)	Reference
Time requirement	Fast (hours)	Moderate (a few hours to a day)	Slow (days to weeks)	Moderate	Gosselink et al. (2004)
Lignin structure preservation	Low (condensation and degradation common)	Organosolv: High acid/base: Moderate	High (enzymes act specifically)	High (non-destructive solvents preserve native structure)	Cai et al. (2023)
Recyclability of reagents	Low (chemicals often disposed)	Moderate (some reuse possible)	Good	High (ILs and DESs designed for recyclability)	Isci and Kaltschmitt (2022)
Industrial examples	Kraft and soda mills, sulfite pulping	Pretreatment in biorefineries, organosolv pilot plants	Fungal bioconversion in waste valorization	Green biorefineries, research in IL/DES valorization	Rodrigues et al. (2018)
Key advantages	Economical, proven at scale	Flexible, cleaner lignin possible	Eco-friendly, selective degradation	Green solvents, high-quality lignin, customizable	Alam et al. (2024)
Key disadvantages	Environmental harm, poor lignin quality	Harsh conditions, waste handling needed	Low yield, slow, expensive	Costly, limited commercial implementation	Bajwa et al. (2019)

TABLE 2.2
Key challenges and solutions of lignin extraction from biomass

Challenge	Solution	Reference
Structural heterogeneity: Makes processing unpredictable; varies with biomass source	Use milder, selective extraction methods and advanced characterization (NMR, FTIR).	Sethupathy et al. (2022)
Low purity: Impurities (e.g., carbohydrates) reduce quality for material or chemical use	Employ membrane filtration, solvent purification, and washing steps.	Mazar et al. (2018)
Recalcitrance (resistance to degradation): Lignin is difficult to depolymerize due to strong C-C and C-O bonds	Catalytic depolymerization (metal catalysts, hydrogenolysis).	Zakzeski et al. (2010)
Inconsistent yield: Varies depending on extraction conditions, affecting economic feasibility	Standardize pretreatment protocols; control temperature and solvent ratios.	Paone et al. (2020)
Chemical inertness: Native lignin lacks reactive sites, limiting direct use in products	Functionalize lignin (oxidation, sulfonation) to introduce reactivity.	Liu et al. (2020)
Environmental impact: Sulfite and kraft processes generate hazardous waste	Adopt green extraction (ionic liquids, deep eutectic solvents, enzymatic extraction).	Tu and Hallett (2018)
Energy-intensive processes: High temperature/pressure extraction methods consume a lot of energy	Develop low-energy processes like organosolv under mild conditions.	Zijlstra et al. (2020)
Economic non-competitiveness: Products from lignin are often more expensive than petroleum-based ones	Integrate lignin valorization into multi-product biorefineries.	Ragauskas et al. (2014)
Scalability issues: Most methods work only at lab scale; industrial scale is challenging	Focus on modular, continuous-flow reactors and scale-up pilot plants.	Beckham et al. (2016)
Variability of biomass feedstock: Different feedstocks (wood, grasses) produce different types of lignin	Develop tailored strategies for different biomass types and feedstock-specific protocols.	Smit et al. (2022)

integrated approach of lignin recovery with biomass pretreatment are considered as the proof-of-concept step for shifting lab-scale success in the direction of commercially viable processes (Ovejero-Pérez et al., 2024).

Demonstration and scale-up of lignin valorization is supposed to happen by 2025–2030 via integrated biorefinery systems based on cellulose, hemicellulose, and lignin streams. However, a number of hurdles exist to achieve this target; for example, solvent toxicity and recovery, corrosion of the reactor, and feedstock variability (Li et al., 2025). An industrial lignin-based economy is expected to rise by 2030–2035, where hardwood and softwood biorefineries have the potential to become zero-waste industries with solvent recycling and energy recovery. Further, hybrid systems associated with the integration of lignin extraction and its upgrading could achieve the direct valorization of lignin fragments into high-value chemicals, adhesives, aromatics, and polymer intermediates. Multi-site, commercial plant operations could begin to supply lignin as an international feedstock for renewable energy resources. Moreover, AI could assist with the execution of processes, maintenance, quality, and consistency (John et al., 2025).

Decentralized lignin extraction subunits close to the biomass generation site could have the potential to reduce transportation costs, carbon emissions, and global certification systems. Existing valorization routes of lignin to additives, pharmaceuticals, catalysts, 3D-printing resins, and carbon fibers could continue to develop into established industries competing with fossil-based aromatics. In the future, lignin could stop being considered as a waste and become a worthy renewable carbon resource pervading the biological infrastructure of a petroleum-based economy (Dixon et al., 2024).

2.7 FUTURE PERSPECTIVES

Future developments in technology, sustainable growth, and industrial integration will probably dictate the advancements in lignin extraction from LCB. Special consideration may be given to green extraction techniques, which use ionic liquids, DES, and supercritical CO_2, as eco-friendly alternatives to existing hazardous methods. The latter methods lower the ecological burden of lignin extraction and are consistent with the current focus on sustainable bioprocessing. Engineering of plants for modified lignin production for targeted applications like bioplastics, carbon fibers, or biomedical materials is another promising approach. Lignin can be commercially exploited and customized to meet certain industrial needs by altering its molecular weight, functional group, or solubility. It is strongly connected with the biorefinery concept, where lignin is no longer considered a waste but a raw material.

Further, catalytic and biological conversions are also receiving more attention due to their significant abilities to depolymerize lignin into monomers and other fine compounds using catalysts and microbial enzymes, respectively. In contrast to thermal degradation, catalytic and biological conversions are selective and are performed under milder conditions, which reduces the cost of the process and scale-up from lab to industry. Novel hybrid extraction methods, such as physico-chemical and physico-biological methods, are also being explored to increase yield and selectivity. These integrated methods can also eliminate the limitations of individual technologies involved and provide flexible and high-throughput lignin recovery for various types of bioproducts. Collectively, these future outlooks could lead to a closer relation between lignin valorization and sustainable bioeconomy, leaving a real impact on the shift from a fossil-based to a bio-based economy.

2.8 CONCLUSION

Lignin, as a kind of natural and underutilized constituent in LCB, has great potential for high-end applications such as bioplastics, biofuels, chemicals, carbon fiber adsorbents, and biomedical materials. Lignin removal is crucial for enhancing the sustainability of bio-based alternatives to oil-based products as well as improving the efficiency of integrated biorefineries. However, the structural heterogeneity, chemical recalcitrance, low purity, and variability of feedstocks are great obstacles to its extensive industrial application. These obstacles require the use of integrated extraction solutions (industrial/chemical, biological, or hybrid) that are adapted for applications and efficient in terms of both yield and purity, while having a low environmental impact. Novel advances in science indicate the significance of green extraction approaches, catalytic depolymerization, and process normalization in addressing these challenges. The future of lignin valorization is integrated with green extraction methods, effect-driven engineering-based design, and industrial-scale implementation. Standardized procedures and downstream process developments and commercialization strategies are required to promote lignin in the bioeconomy. Lignin has the potential to provide a substantial contribution to a sustainable, circular, and carbon-neutral future after the execution of advanced technological innovation in extraction and valorization processes.

ACKNOWLEDGMENTS

The authors gratefully acknowledge Guru Nanak Dev University, Amritsar, India, for providing research facilities. Meenu Hans is also grateful to DST-SERB for providing funds to carry out the necessary research work via grant no. EEQ/2023/000427. Suman Salhotra (Registration No. 2014J/A.1794) and Pragati Bhardwaj (Registration No. 2019.ZOO/A.25) are thankful to Guru Nanak Dev University, Amritsar, India, for the full-time PhD in the Department of Biotechnology.

REFERENCES

Abu-Omar, M. M., K. Barta, G. T. Beckham, et al. 2021. Guidelines for performing lignin-first biorefining. *Energy and Environmental Science* 14 (1): 262–92. https://doi.org/10.1039/D0EE02870C

Adebayo, A. J., A. Olanrewaju. 2025. A complete review of lignin's extraction, analysis, applications, and future outlook. *Chemistry Africa* 1–31. https://doi.org/10.1007/s42250-025-01323-y

Adewuyi, S. O., H. A. Ahmed, H. M. Ahmed. 2020. Methods of ore pretreatment for comminution energy reduction. *Minerals* 10 (5): 423. https://doi.org/10.3390/min10050423

Akpan, E. I. 2019. Chemistry and structure of lignin. In *Sustainable lignin for carbon fibers: Principles, techniques, and applications*, ed. E. I. Akpan and S. O. Adeosun, 1–50. Cham: Springer International Publishing. https://doi.org/10.1007/978-3-030-18792-7_1

Alam, M. M., A. Greco, Z. Rajabimashhadi, C. E. Corcione. 2024. Efficient and environmentally friendly techniques for extracting lignin from lignocellulose biomass and subsequent uses: A review. *Cleaner Materials* 13: 100253. https://doi.org/10.1016/j.clema.2024.100253

Ali, S., A. Rani, M. A. Dar, et al. 2024. Recent advances in characterization and valorization of lignin and its value-added products: Challenges and future perspectives. *Biomass* 4 (3): 947–77. https://doi.org/10.3390/biomass4030053

Bajwa, D. S., G. Pourhashem, A. H. Ullah, S. G. Bajwa. 2019. A concise review of current lignin production, applications, products and their environmental impact. *Industrial Crops and Products* 139: 111526. https://doi.org/10.1016/j.indcrop.2019.111526

Beckham, G. T., C. W. Johnson, E. M. Karp, D. Salvachúa, D. R. Vardon. 2016. Opportunities and challenges in biological lignin valorization. *Current Opinion in Biotechnology* 42: 40–53. https://doi.org/10.1016/j.copbio.2016.02.030

Brienza, F., D. Cannella, D. Montesdeoca, I. Cybulska, D. P. Debecker. 2024. A guide to lignin valorization in biorefineries: Traditional, recent, and forthcoming approaches to convert raw lignocellulose into valuable materials and chemicals. *RSC Sustainability* 2 (1): 37–90. https://doi.org/10.1039/D3SU00140G

Cai, C., C. Zhang, N. Li, et al. 2023. Changing the role of lignin in enzymatic hydrolysis for a sustainable and efficient sugar platform. *Renewable and Sustainable Energy Reviews* 183: 113445. https://doi.org/10.1016/j.rser.2023.113445

Capolupo, L., V. Faraco. 2016. Green methods of lignocellulose pretreatment for biorefinery development. *Applied Microbiology and Biotechnology* 100 (22): 9451–67. https://doi.org/10.1007/s00253-016-7884-y

Carvajal, J. C., Á. Gómez, C. A. Cardona. 2016. Comparison of lignin extraction processes: Economic and environmental assessment. *Bioresource Technology* 214: 468–76. https://doi.org/10.1016/j.biortech.2016.04.103

Chakar, F. S., A. J. Ragauskas. 2004. Review of current and future softwood kraft lignin process chemistry. *Industrial Crops and Products* 20 (2): 131–41. https://doi.org/10.1016/j.indcrop.2004.04.016

Chauhan, S., P. Solanki, C. Putatunda, et al. 2025. Recent advancements in biomass to bioenergy management and carbon capture through artificial intelligence integrated technologies to achieve carbon neutrality. *Sustainable Energy Technologies and Assessments* 73: 104123. https://doi.org/10.1016/j.seta.2024.104123

Cousins, W. J. 1976. Elastic modulus of lignin as related to moisture content. *Wood Science and Technology* 10 (1): 9–17. https://doi.org/10.1007/BF00376380

Dixon, R. A., A. Puente-Urbina, G. T. Beckham, Y. Román-Leshkov. 2024. Enabling lignin valorization through integrated advances in plant biology and biorefining. *Annual Review of Plant Biology* 75 (1): 239–63. https://doi.org/10.1146/annurev-arplant-062923-022602

Figueiredo, P., K. Lintinen, J. T. Hirvonen, M. A. Kostiainen, H. A. Santos. 2018. Properties and chemical modifications of lignin: Towards lignin-based nanomaterials for biomedical applications. *Progress in Materials Science* 93: 233–69. https://doi.org/10.1016/j.pmatsci.2017.12.001

Gosselink, R. J. A., A. Abächerli, H. Semke, et al. 2004. Analytical protocols for characterisation of sulphur-free lignin. *Industrial Crops and Products* 19 (3): 271–81. https://doi.org/10.1016/j.indcrop.2003.10.008

Gui, C., L. Wang, G. Liu, A. T. Ogunbiyi, W. Li. 2025. The catalytic valorization of lignin from biomass for the production of liquid fuels. *Energies* 18 (6): 1478. https://doi.org/10.3390/en18061478

Haykir, N. I., J. Viell. 2024. Challenges and opportunities for gamma-valerolactone in biomass pretreatment. *Industrial & Engineering Chemistry Research* 63 (28): 12251–64. https://doi.org/10.1021/acs.iecr.4c01277?urlappend=%3Fref%3DPDF&jav=VoR&rel=cite-as

Hoang, A. T., X. P. Nguyen, X. Q. Duong, et al. 2023. Steam explosion as sustainable biomass pretreatment technique for biofuel production: Characteristics and challenges. *Bioresource Technology* 385: 129398. https://doi.org/10.1016/j.biortech.2023.129398

Hu, X., Y. Xia, H. Wang, et al. 2024. Holistic investigation of structural evolution in corn stover lignin under pretreatment with varying γ-valerolactone concentrations. *Biomass and Bioenergy* 183: 107155. https://doi.org/10.1016/j.biombioe.2024.107155

Isci, A., M. Kaltschmitt, 2022. Recovery and recycling of deep eutectic solvents in biomass conversions: A review. *Biomass Conversion and Biorefinery* 12 (1): 197–26. https://doi.org/10.1007/s13399-021-01860-9

Jindal, M., P. Uniyal, B. Thallada. 2023. Reductive catalytic fractionation as a novel pretreatment/lignin-first approach for lignocellulosic biomass valorization: A review. *Bioresource Technology* 385: 129396. https://doi.org/10.1016/j.biortech.2023.129396

John, S. U., C. E. Onu, C. M. J. Ezechukwu, I. C. Nwokedi, C. N. Onyenanu. 2025. Multi-product biorefineries for biofuels and value-added products: Advances and future perspectives. *Academia Green Energy* 2 (1): 1–39. https://doi.org/10.20935/AcadEnergy7605

Kumar, V., P. Pallavi, S. K. Sen, S. Raut. 2024. Harnessing the potential of white rot fungi and ligninolytic enzymes for efficient textile dye degradation: A comprehensive review. *Water Environment Research* 96 (1): 10959. https://doi.org/10.1002/wer.10959

Kumari, D., G. S. Maurya, E. Tarif, Y. Jain. 2025. Sustainable biohythane production from lignocellulosic biomass and industrial by-products: A review of pretreatment technologies and economic feasibility. *Journal of Environmental Chemical Engineering* 13(6): 119936. https://doi.org/10.1016/j.jece.2025.119936

Kumari, D., R. Singh. 2018. Pretreatment of lignocellulosic wastes for biofuel production: A critical review. *Renewable Sustainable Energy Reviews* 90: 877–91. https://doi.org/10.1016/j.rser.2018.03.111

Lehto, J., R. Alén. 2015. Chemical pretreatments of wood chips prior to alkaline pulping: A review of pretreatment alternatives, chemical aspects of the resulting liquors, and pulping outcomes. *BioResources* 10 (4): 8604–56. https://doi.org/10.15376/biores.10.4.lehto

Li, X., J. Yang, W. He, et al. 2025. A review: Using ionic liquids for lignin extraction from lignocellulose and high-value utilization. *Molecules* 30 (12): 2514. https://doi.org/10.3390/molecules30122514

Liu, R., L. Dai, C. Xu, K. Wang, C. Zheng, C. Si. 2020. Lignin based micro and nano materials and their composites in biomedical applications. *ChemSusChem* 13 (17): 4266–83. https://doi.org/10.1002/cssc.202000783

Lobato-Peralta, D. R., E. Duque-Brito, H. I. Villafan-Vidales, et al. 2021. A review on trends in lignin extraction and valorization of lignocellulosic biomass for energy applications. *Journal of Cleaner Production* 293: 126123. https://doi.org/10.1016/j.jclepro.2021.126123

Luo, H., E. P. Weeda, M. Alherech, et al. 2021. Oxidative catalytic fractionation of lignocellulosic biomass under non-alkaline conditions. *Journal of the American Chemical Society* 143 (37): 15462–70. https://doi.org/10.1021/jacs.1c08635

Mazar, A., N. Jemaa, W. W. Al Dajani, M. Marinova, M. Perrier. 2018. Influence of membrane filtration on extraction and characteristics of lignin from a kraft dissolving pulp mill pre-hydrolysate. *Industrial Crops and Products* 124: 726–34. https://doi.org/10.1016/j.indcrop.2018.08.046

Morais, E. S., A. M. D. C. Lopes, M. G. Freire, C. S. Freire, J. A. Coutinho, A. J. Silvestre. 2020. Use of ionic liquids and deep eutectic solvents in polysaccharides dissolution and extraction processes towards sustainable biomass valorization. *Molecules* 25 (16): 3652. https://doi.org/10.3390/molecules25163652

Norgren, M., H. Edlund. 2014. Lignin: Recent advances and emerging applications. *Current Opinion in Colloid & Interface Science* 19 (5): 409–16. https://doi.org/10.1016/j.cocis.2014.08.004

Ovejero-Pérez, A., P. Y. Nakasu, C. Hopson, J. M. Costa, J. P. Hallett. 2024. Challenges and opportunities on the utilisation of ionic liquid for biomass pretreatment and valorisation. *npj Materials Sustainability* 2 (1): 7. https://doi.org/10.1038/s44296-024-00015-x

Paone, E., T. Tabanelli, F. Mauriello. 2020. The rise of lignin biorefinery. *Current Opinion in Green and Sustainable Chemistry* 24: 1–6. https://doi.org/10.1016/j.cogsc.2019.11.004

Ragauskas, A. J., G. T. Beckham, M. J. Biddy, et al. 2014. Lignin valorization: Improving lignin processing in the biorefinery. *Science* 344 (6185): 1246843. https://doi.org/10.1126/science.1246843

Rathour, R. K., M. Behl, D. Sakhuja, et al. 2025. Recent updates in biomass pretreatment techniques for a sustainable biofuel industry: A comprehensive review. *Waste and Biomass Valorization* 1–26. https://doi.org/10.1007/s12649-025-03016-6

Renders, T., G. Van den Bossche, T. Vangeel, K. Van Aelst, B. Sels. 2019. Reductive catalytic fractionation: State of the art of the lignin-first biorefinery. *Current Opinion in Biotechnology* 56: 193–01. https://doi.org/10.1016/j.copbio.2018.12.005

Rodrigues, A. E., P. C. D. O. R. Pinto, M. F. Barreiro, C. A. Esteves da Costa, M. I. Ferreira da Mota, I. Fernandes. 2018. Chemical pulp mills as biorefineries. In An integrated approach for added-value products from lignocellulosic biorefineries: Vanillin, syringaldehyde, polyphenols and polyurethane, ed. A. E. Rodrigues, P. C. D. O. R. Pinto, M. F. Barreiro, C. A. Esteves da Costa, M. I. Ferreira da Mota, and I. Fernandes, 1–51. Cham: Springer International Publishing. https://doi.org/10.1007/978-3-319-99313-3_1

Saadan, R., C. H. Alaoui, A. Ihammi, M. Chigr, A. Fatimi. 2024. A brief overview of lignin extraction and isolation processes: From lignocellulosic biomass to added-value biomaterials. *Environmental and Earth Sciences Proceedings* 31 (1): 3. https://doi.org/10.3390/eesp2024031003

Sethupathy, S., G. M. Morales, L. Gao, et al. 2022. Lignin valorization: Status, challenges and opportunities. *Bioresource Technology* 347: 126696. https://doi.org/10.1016/j.biortech.2022.126696

Singh, K., S. Mehra, A. Kumar. 2024. Recent advances in catalytic conversion of lignin to value-added chemicals using ionic liquids and deep eutectic solvents: A critical review. *Green Chemistry* 26(3): 1062–91. https://doi.org/10.1039/D3GC03439A

Smit, A. T., A. van Zomeren, K. Dussan, et al. 2022. Biomass pre-extraction as a versatile strategy to improve biorefinery feedstock flexibility, sugar yields, and lignin purity. *ACS Sustainable Chemistry and Engineering* 10 (18): 6012–22. https://doi.org/10.1021/acssuschemeng.2c00838

Tarabanko, V. E., K. L. Kaygorodov, E. A. Skiba, et al. 2017. Processing pine wood into vanillin and glucose by sequential catalytic oxidation and enzymatic hydrolysis. *Journal of Wood Chemistry and Technology* 37 (1): 43–51. https://doi.org/10.1080/02773813.2016.1235583

Tu, W. C., J. P. Hallett. 2018. Ionic liquids for biomass processing. In *Sustainable catalysis in ionic liquids*, ed. P. Lozano, 231–64. Boca Raton: CRC Press. https://www.taylorfrancis.com/chapters/edit/10.1201/9781315148694-10/ionic-liquids-biomass-processing-wei-chien-tu-jason-hallett

Verdía Barbará, P., H. Choudhary, P. S. Nakasu, et al. 2025. Recent advances in the use of ionic liquids and deep eutectic solvents for lignocellulosic biorefineries and biobased chemical and material production. *Chemical Reviews* 125 (12): 5461–83. https://doi.org/10.1021/acs.chemrev.4c00754

Wang, L., G. Li, X. Chen, Y. Yang, et al. 2024. Extraction strategies for lignin, cellulose, and hemicellulose to obtain valuable products from biomass. *Advanced Composites and Hybrid Materials* 7 (6): 219. https://doi.org/10.1007/s42114-024-01009-y

Zakzeski, J., P. C. Bruijnincx, A. L. Jongerius, B. M. Weckhuysen. 2010. The catalytic valorization of lignin for the production of renewable chemicals. *Chemical Reviews* 110 (6): 3552–99. https://doi.org/10.1021/cr900354u

Zhuo, R., F. Fan. 2021. A comprehensive insight into the application of white rot fungi and their lignocellulolytic enzymes in the removal of organic pollutants. *Science of the Total Environment* 778: 146132. https://doi.org/10.1016/j.scitotenv.2021.146132

Zijlstra, D. S., C. W. Lahive, C. A. Analbers, et al. 2020. Mild organosolv lignin extraction with alcohols: The importance of benzylic alkoxylation. *ACS Sustainable Chemistry and Engineering* 8 (13): 5119–31. https://doi.org/10.1021/acssuschemeng.9b07222

3 Lignin structure and analytics

Amisha Patel, Mateo Bello Villarino, Payam Ghiaci, Mattias Hedenström, and Leif J. Jönsson

3.1 INTRODUCTION

3.1.1 Occurrence and biological function of lignin

Lignin is a polymeric, aromatic substance consisting of phenylpropane units, which are interconnected by a variety of carbon-carbon and ether bonds, of which the β-O-4 aryl ether bond is the most common (Ralph et al., 2004, 2019). As schematically shown in Figure 3.1, the carbon atoms in the propyl side chain are referred to as α, β, and γ (with α referring to the benzylic carbon atom), whereas the carbon atoms in the aromatic ring are referred to using numerals. Like other phenylpropanoids, lignin is a biosynthetic derivative of aromatic amino acids, primarily phenylalanine (Kumari et al., 2023). As lignin is by definition a polymer, monomeric, dimeric, and oligomeric phenylpropanoids or other non-polymeric lignin-like aromatic substances do not qualify as lignin, although fragmentation of lignin may result in the formation of such substances. Similarly, if a substance is not predominantly a phenylpropanoid, it does not qualify as lignin. Lignin is a polyphenolic substance, although it is noteworthy that the majority of phenylpropane units in lignin are non-phenolic rather than phenolic. In contrast, the three main monolignols (i.e., *p*-coumaryl alcohol, coniferyl alcohol, and sinapyl alcohol) (Ralph et al., 2004, 2019; Kumari et al., 2023) that after oxidation serve as building blocks for lignin are all phenolic substances, and another difference is that they are phenylpropenes rather than phenylpropanes. Many terrestrial environments are dominated by vascular plants (tracheophytes), which include plants such as ferns, fern allies, gymnosperms [such as conifers (a.k.a. softwoods), ginkgo, cycads, and gnetophytes], and angiosperms [flowering plants, including magnoliids, eudicots (which include, for example, the majority of broad-leaved trees, a.k.a. hardwoods), and monocots (such as grasses, which include sugar cane, bamboos, reeds, and palms)] (Taiz et al., 2015). Owing to their lignified vascular tissues transporting water and mineral salts upwards from the roots and photosynthate downwards from leaves and needles, vascular plants may reach heights of more than one hundred meters. In contrast, non-vascular terrestrial plants (bryophytes such as mosses, hornworts, and liverworts) and most algae typically do not contain lignified tissues, although the occurrence of very small amounts of lignin in the red alga

 DOI: 10.1201/9781003653189-3

Calliarthron cheilosporioides has been reported (Martone et al., 2009). The hydrophobic properties of lignin result in improved resistance to water, and lignification also results in increased strength and improved resistance to herbivores and microorganisms. Lignin is especially prevalent in secondary cell walls (SCWs) and is tightly associated with the main polysaccharides of lignocellulosic materials, i.e., cellulose and hemicelluloses. This is substantiated by the existence of lignin-carbohydrate complexes (LCCs) (Yuan et al., 2011), in which lignin forms covalent bonds (such as benzyl ether bonds, phenyl glycoside bonds, and γ-ester bonds) with carbohydrates, primarily hemicelluloses. The tight association with other cell wall constituents makes it difficult to extract lignin in its native form, and lignin preparations are therefore more or less modified compared to native lignin in plants.

3.1.2 Importance of studying lignin

The study of lignin is important across several scientific and technical areas, including plant biology, biochemistry, environmental science, chemical engineering, and bioengineering. Lignin, which is already widely accessible as a by-product of chemical pulping, is increasingly recognized as a valuable renewable feedstock to produce biofuels, bioplastics, and aromatic chemicals. Furthermore, lignin is utilized in (nano) composites, nanoparticles, controlled-release fertilizers in agriculture, and electrochemical applications (Lu et al., 2022). To create practical uses for lignin, precise analytical tools and techniques are needed to evaluate both qualitative and quantitative aspects of the material, such as the structure or the proportion of lignin in the cellular makeup of a given feedstock. Over the past decade, diverse strategies such as advanced nuclear magnetic resonance (NMR) spectroscopy, vibrational spectroscopy, linkage analysis, and refined high-throughput methods have been used to study lignin (Lupoi et al., 2015).

3.2 LIGNIN STRUCTURE

3.2.1 Biosynthesis of lignin

The shikimate pathway can be considered the starting point of lignin biosynthesis. The convergence of phosphoenolpyruvate from glycolysis and erythrose 4-phosphate from the pentose phosphate pathway results in the formation of chorismate through a seven-reaction process, which in plants is catalyzed by six enzymes (Shende et al., 2024). Chorismate is used to generate aromatic amino acids in a branched pathway that determines its ultimate conversion into either L-tryptophan or L-tyrosine and L-phenylalanine (Yokoyama, 2024). The cytosolic phenylpropanoid pathway involves a series of enzymatic reactions, which, starting from L-phenylalanine, results in its deamination by an ammonia lyase into cinnamic acid, which is in turn converted into *p*-coumaric acid by a hydroxylase (de Vries et al., 2021). However, grasses can deaminate L-tyrosine directly into *p*-coumaric acid (Martarello et al., 2023). From *p*-coumaric acid, successive hydroxylations and/or methoxylations of the aromatic ring, together with reduction of the side chain, ultimately lead to the generation of cinnamyl alcohols, also known as monolignols. Monolignols are usually divided into traditional (*p*-coumaryl alcohol, coniferyl alcohol, and sinapyl alcohol) and nontraditional. These traditional monolignols give rise to lignin units H (*p*-**h**ydroxyphenyl), G (**g**uaiacyl), and S (**s**yringyl), respectively (Figure 3.1). Nontraditional monolignols can also be naturally incorporated into lignin, such as C-lignin from caffeyl alcohol in seed coats (Chen et al., 2012) and 5HG (5-hydroxyguaiacyl) unit lignin in some cacti species (Chen et al., 2013). Nontraditional components can be engineered into lignin, but often in a minor fraction. Oxidative polymerization of monolignols to initiate and expand lignin is catalyzed by two enzymes: laccase (EC 1.10.3.2), which uses molecular oxygen to catalyze oxidation reactions, and peroxidase (EC 1.11.1.7), which uses hydrogen peroxide to catalyze oxidation reactions. These have been found to co-localize during plant development (Hoffmann et al., 2020), suggesting a conjoined action in polymerization. It is likely that different sets of laccases and peroxidases are the primary drivers of lignification depending on cell

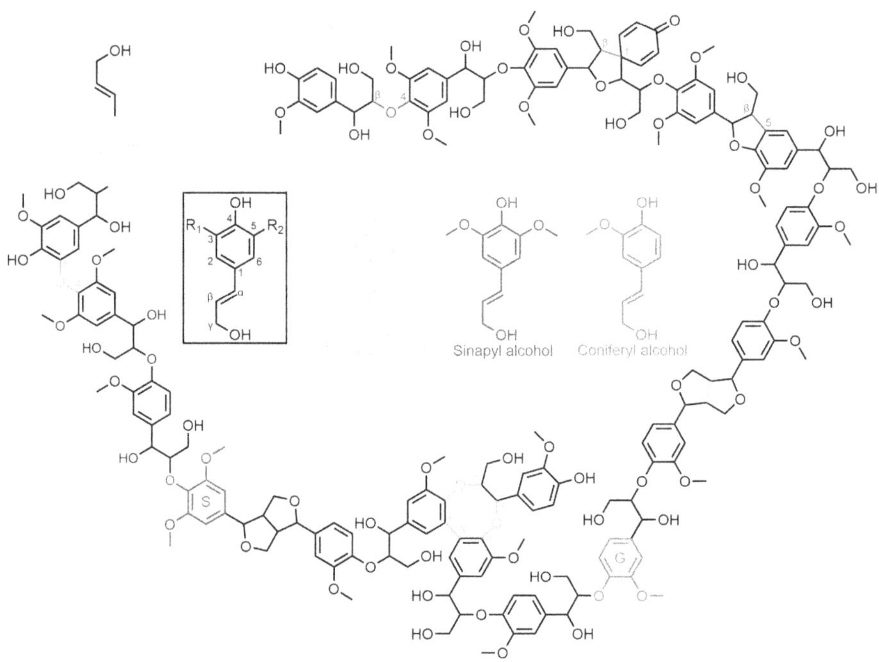

FIGURE 3.1 Schematic representation of lignin, highlighting interunit linkages and subunits that may not be present in the depicted proportions in nature. The scaffold structure of a monolignol and its consensus numbering system is shown inside a box: R_1 and R_2 can be either H atoms, O-CH_3, or a combination of both, depending on the monolignol. The traditional lignin subunits and their corresponding monolignols are shown in blue (sinapyl alcohol; S/syringyl unit), red (coniferyl alcohol; G/guaiacyl unit), and yellow (*p*-coumaryl alcohol; H/*p*-hydroxyphenyl unit). R_1 can also be an OH group (R_2=H) in caffeyl alcohol, which gives rise to C units; if R_1=O-CH_3 and R_2=OH, then the monolignol is 5-hydroxyconiferyl alcohol, which results in 5HG units (neither is depicted). Common linkages in between subunits are also highlighted: β-O-4 (aryl ether), brown; β-β (resinol), cyan; 4-O-5 (diaryl ether), orange; β-1 (spirodienone), magenta; β-5 (phenylcoumaran), purple; 5-5/β-O-4/α-O-4 (dibenzodioxocin), light green.

and tissue type (Choi et al., 2023). Exactly how monolignol synthesis and lignin polymerization are linked in an all-encompassing model to describe lignin biosynthesis starting from its monomers is still an area of active research. The current hypothesis involves monolignol synthesis in the cytoplasm through the phenylpropanoid pathway, followed by the exiting of these monolignols outwards through the plasma membrane into the apoplast and ultimately their oxidative polymerization into the lignin macromolecule taking place in the SCW. Major foci of study include how monolignols travel to the apoplast (whether this happens through passive diffusion or if specialized mechanisms can actively transport them out of the cytoplasm) and whether the initial dimerization occurs in the cytoplasm before cytosolic exit or once monolignols are in the apoplast (Dixon & Barros, 2019). The ending of polymerization has not been deciphered either, and it is unclear what stops the reaction. Although the metabolic pathways in which lignin polymerization happens have not been completely described, no evidence has been shown for strict control of lignin primary structure through an organized scaffolding process. Polymer structure is instead primarily defined by monolignol availability and the metabolic regulation of the pathways that lead to their synthesis, as well as plant capabilities to generate radicals and the cellular and environmental conditions of the SCW in connection with combinatorial radical coupling resulting in the formation of lignin (Ralph et al., 2004). Lignin deposition entails an irreversible sinking of carbon from the plant's perspective, and although the oxidative polymerization process can appear random owing to its great variability, it is logically tightly regulated (Cesarino, 2019). As paradoxical as it may seem, the plant has control over which structural components become part of the polymer and the proportion in which they do so, but due to the combinatorial nature of the radical coupling, the precise chemical structure of the resulting lignin molecule cannot be predicted.

3.2.2 Natural lignins

Lignin composition varies between plant species, tissues, and developmental stages. However, certain generalizations can be made for a basic understanding of lignin profiles in different plant types. Softwood lignins predominantly contain G units, with only minor amounts of H units and no or very small amounts of S units, whereas lignins from hardwoods contain both G and S units (monocots showing a greater fraction of S units and dicots having varying S/G ratios), with no H units (Boerjan et al., 2003; Mansfield et al., 2012). Although also monocots, grasses exhibit a more heterogeneous lignin composition, as they incorporate significant fractions of all three units (Ralph et al., 2004, 2019).

Interunit linkages greatly influence lignin reactivity and functionality. In natural lignins, the β-O-4 aryl ether bond is the most prevalent linkage, particularly in hardwoods. In contrast, softwood lignins contain more condensed structures, with lower proportions of β-O-4 and more β-5 (phenylcoumaran) and 5-5 (biphenyl) carbon-carbon bonds, which increases their resistance to degradation (Vanholme et al., 2010). The interunit substructures also include resinol (β-β), dibenzodioxocin, spirodienone, and 4-O-5 diaryl ether linkages (Figure 3.1). The relative abundance of phenolic hydroxyl groups also varies, with softwood lignins typically possessing a higher degree of free phenolic hydroxyl groups compared to hardwoods, which exhibit a lower fraction (Lai & Guo, 1991). Due to their significant fraction of H units and the presence of ester-linked ferulic acid, grasses present more variable phenolic hydroxyl content (Lapierre, 1993).

3.2.3 Engineered lignins

Monolignols that can be substrates for oxidation and further combinatorial polymerization are synthesized through metabolic routes under strict enzymatic control. Biotechnological techniques have been used to modify or disrupt the enzymes involved in different steps of lignin biosynthesis to learn more about how the process works through the effects of said changes. The effect of plant mutants on lignin composition was discussed with relevant examples by Martarello et al. (2023) and by De Meester et al. (2022) for heterologous expression of enzymes involved in the phenylpropanoid pathway. Genetic engineering also offers the possibility of generating plants with synthetically modified lignin for specific purposes. The most prominent example is zip lignin, which incorporates substructures that are less recalcitrant to degradation, such as those more prone to establish ester linkages, making them easier to break down during biorefining processes (Ralph et al., 2019).

3.3 LIGNIN PREPARATIONS

3.3.1 Industrial lignins

Industrial or technical lignins are by-products produced during chemical pulping and some types of industrial biorefining of lignocellulosic biomass, such as biochemical conversion. Industrial lignins are more chemically and structurally diverse than native lignins, containing many minor structural components (Berlin & Balakshin, 2014). Industrial lignins can be classified into different types based on traditional pulp and paper processes and emerging biomass biorefining methods, as listed in Table 3.1. It is noteworthy in this context that among the traditional pulping processes, it is in chemical pulping (kraft, sulfite, soda, and organosolv processes) that lignin is targeted and fractionated, whereas in the mechanical pulping [groundwood and thermo-mechanical pulp (TMP) processes] and in the chemithermomechanical pulp (CTMP) process, lignin remains as a main constituent of the pulp (Mboowa, 2024). Similarly, among different alternative and emerging biorefinery approaches (Liu et al., 2025), industrial lignins are obtained in biochemical conversion and potentially also in so-called lignin-first processes, which are designed to target lignin, as the conventional chemical pulping processes are.

3.3.1.1 Kraft lignin

The kraft process, developed in 1879, is the dominant method for lignin extraction from lignocellulosic biomass and the main source of industrial lignin (Mboowa, 2024). It is an alkaline process based on cooking liquor with sodium hydroxide and hydrosulfide ions. Two key reactions take place: (1) depolymerization, where ether linkages are cleaved to yield alkali-soluble lignin fragments in black liquor, and (2) condensation, where reactive intermediates form high-molecular-weight structures, potentially leading to precipitation (Melro et al., 2021). In the LignoBoost

TABLE 3.1
Comparison of different industrial (technical) lignins

Parameters	Kraft Lignin	Soda Lignin	Lignosulfonates	Organosolv Lignin	Hydrolysis Lignin
Ash, %	0.5–3.0	0.7–2.3	4.0–8.0	1.7	1.0–3.0
Moisture, %	3.0–6.0	2.5–5.0	5.8	7.5	4.0–9.0
Carbohydrates, %	1.0–2.3	1.5–3.0	NA	1.0–3.0	10.0–22.4
Nitrogen, %	0.05	0.2–1.0	0.02	0–0.3	0.5–1.4
Sulfur, %	1.0–3.0	0	3.5–8.0	0	0–1.0
Acid-soluble lignin, %	1.0–4.9	1.0–11	NA	1.9	2.9
Molecular weight, Da	1,000–3,000, up to 15,000	1,000–3,000 up to 15,000	10,000–40,000, up to 100,000	500–5,000	5,000–10,000
Polydispersity	2.5–3.5	2.5–3.5	4.2–7.0	1.5	4.0–11.0
Commercial technology	LignoBoost technology for upgrading	–	Borregaard Domsjö Fabriker, Tembec, LignoTech, MeadWestvaco	Alcell, Organocell, Milox, Biolignin	–

Source: Liao et al. (2020) and Berlin and Balakshin (2014).

process, the kraft lignin in black liquor is purified using precipitation with carbon dioxide followed by filtration (Zhu & Theliander, 2015).

3.3.1.2 Lignosulfonates

Lignosulfonates are co-products of the sulfite process, which was first developed in 1866 (Mboowa, 2024). It is based on bisulfite ions, and the pH is typically acidic or close to neutral. The base might be calcium, ammonium, magnesium, or sodium. In contrast to most other industrial lignins, lignosulfonates are soluble in water even at neutral or acidic pH due to the presence of sulfonic acid groups, which are deprotonated over a large pH interval.

3.3.1.3 Organosolv lignin

The organosolv process uses aqueous organic solvents (commonly ethanol, butanol, formic acid, or acetic acid) as the cooking medium to solubilize and extract lignin. This process yields relatively pure lignin with higher homogeneity than lignosulfonates or alkali lignins. Despite the ease of solvent recovery due to low boiling points, its high recovery cost, equipment corrosion, and use of costly solvents at elevated temperatures and pressures make it expensive. Organosolv lignins are hydrophobic and poorly soluble in water (Liao et al., 2020).

3.3.1.4 Soda lignin

The soda process was developed already in the 1850s (Mboowa, 2024). Soda lignin is derived from soda or soda-anthraquinone pulping, which is nowadays typically applied to low-lignin agricultural residues like bagasse, straw, kenaf, and hemp (Melro et al., 2021). It is an alkaline process that uses sodium hydroxide under high pressure (140–170°C) without Na_2S, resulting in a milder, sulfur-free delignification. Compared to kraft pulping, it yields purer lignin but with less efficient depolymerization. However, lignin recovery is challenging due to its high carboxylic acid content (Liao et al., 2020).

3.3.1.5 Hydrolysis lignin

Hydrolysis lignins are obtained from processes in which acids and/or enzymes are used to hydrolyze and solubilize hemicelluloses and cellulose. This leaves a lignin-rich solid fraction known as hydrolysis lignin. The feedstock might be woody or non-woody biomass (Melro et al., 2021). As crystalline cellulose is sometimes difficult to hydrolyze completely, it is often present as a contaminant in hydrolysis lignin. If high temperatures and acidic conditions are used, hydrolysis lignin might also be contaminated with pseudo-lignin (Shinde et al., 2018; Chen et al., 2024). As the name implies, pseudo-lignin is not truly lignin. Instead, it is formed from carbohydrates under harsh conditions. It only has a superficial resemblance to lignin with respect to aromaticity and solubility properties in acid. Pseudo-lignin might appear as droplets or microspheres on biomass surfaces. It is misidentified as Klason lignin in compositional analysis, as it is not solubilized by the sulfuric acid that is used to degrade hemicelluloses and cellulose to sugars.

3.3.2 Analytical and Synthetic Lignins

Besides the industrial lignins discussed in Section 3.3.1, there are also other lignin preparations, which have so far primarily been studied in laboratory settings. These include milled wood lignin (MWL), also known as Björkman lignin, which is typically prepared on a laboratory scale and used for analytical studies. MWL has provided important insights into the structure and properties of lignin from woody plants. Among synthetic lignins, dehydrogenation polymer (DHP) lignin has provided important information on the biosynthesis of lignin. Another form of synthetic lignin is lignin nanoparticles (LNPs), which are artificially produced nanoparticles that can be prepared from various lignin preparations. MWL is a preparation of native lignin made through extensive milling and extraction with dioxane, first described in 1956 (Björkman, 1956). Both softwood and hardwood can be used to obtain MWL, and although it has been suggested that the resulting structure is that of a non-branched, linear oligomer (Crestini et al., 2011) rather than the natural lignin polymer, this is still a contested topic of debate (Balakshin et al., 2020). DHP lignin is polymerized *in vitro* from monolignols through the action of laccases or peroxidases (Glasser et al., 1999). In the 1950s, Karl Freudenberg used newly available coniferyl alcohol and mushroom laccase or horseradish peroxidase as catalysts to synthesize polymerization products. More recently, it has been assessed how reaction conditions, such as pH, enzyme loading, reaction scale, and the presence of organic solvent, affect the resulting polymer (Tokunaga & Watanabe, 2023). Given the great variability and the difficulty of isolating natural lignins, DHP lignin has emerged as a useful model substance in studies of lignin polymerization and lignin reactivity. The amphiphilic nature of lignin molecules can be utilized for generating spherical assemblies, LNPs, with hydrophobic aromatic groups on the inside of the nanoparticle and hydrophilic hydroxyl and potential carboxylic groups pointing outwards. First synthesized in 2012 using acidic precipitation, the average LNP size ranged from 100 nm to 2 mm (Frangville et al., 2012), presenting irregular shapes, although sizes as low as 15–20 nm have been reported for spherical LNPs (ur Rahman et al., 2018). The poor solubility of lignin in water has been a limiting factor in industrial and commercial contexts. LNPs overcome this issue by facilitating aqueous dispersion, unlocking their potential capabilities as molecule encapsulators, bioremediators, and antimicrobials, among others (Schneider et al., 2021).

3.4 ANALYTICAL TECHNIQUES FOR CHARACTERIZATION OF LIGNIN

3.4.1 Lignin content

Various methods exist to quantify lignin, with factors like speed, sample size, and ecotoxic by-products being crucial, especially for high-throughput phenotyping (Table 3.2).

3.4.1.1 Acid-insoluble and acid-soluble lignin

The Klason method, developed in the early 1900s by Peter Klason and others, is a gravimetric procedure for quantification of lignin. It is based on solubilization of cell wall polysaccharides through hydrolysis using sulfuric acid. The hydrolysis is today typically performed in two steps, first with 72% sulfuric acid at low temperature (such as 30°C), and then with 4% sulfuric acid at elevated temperature (such as 121°C) to complete the hydrolysis. The insoluble lignin residue thus obtained is filtered, dried, and weighed (Bozek et al., 2025). The Klason method requires approximately one gram of sample, specialized glassware, and up to two days for completion, which limits its usefulness in high-throughput applications (Hatfield & Fukushima, 2005). Several protocols are available, such as KCL no. 115b:82, TAPPI T222, and NREL TP-510-42618. Structural changes in lignin may occur during acid hydrolysis (Melro et al., 2021). Limitations include that hydrolysis conditions might be necessary to adjust depending on the material being analyzed, as too mild or too harsh hydrolysis conditions will result in erroneous results, and that the potential presence of other substances than lignin that are not solubilized by acid (such as pseudo-lignin) complicates analysis. Acid-soluble lignin (ASL) can be quantified by measuring the UV absorbance of the hydrolysate from the second stage of the Klason procedure, providing total lignin content (insoluble + soluble). A small fraction of the lignin is typically solubilized in dilute acid (Melro et al., 2021). Limitations include that different lignin degradation products have different molar absorptivity (ε) and that UV absorbance at wavelengths commonly used to determine ASL is in no way limited to lignin degradation products. For instance, carbohydrate degradation products, such as furfural and 5-hydroxymethylfurfural, also absorb strongly in UV. Thus, caution is necessary when interpreting ASL values (Hatfield & Fukushima, 2005).

TABLE 3.2
Comparison of methods used to quantify lignin content*

Parameters	Klason Method	Acid-Soluble Lignin	Acetyl Bromide	CASA Method	Pyrolysis-GC/MS	Kappa Number
Process time	Long (2 days)	Medium (~ 30 min plus previous Klason steps)	Medium (1 day)	Quick (half day)	Quick (15 min-1 h)	Quick (1–2 h)
Sample size	High (>1 g)	Low (depends on Klason method)	Low (5–25 mg)	Low (30 mg)	Low (in mg)	High (>1 g)
Accuracy	High	Moderate	High	High	High	Moderate
Ecotoxic by-products	Yes (concentrated acids)	Yes (from Klason hydrolysis)	Yes (acetyl bromide, corrosive)	Yes (acids)	Minimal	Yes ($KMnO_4$, acidic waste)
Instrumentation	Basic lab glassware, oven	UV-Vis spectro-photometer	UV-Vis spectro-photometer	Spectro-photometer, fume hood	Pyrolyzer and GC/MS	Titration set up
High-throughput suitability	Very low	Medium to high (limited mainly by the Klason step)	Medium	Medium to high	Medium to high if advanced instrumentation is available	Low
Cost	Inexpensive	Inexpensive	Medium	Medium	Low compared to other mass spectrometry	Inexpensive
Limitations	Time-consuming; overestimation due to co-precipitation; underestimation of syringyl-rich lignin lost in acid	Molar absorptivity variability; spectral interference; overestimation risk	Sensitive to polyphenols or cell residues; require well-characterized lignin standard	Tested on limited wood samples	Limited to detecting mostly monomers and small dimers	Influenced by pulp composition

Source: Bozek et al. (2025), Melro et al. (2021), Dence (1992), Hatfield and Fukushima (2005), Letourneau and Volmer (2023) and Burkhardt (2018).

3.4.1.2 Acetyl bromide

Originally developed by Johnson et al. (1961), the acetyl bromide soluble lignin (ABSL) method is a spectrophotometric technique in which lignin is solubilized by bromination at benzylic positions and acetylation of hydroxyl groups. The reaction is terminated with additional reagents, and the absorbance is measured at 280 nm (Bozek et al., 2025). It has been widely used and modified for nonwoody plant samples. The method is rapid, simple, does not require correction for ASL, is suitable for small samples (5–25 mg), provides accurate absorbance values, and exhibits minimal interference from non-lignin compounds. A key limitation is the need for a well-characterized lignin standard for accurate calibration (Hatfield & Fukushima, 2005).

3.4.1.3 CASA method

The recently developed cysteine-assisted sulfuric acid (CASA) method by Lu et al. (2021) is a spectrophotometric technique that enables faster reaction times by enhancing lignin dissolution in sulfuric acid through cysteine. The molar absorptivity (ε) has been estimated for woody samples, which typically contain higher levels of lignin as compared to fibrous annual plants. Further validation is required for low-lignin biomass, where low H monomer content may introduce bias. Bozek et al. (2025) validated CASA for fibrous plants and low-lignin samples. However, ε varies with monomer ratio, ranging from < 10 to > 20 L/g/cm. As compared to ABSL, CASA yields more precise results, with a relative difference of −8% to +9% from Klason lignin values. Prior to CASA quantification, monomer ratios can potentially be assessed using nondestructive methods, such as quantitative NMR, to select an appropriate ε value.

3.4.1.4 Pyrolysis-GC/MS

Pyrolysis-gas chromatography/mass spectrometry (Py-GC/MS) is a commonly applied method in lignin analytics that thermally decomposes lignin into smaller, volatile compounds, enabling the identification and semi-quantitative profiling of its monomeric units through GC-MS. This approach offers rapid insights into the H, G, and S unit composition without the need for prior lignin extraction.

The strength of Py-GC/MS lies in its broad applicability, affordability, and compatibility with various biomass types. Despite its utility, Py-GC/MS is limited to detecting mostly monomers and small dimers due to the extent of thermal fragmentation. The analytical sensitivity and specificity have been improved through temperature optimization, derivatization strategies, and the inclusion of catalysts such as zeolites. Enhanced formats like Py-MBMS (molecular beam mass spectrometry) and 2D GC-TOF-MS allow high-throughput screening and more detailed compositional analysis, supporting applications in structural elucidation and chemotaxonomic classification using advanced statistical tools (Letourneau & Volmer, 2023).

3.4.1.5 Kappa number

The Kappa number method, developed in the 1950s, is a redox titration technique used to estimate residual lignin content in pulp by measuring the consumption of potassium permanganate under controlled conditions. It is widely applied in the pulp and paper industry due to its speed, simplicity, and effectiveness for routine monitoring. Typically used for chemical pulps, it provides an indirect measure of lignin, with a standard conversion factor of approximately 0.15 for kraft pulps (Dence, 1992). While effective for samples with moderate lignin content, its accuracy diminishes in bleached or low-lignin pulps and can be influenced by other oxidizable substances, such as hexenuronic acids. Additionally, differences in lignin composition across biomass sources may affect reactivity with permanganate, limiting precision. Nonetheless, the method remains a practical tool for assessing delignification progress and supporting operational control during pulp processing (Burkhardt, 2018).

3.4.2 Lignin subunits and functional groups

3.4.2.1 Nuclear Magnetic Resonance (NMR) spectroscopy

Nuclear magnetic resonance (NMR) spectroscopy has been used extensively in lignin analysis for decades, both in the solid state and in solution. It provides detailed structural information and can be used to characterize a wide range of lignin preparations. Structural information on an atomic level is obtained as each observable nucleus (e.g., ^{1}H, ^{13}C, and ^{31}P) with a unique chemical environment yields a unique resonance frequency (chemical shift). NMR spectra of complex molecules, such as lignin, will thus contain hundreds of unique signals, resulting in very information-rich spectra. Another key feature of NMR spectroscopy is its inherent quantitative nature that allows for molecular ratios to be determined directly from peak areas and, in the case of lignin, enables quantification of different lignin subunits and interunit linkages. However, not all types of NMR experiments are quantitative, as discussed below.

3.4.2.2 Solid-state NMR

A major benefit of analyzing lignin using solid-state rather than solution-state NMR is the simple sample preparation that preserves the native structure. Poor solubility of many lignocellulosic compounds also makes solid-state approaches attractive. ^{13}C Cross-Polarization Magic Angle Spinning (CPMAS) is by far the most commonly employed technique in solid-state and has been used, for example, to determine the S/G ratios in wood (Manders, 1987; Wikberg & Maunu, 2004) and to study changes in lignin structure resulting from various processes, such as decomposition (Baldock et al., 1997). ^{13}C CPMAS NMR continues to be a popular method for analysis of lignocellulosic materials, but broad lines and poor sensitivity prohibit detailed structural information from being extracted. Also, because the effectiveness of polarization to ^{13}C from adjacent ^{1}H nuclei varies depending on the molecular structure, ^{13}C CPMAS spectra are not quantitative. ^{13}C-labeling increases sensitivity and enables multidimensional NMR that could, e.g., detect interaction between lignin and other cell wall components (Dupree et al., 2015). Even higher sensitivity gains can be reached using Dynamic Nuclear Polarization (DNP) NMR, where the much higher Boltzmann polarization of electrons compared to nuclei can be utilized to boost the NMR signal dramatically (Perras et al., 2017). Although reliant on highly specialized equipment, DNP NMR is becoming increasingly popular.

3.4.2.3 Solution-state NMR

NMR of lignin and lignocelluloses in solution exhibits greater sensitivity and resolution compared to solid-state NMR and is a very versatile analytical approach for a wide range of lignin preparations, both for isolated lignin and for lignocellulosic materials. One-dimensional ^{1}H and ^{13}C NMR has been used for a very long time to characterize lignin from various sources and continues to be of great value because of the quantitative nature of these experiments (Lundquist, 1980; Gellerstedt et al., 1987; Balakshin & Capanema, 2015). More recently, ^{31}P NMR has emerged as a useful technique to characterize and quantify lignin hydroxyl groups (Meng et al., 2019). Compared to other analytical techniques that quantify the total amount of aromatic hydroxyl groups, ^{31}P NMR can discriminate between hydroxyl groups on different subunits, all in a single experiment. To achieve quantitative results from 1D spectra (irrespective of the observed nucleus), the delay between consecutive scans should be longer than five times the longitudinal relaxation time (T_1). Two-dimensional NMR can be used to dramatically increase resolution in the spectrum by separating overlapping peaks in a second dimension. In particular, ^{1}H-^{13}C Heteronuclear Single Quantum Coherence (HSQC) NMR (Figure 3.2) has become very popular and is now used extensively in lignin (both for industrial lignin and lignocellulosics) analysis and can be used for detailed structural characterization, including detection of low abundance interunit linkages and incorporation of unusual subunits (Kim et al., 2008;

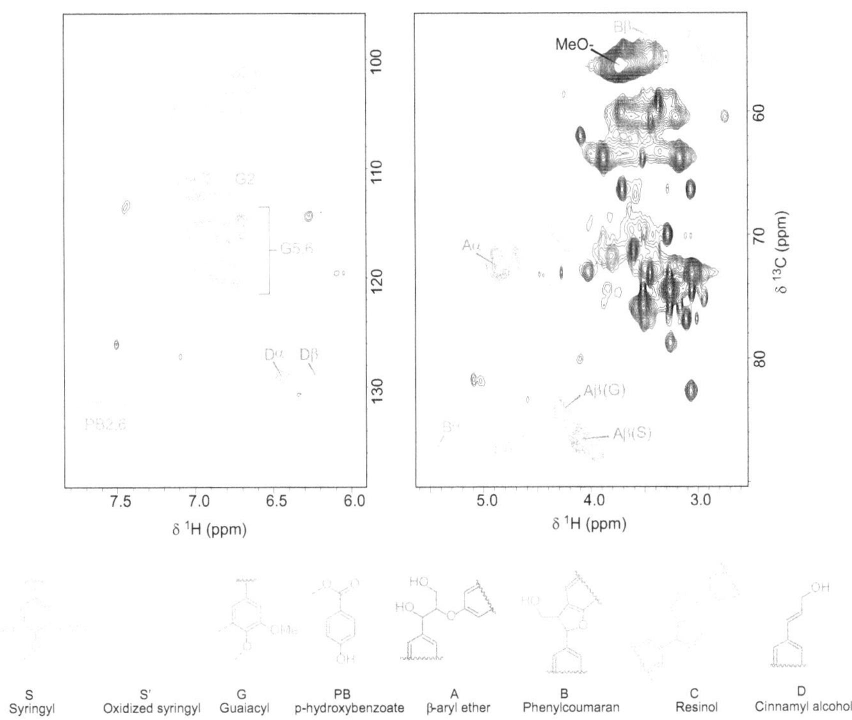

FIGURE 3.2 2D ^{1}H-^{13}C HSQC spectrum of aspen MWL. The left panel shows the aromatic spectral region, and the right panel the interunit linkage region.

Kim & Ralph, 2010; Mansfield et al., 2012). It is important to note that the standard 2D HSQC setup does not yield a quantitative spectrum because of differences in $^{1}J_{CH}$ coupling constants and relaxation rates for different structural moieties of the lignin molecules. However, it can still be used in a semi-quantitative manner to compare the relative abundance of, e.g., interunit linkages between samples. Considerable efforts have been put into developing methods that would allow reliable quantification from HSQC spectra, either by combining HSQC with quantitative ^{13}C spectra (Yuan et al., 2011) or by novel HSQC pulse sequences (Heikkinen et al., 2003; Sette et al., 2011; Bourmaud et al., 2024).

3.4.2.4 Infrared spectroscopy

Infrared spectroscopy is a powerful tool for analyzing phenolic compounds, including lignin, in plant cell walls. It enables identification of specific bond vibrations (e.g., C–O, C=O, O–H), yielding extensive structural information from minimal sample quantities. Although spectral overlap can complicate analysis, certain regions offer lignin-specific signals useful for quantification (Hatfield & Fukushima, 2005). Techniques like Fourier transform infrared (FTIR) spectroscopy, particularly in solid-state modes such as attenuated total reflectance (ATR), diffuse reflectance (DRIFT), and photoacoustic (PAS), enhance applicability to wood, pulp, and paper. FTIR spectra typically show two key regions: 3,700–2,750 cm^{-1} (-OH and methyl groups) and 1,800–900 cm^{-1} (fingerprint region), with peaks indicating guaiacyl and syringyl units, as well as carbonyl and ether groups near 1,710–1,600 cm^{-1} (Melro et al., 2021). While structural and compositional comparisons are straightforward, correlating spectra with total lignin content remains challenging. Recently, near-infrared spectroscopy (NIRS), covering wavelengths between visible and infrared light, has also been employed for lignin quantification (Wang et al., 2025).

3.4.2.5 Thioacidolysis

Thioacidolysis can be used for investigating the occurrence of lignin and for estimating the substructure composition, i.e., the fractions of H, G, and S units. It is based on the hydrolysis of β-O-4 aryl ether linkages using solvolysis in dioxane-ethanethiol with boron trifluoride etherate, followed by separation and quantitation of products using GC-MS (Lapierre et al., 1991). It is a widely used method in lignin chemistry due to the fact that it provides unambiguous evidence for the occurrence of the arylglycerol aryl ether structures that are indicative of lignin. Although advantageous compared to several other methods, it has potential drawbacks, including side reactions occurring at high temperature (100°C) and the use of foul-smelling mercaptan reagent.

3.4.2.6 DFRC method

DFRC (derivatization followed by reductive cleavage) is an alternative to the thioacidolysis method for investigating the occurrence of lignin and estimating the substructure composition of lignin (Martone et al., 2009). It is based on derivatization by acetylation with acetyl bromide followed by reductive cleavage of β-O-4 aryl ether linkages using zinc and separation and quantitation of resulting products using GC (Lu & Ralph, 1997). Drawbacks include incomplete degradation of lignin, whereas advantages compared to thioacidolysis include fewer side reactions using lower temperature (such as 50°C) and that it is less malodorous.

3.4.2.7 Phenolic hydroxyl groups

The content of phenolic hydroxyl groups in lignins is important with regard to reactivity as well as solubility in polar solvents. Deprotonation of phenolic hydroxyl groups in aqueous solutions under alkaline conditions creates phenolate ions, which facilitates extraction of lignin, for example, during kraft pulping and soda pulping. As mentioned in the section on solution-state NMR spectroscopy, ^{31}P NMR has emerged as a standard method to investigate hydroxyl groups in lignin (Meng et al., 2019). More recently, PFP (pentafluoropyridine) ^{19}F NMR spectroscopy has been proposed as an alternative, as it was found that PFP exhibits high selectivity for reactions with phenolic hydroxyl groups relative to aliphatic alcohols (Kenny et al., 2023). Other approaches include the periodate oxidation method (Lai et al., 1990), in which periodate targets vicinal diols in lignin, and methanol that is released during the reaction is quantitated. Furthermore, the spectrophotometric Folin-Ciocalteu method has been adapted for determination of the content of phenolic hydroxyl groups in pulps (de Sousa et al., 2001). Simultaneous conductometric and pH-metric titrations in aqueous medium have also been proposed for determination of the content of phenolic hydroxyl groups (Serrano et al., 2018).

3.4.3 Other analytical techniques

3.4.3.1 Electron microscopy

Electron microscopy is a powerful technique for analyzing the physical changes in lignocellulosic biomass, especially those associated with lignin removal and structural reorganization. Scanning electron microscopy (SEM) enables detailed visualization of surface morphology, revealing key lignin-related transformations, such as cell wall disruption, pore formation, and fiber delamination after pretreatment. These morphological changes often correlate with lignin solubilization or redistribution. Transmission electron microscopy (TEM), though more technically demanding, offers higher resolution and can detect nanoscale lignin domains and their association with other cell wall polymers. Both SEM and TEM provide visual confirmation of lignin removal and are essential for linking chemical delignification efficiency with structural outcomes. Such insights are crucial for optimizing pretreatment strategies and understanding the role of lignin in biomass recalcitrance (Lupoi et al., 2015).

3.4.3.2 Size exclusion chromatography

Size exclusion chromatography (SEC), often referred to as gel permeation chromatography (GPC) or gel filtration chromatography (GFC), is the most common method for determining lignin molecular weight (MW). It provides number-average (Mn), weight-average (Mw), and polydispersity index (PDI = Mw/Mn) values. However, results vary widely due to differences in lignin pretreatment, SEC setup, and data handling. The heterogeneous structure of lignin causes non-size-based interactions in columns, making separation complicated (Andrianova et al., 2018). The calibration is typically done using standards like polystyrene, which poorly represent lignin, leading to inaccurate MW estimates. Specific lignin standards are not available (Liao et al., 2020). Poor lignin solubility in most solvents often requires acetylation, which can artificially raise MW depending on the hydroxyl group content. The results also depend on column type, eluent, and detector setup. Universal calibration using hydrodynamic volume has been proposed, but challenges remain. Polymer association and fluorescence can interfere with advanced detectors like multi-angle light scattering (Andrianova et al., 2018). Despite its limitations, SEC is valuable for comparing lignin samples and studying structural changes under various processing conditions. It has largely replaced older methods like ultracentrifugation, light scattering, and vapor pressure osmometry (Melro et al., 2021).

3.5 APPLICATIONS OF LIGNIN ANALYSIS

3.5.1 Industrial applications

Lignin analysis is fundamental in industries such as pulp and paper, biofuels, and polymer manufacturing. In the pulp and paper sector, methods like the Kappa number and UV spectroscopy are used to monitor lignin removal during chemical pulping and bleaching (Burkhardt, 2018). In bioenergy and biorefinery operations, accurate quantification and structural assessment of lignin pave the direction for pretreatment conditions, enzymatic hydrolysis, and downstream valorization. High-resolution techniques such as NMR, FTIR, and SEC provide insight into functional groups, monomeric composition, and molecular weight, facilitating the development of lignin-derived adhesives, resins, and composites (Adebayo & Olanrewaju, 2025). The understanding of lignin variability across biomass types supports the production of uniform and high-value products at industrial scales.

3.5.2 Environmental impact studies

The stability and structural complexity of lignin make it a key marker in environmental studies of soil carbon cycling and organic matter decomposition. Analytical methods are used to track lignin-derived phenols and monomers in aquatic and terrestrial systems, helping to assess lignin degradation, transport, and persistence. For example, the Py-GC/MS method allows researchers to quantify lignin-derived biomarkers and monitor degradation pathways under different environmental conditions. These insights contribute to understanding the role of lignin in carbon sequestration, pollutant binding, and nutrient cycling, especially in ecosystems affected by land-use change or climate variability.

3.5.3 Biotechnological applications

In biotechnology, lignin analysis supports both microbial engineering and plant trait optimization. The structural profiling of lignin informs enzyme development for more efficient lignin breakdown, a key step in converting biomass into value-added chemicals. In plant biotechnology, compositional analysis helps in screening genetically modified lines or naturally occurring variants with altered lignin content or structure, aiding in the development of bioenergy crops with reduced recalcitrance. Emerging high-throughput and spectroscopic techniques, including NIRS, CASA, and electrophoresis, are especially useful in large-scale phenotyping and strain selection. By integrating detailed lignin characterization, biotechnological applications are increasingly able to target sustainable, efficient biomass utilization.

3.6 HIGH-THROUGHPUT TECHNIQUES

What often impedes a process from being applied in a high-throughput format, particularly if robotized, are multiple or complex sample preparation steps, including special treatments, such as extreme temperature and pressure, phase extraction (especially if solid), and filtration. The definition of high throughput is, however, a matter of discussion. With respect to the quantification and characterization of lignin, there have been recent reports on fast-paced, scaled-down methods relative to the conventional. This has been achieved either by obviating solvent extraction steps in combination with chromatographic detection (Chen et al., 2021) or by applying methods based on spectroscopy (Lu et al., 2021) and gel electrophoresis (Holt et al., 2020). Chen et al. (2021) demonstrated a modified thioacidolysis method where the solvent-solvent extraction step, iterations of mixing-centrifugation-separation, was removed, leading to an 80-sample per day-person throughput for lignin composition characterization and tricin content quantification by GC-MS and LC-MS. The results also showed the possibility of mitigating the adverse effect of contaminants in samples by parameter adjustment of analytical instruments.

As outlined in Section 3.4.1, Lu et al. (2021) presented the CASA method, which is based on mild reaction conditions, such as dissolution at 24°C, and detection using UV absorbance at 283 nm. Using three simple steps, i.e., biomass dissolution, dilution, and UV absorbance measurement, the method was found to be useful for the quantification of the lignin content of various types of biomass, which was verified using the Klason method. Gel electrophoresis is yet another novel approach that offers the advantage of fast separation of many samples at once (Holt et al., 2020). By applying an electric field across an agarose gel, lignin preparations (including kraft, soda, and organosolv lignins) were separated by charge, shape, and molecular weight. This was cross-validated by GPC, ^{1}H NMR, and Py-GC/MS. Given recent reports and the fast development of robotic platforms, the maturation of fast-paced methods into real high-throughput procedures seems definite.

3.7 CONCLUSIONS

Despite significant advances in lignin analytics, current methodologies still face challenges in precision, standardization, and adaptability across diverse types of lignin. Although most classical methods are informative, they rely heavily on indirect estimation and are often sensitive to variations in sample preparation, chemical complexity, and source material. Future work needs to prioritize the development of robotized high-throughput, minimally destructive, and quantitative methods capable of handling heterogeneous lignin matrices. More comprehensive structural insights can be achieved by integration of multi-modal approaches like combining spectroscopic, chromatographic, and imaging tools. Electrophoretic methods offer unexplored potential for molecular-weight profiling and merit further investigation due to their rapid and accessible format, although they are still in early development. The development of universal calibration standards, especially for spectrophotometric and chromatographic techniques, is also important. Finally, robust and scalable analytical workflows will be essential for unlocking the full potential of lignin and its valorization, particularly in materials science, biorefinery, and environmental applications.

REFERENCES

Adebayo, A. J., A. Olanrewaju. 2025. A complete review of lignin's extraction, analysis, applications, and future outlook. *Chem. Afr.* 1–31. https://doi.org/10.1007/s42250-025-01323-y

Andrianova, A. A., N. A. Yeudakimenka, S. L. Lilak, et al. 2018. Size exclusion chromatography of lignin: The mechanistic aspects and elimination of undesired secondary interactions. *J. Chromatogr. A.* 1534: 101–10. https://doi.org/10.1016/j.chroma.2017.12.051

Balakshin, M. Y., E. A. Capanema. 2015. Comprehensive structural analysis of biorefinery lignins with a quantitative ^{13}C NMR approach. *RSC Adv.* 5 (106): 87187–99. https://doi.org/10.1039/C5RA16649G

Balakshin, M., E. A. Capanema, X. Zhu, et al. 2020. Spruce milled wood lignin: Linear, branched or cross-linked? *Green Chem.* 22 (13): 3985–01. https://doi.org/10.1039/D0GC00926A

Baldock, J. A., J. M. Oades, P. N. Nelson, T. M. Skene, A. Golchin, P. Clarke. 1997. Assessing the extent of decomposition of natural organic materials using solid-state C-13 NMR spectroscopy. *Aust. J. Soil Res.* 35 (5): 1061–83. https://doi.org/10.1071/S97004

Berlin, A., M. Balakshin. 2014. Industrial lignins: Analysis, properties, and applications. In *Bioenergy research: Advances and applications*, ed. Kusumakar, 315–36. Amsterdam: Elsevier.

Björkman, A. 1956. Studies on finely divided wood. Part 1. Extraction of lignin with neutral Solvents. *Sven. Papperstid.* 59 (13): 477–85. https://doi.org/10.1038/1741057a0

Boerjan, W., J. Ralph, M. Baucher. 2003. Lignin biosynthesis. *Annu. Rev. Plant Biol.* 54 (1): 519–46. https://doi.org/10.1146/annurev.arplant.54.031902.134938

Bourmaud, C. L., S. Bertella, A. Bosch Rico, S. D. Karlen, J. Ralph, J. S. Luterbacher. 2024. Quantification of native lignin structural features with gel-phase 2D-$HSQC_0$ reveals lignin structural changes during extraction. *Angew. Chem.* 136 (31): e202404442. https://doi.org/10.1002/anie.202404442

Bozek, A., F. Mathis, L. Guillaumat, et al. 2025. Determination of a suitable molar absorption coefficient (ε) for lignin analysis of fibrous plants using the CASA method. *Int. J. Biol. Macromol.* 308: 142324. https://doi.org/10.1016/j.ijbiomac.2025.142324

Burkhardt, S. 2018. Does the kappa number method accurately reflect lignin content in nonwood pulps? *TAPPI J* 17 (11): 611–17. https://api.semanticscholar.org/CorpusID:104433411

Cesarino, I. 2019. Structural features and regulation of lignin deposited upon biotic and abiotic stresses. *Curr. Opin. Biotechnol.* 56: 209–14. https://doi.org/10.1016/j.copbio.2018.12.012

Chen, N., K. Jiang, M. Zhao, C. Zhang, Y. Jin, W. Wu. 2024. Pretreatment process of lignocellulosic biomass: A review of pseudo-lignin formation. *Biomass Bioenergy* 188: 107339. https://doi.org/10.1016/j.biombioe.2024.107339

Chen, F., Y. Tobimatsu, D. Havkin-Frenkel, R. A. Dixon, J. Ralph. 2012. A polymer of caffeyl alcohol in plant seeds. *Proc. Natl. Acad. Sci. USA* 109 (5): 1772–77. https://doi.org/10.1073/pnas.1120992109

Chen, F., Y. Tobimatsu, L. Jackson, J. Nakashima, J. Ralph, R. A. Dixon. 2013. Novel seed coat lignins in the Cactaceae: Structure, distribution and implications for the evolution of lignin diversity. *Plant J.* 73 (2): 201–11. https://doi.org/10.1111/tpj.12012

Chen, F., C. Zhuo, X. Xiao, T. H. Pendergast, K. M. Devos. 2021. A rapid thioacidolysis method for biomass lignin composition and tricin analysis. *Biotechnol Biofuels* 14: 1–9. https://doi.org/10.1186/s13068-020-01865-y

Choi, S. J., Z. Lee, S. Kim, E. Jeong, J. S. Shim. 2023. Modulation of lignin biosynthesis for drought tolerance in plants. *Front. Plant Sci.* 14: 1116426. https://doi.org/10.3389/fpls.2023.1116426

Crestini, C., F. Melone, M. Sette, R. Saladino. 2011. Milled wood lignin: A linear oligomer. *Biomacromolecules* 12 (11): 3928–35. https://doi.org/10.1021/bm200948r

De Meester, B., R. Vanholme, T. Mota, W. Boerjan. 2022. Lignin engineering in forest trees: From gene discovery to field trials. *Plant Commun.* 3 (6): 100465. https://doi.org/10.1016/j.xplc.2022.100465

de Sousa, F., A. Reimann, M. Björklund Jansson, N. O. Nilvebrant. 2001. Estimating the amount of phenolic hydroxyl groups in lignins. In *Proceedings of the 11th International Symposium on Wood and Pulping Chemistry (ISWPC)*, Nice, France 3: 649–53.

de Vries, S., J. M. Fürst-Jansen, I. Irisarri, et al. 2021. The evolution of the phenylpropanoid pathway entailed pronounced radiations and divergences of enzyme families. *Plant J.* 107 (4): 975–02. https://doi.org/10.1111/tpj.15387

Dence, C. W. 1992. The determination of lignin. In *Methods in lignin chemistry,* ed. S. Y. Lin and C. W. Dence, 33–61. Berlin and Heidelberg: Springer.

Dixon, R. A., J. Barros. 2019. Lignin biosynthesis: Old roads revisited and new roads explored. *Open Biol.* 9 (12): 190215. https://doi.org/10.1098/rsob.190215

Dupree, R., T. J. Simmons, J. C. Mortimer, et al. 2015. Probing the molecular architecture of *Arabidopsis thaliana* secondary cell walls using two-and three-dimensional ^{13}C solid state nuclear magnetic resonance spectroscopy. *Biochem.* 54 (14): 2335–45. https://doi.org/10.1021/bi501552k

Frangville, C., M. Rutkevičius, A. P. Richter, O. D. Velev, S. D. Stoyanov, V. N. Paunov. 2012. Fabrication of environmentally biodegradable lignin nanoparticles. *ChemPhysChem.* 13 (18): 4235–43. https://doi.org/10.1002/cphc.201200537

Gellerstedt, G., D. Robert, V. Parker, M. Oivanen, L. Eberson. 1987. Quantitative ^{13}C NMR analysis of kraft lignins. *Acta Chem. Scand., B.* 41 (7): 541–46. https://doi.org/10.3891/acta.chem.scand.41b-0541

Glasser, W. G., R. A. Northey, T. P. Schultz, (Eds.). 1999. *Lignin: Historical, biological, and materials perspectives.* Washington, DC: American Chemical Society.

Hatfield, R., R. S. Fukushima. 2005. Can lignin be accurately measured? *Crop Sci.* 45 (3): 832–39. https://doi.org/10.2135/cropsci2004.0238

Heikkinen, S., M. M. Toikka, P. T. Karhunen, I. A. Kilpeläinen. 2003. Quantitative 2D HSQC (Q-HSQC) via suppression of J-dependence of polarization transfer in NMR spectroscopy: Application to wood lignin. *J. Am. Chem. Soc.* 125 (14): 4362–67. https://doi.org/10.1021/ja029035k

Hoffmann, N., A. Benske, H. Betz, M. Schuetz, A. L. Samuels. 2020. Laccases and peroxidases co-localize in lignified secondary cell walls throughout stem development. *J. Plant Physiol.* 184 (2): 806–22. https://doi.org/10.1104/pp.20.00473

Holt, C. A., B. Cottyn, S. Baumberger, K. Kovacs-Schreiner, A. J. Blacker. 2020. High-throughput analysis of lignin by agarose gel electrophoresis. *J. Agric. Food Chem.* 68 (48): 14297–06. https://doi.org/10.1021/acs.jafc.0c06308

Johnson, D. B., W. E. Moore, L. C. Zank. 1961. The spectrophotometric determination of lignin in small wood samples. *TAPPI J.* 44 (11): 793–98.

Kenny, J. K., J. W. Medlin, G. T. Beckham. 2023. Quantification of phenolic hydroxyl groups in lignin via ^{19}F NMR spectroscopy. *ACS Sustain. Chem. Eng.* 11 (14): 5644–55. https://doi.org/10.1021/acssuschemeng.3c00115

Kim, H., J. Ralph. 2010. Solution-state 2D NMR of ball-milled plant cell wall gels in DMSO-d6/pyridine-d5. *Org. Biomol. Chem.* 8 (3): 576–91. https://doi.org/10.1039/B916070A
Kim, H., J. Ralph, T. Akiyama. 2008. Solution-state 2D NMR of ball-milled plant cell wall gels in DMSO-d6. *Bioenergy Res.* 1: 56–66. https://doi.org/10.1007/s12155-008-9004-z
Kumari, A., V. Kumar, R. Ovadia, M. Oren-Shamir. 2023. Phenylalanine in motion: A tale of an essential molecule with many faces. *Biotechnol. Adv.* 68: 108246. https://doi.org/10.1016/j.biotechadv.2023.108246
Lai, Y. Z., X. P. Guo. 1991. Variation of the phenolic hydroxyl group content in wood lignins. *Wood Sci. Technol.* 25: 467–72. https://doi.org/10.1007/BF00225239
Lai, Y. Z., X. P. Guo, W. Situ. 1990. Estimation of phenolic hydroxyl groups in wood by a periodate oxidation method. *J. Wood Chem. Technol.* 10 (3): 365–77. https://doi.org/10.1080/02773819008050245
Lapierre, C. 1993. Application of new methods for the investigation of lignin structure. In *Forage cell wall structure and digestibility*, ed. H. G. Jung, D. R. Buxton, R. D. Hatfield, and J. Ralph, 133–66. Madison, WI: American Society of Agronomy.
Lapierre, C., B. Pollet, B. Monties, C. Rolando. 1991. Thioacidolysis of spruce lignin: GC-MS analysis of the main dimers recovered after Raney nickel desulphuration. *Holzforschung* 45 (1): 61–68. https://doi.org/10.1515/hfsg.1991.45.1.61
Letourneau, D. R., D. A. Volmer. 2023. Mass spectrometry-based methods for the advanced characterization and structural analysis of lignin: A review. *Mass Spectrom. Rev.* 42 (1): 144–88. https://doi.org/10.1002/mas.21716
Liao, J. J., N. H. Abd Latif, D. Trache, N. Brosse, M. H. Hussin. 2020. Current advancement on the isolation, characterization and application of lignin. *Int. J. Biol. Macromol.* 162: 985–24. https://doi.org/10.1016/j.ijbiomac.2020.06.168
Liu, Z.-H., H. Liu, T. Xu, et al. 2025. Lignin valorization reshapes sustainable biomass refining. *Renew. Sustain. Energy Rev.* 211: 115296. https://doi.org/10.1016/j.rser.2024.115296
Lu, F., J. Ralph. 1997. Derivatization followed by reductive cleavage (DFRC method), a new method for lignin analysis: Protocol for analysis of DFRC monomers. *J. Agric. Food Chem.* 45 (7): 2590–92. https://doi.org/10.1021/jf970258h
Lu, F., C. Wang, M. Chen, F. Yue, J. Ralph. 2021. A facile spectroscopic method for measuring lignin content in lignocellulosic biomass. *Green Chem.* 23 (14): 5106–12. https://doi.org/10.1039/D1GC01507A
Lu, J., M. Cheng, C. Zhao, et al. 2022. Application of lignin in preparation of slow-release fertilizer: Current status and future perspectives. *Ind. Crop. Prod.* 176: 114267. https://doi.org/10.1016/j.indcrop.2021.114267
Lundquist, K. 1980. NMR studies of lignins. 4. Investigation of spruce lignin by ^{1}H NMR spectroscopy. *Acta Chem. Scand., Ser. B.* 34 (1): 21–26. https://doi.org/10.3183/npprj-1989-04-03-p210-213
Lupoi, J. S., S. Singh, R. Parthasarathi, B. A. Simmons, R. J. Henry. 2015. Recent innovations in analytical methods for the qualitative and quantitative assessment of lignin. *Renew. Sustain. Energy Rev.* 49: 871–06. https://doi.org/10.1016/j.rser.2015.04.091
Manders, W. F. 1987. Solid-state ^{13}C NMR determination of the syringyl/guaiacyl ratio in hardwoods. *Holzforschung* 41 (1): 13–18. https://doi.org/10.1515/hfsg.1987.41.1.13
Mansfield, S. D., H. Kim, F. Lu, J. Ralph. 2012. Whole plant cell wall characterization using solution-state 2D NMR. *Nat. Protoc.* 7 (9): 1579–89. https://doi.org/10.1038/nprot.2012.064
Martarello, D. C., A. M. Almeida, R. C. Sinzker, et al. 2023. The known unknowns in lignin biosynthesis and its engineering to improve lignocellulosic saccharification efficiency. *Biomass Convers. Biorefin.* 13 (4): 2497–15. https://doi.org/10.1007/s13399-021-01291-6

Martone, P. T., J. M. Estevez, F. Lu, et al. 2009. Discovery of lignin in seaweed reveals convergent evolution of cell-wall architecture. *Curr. Biol.* 19 (2): 169–75. https://doi.org/10.1016/j.cub.2008.12.031

Mboowa, D. 2024. A review of the traditional pulping methods and the recent improvements in the pulping processes. *Biomass Convers. Biorefin.* 14 (1): 1–12. https://doi.org/10.1007/s13399-020-01243-6

Melro, E., A. Filipe, D. Sousa, B. Medronho, A. Romano. 2021. Revisiting lignin: A tour through its structural features, characterization methods and applications. *New J. Chem.* 45 (16): 6986–13. https://doi.org/10.1039/d0nj06234k

Meng, X., C. Crestini, H. Ben, et al. 2019. Determination of hydroxyl groups in biorefinery resources via quantitative ^{31}P NMR spectroscopy. *Nat. Protoc.* 14 (9): 2627–47. https://doi.org/10.1038/s41596-019-0191-1

Perras, F. A., H. Luo, X. Zhang, N. S. Mosier, M. Pruski, M. M. Abu-Omar. 2017. Atomic-level structure characterization of biomass pre-and post-lignin treatment by dynamic nuclear polarization-enhanced solid-state NMR. *J. Phys. Chem. A.* 121 (3): 623–30. https://doi.org/10.1021/acs.jpca.6b11121

Ralph, J., C. Lapierre, W. Boerjan. 2019. Lignin structure and its engineering. *Curr. Opin. Biotechnol.* 56: 240–49. https://doi.org/10.1016/j.copbio.2019.02.019

Ralph, J., K. Lundquist, G. Brunow, et al. 2004. Lignins: Natural polymers from oxidative coupling of 4-hydroxyphenyl-propanoids. *Phytochem. Rev.* 3 (1–2): 29–60. https://doi.org/10.1023/B:PHYT.0000047809.65444.a4

Schneider, W. D. H., A. J. P. Dillon, M. Camassola. 2021. Lignin nanoparticles enter the scene: A promising versatile green tool for multiple applications. *Biotechnol. Adv.* 47: 107685. https://doi.org/10.1016/j.biotechadv.2020.107685

Serrano, L., E. S. Esakkimuthu, N. Marlin, M. C. Brochier-Salon, G. Mortha, F. Bertaud. 2018. Fast, easy, and economical quantification of lignin phenolic hydroxyl groups: comparison with classical techniques. *Energy Fuels* 32 (5): 5969–77. https://doi.org/10.1021/acs.energyfuels.8b00383

Sette, M., R. Wechselberger, C. Crestini. 2011. Elucidation of lignin structure by quantitative 2D NMR. *Chem. Eur. J.* 17 (34): 9529–35. https://doi.org/10.1002/chem.201003045

Shende, V. V., K. D. Bauman, B. S. Moore. 2024. The shikimate pathway: Gateway to metabolic diversity. *Nat. Prod. Rep.* 41 (4): 604–48. https://doi.org/10.1039/D3NP00037K

Shinde, S. D., X. Meng, R. Kumar, A. J. Ragauskas. 2018. Recent advances in understanding the pseudo-lignin formation in a lignocellulosic biorefinery. *Green Chem.* 20 (10): 2192–05. https://doi.org/10.1039/c8gc00353j

Taiz, L., E. Zeiger, I. M. Møller, A. Murphy. 2015. *Plant physiology and development*. Sunderland, MA: Sinauer Associates.

Tokunaga, Y., T. Watanabe. 2023. An investigation of the factors controlling the chemical structures of lignin dehydrogenation polymers. *Holzforschung* 77 (1): 51–62. https://doi.org/10.1515/hf-2022-0129

ur Rahman, O., S. Shi, J. Ding, D. Wang, S. Ahmad, H. Yu. 2018. Lignin nanoparticles: Synthesis, characterization and corrosion protection performance. *New J. Chem.* 42 (5): 3415–25. https://doi.org/10.1039/c7nj04103a

Vanholme, R., B. Demedts, K. Morreel, J. Ralph, W. Boerjan. 2010. Lignin biosynthesis and structure. *Plant Physiol.* 153 (3): 895–05. https://doi.org/10.1104/pp.110.155119

Wang, L., C. Lu, J. Wang, C. Wang, C. Li. 2025. Rapid determination of cellulose, hemicellulose and lignin content in hemp fibers using near-infrared spectroscopy. *Fibers Polym.* 1–11. https://doi.org/10.1007/s12221-024-00834-8

Wikberg, H., S. L. Maunu. 2004. Characterisation of thermally modified hard-and softwoods by ^{13}C CPMAS NMR. *Carbohydr. Polym.* 58 (4): 461–66. https://doi.org/10.1016/j.carbpol.2004.08.008

Yokoyama, R. 2024. Evolution of aromatic amino acid metabolism in plants: A key driving force behind plant chemical diversity in aromatic natural products. *Philos. Trans. R. Soc. B.* 379 (1914): 20230352. https://doi.org/10.1098/rstb.2023.0352

Yuan, T. Q., S. N. Sun, F. Xu, R. C. Sun. 2011. Characterization of lignin structures and lignin–carbohydrate complex (LCC) linkages by quantitative ^{13}C and 2D HSQC NMR spectroscopy. *J. Agric. Food Chem.* 59 (19): 10604–14. https://doi.org/10.1021/jf2031549

Zhu, W., H. Theliander. 2015. Precipitation of lignin from softwood black liquor: An investigation of the equilibrium and molecular properties of lignin. *BioResources* 10 (1): 1696–14. https://doi.org/10.15376/biores.10.1.1696-1715

4 Depolymerization of lignin and conversion into biopolymer resins

Gayathri Balaji Venkata, Glen Cletus Dsouza, and Fatemeh Dodangeh

4.1 INTRODUCTION

The continued reliance on fossil fuels, currently supplying approximately 80% of global energy, has driven unprecedented greenhouse gas emissions and contributed to climate change, while the finite nature of petroleum and natural gas reserves highlights the urgency of developing renewable alternatives (Kumar et al., 2020). Biomass, and in particular lignocellulosic feedstocks, offers a versatile platform for sustainable production of both fuels and materials (Kariim et al., 2025). Lignocellulosic biomass consists predominantly of cellulose (40–60%), hemicellulose (10–40%), and lignin (10–30%). Although cellulose and hemicellulose have been successfully converted to bioethanol and other platform chemicals, lignin remains underexploited and is largely combusted for process heat (Seidi et al., 2025). Lignin is a complex, three-dimensional aromatic polymer formed by radical coupling of p-coumaryl, coniferyl, and sinapyl alcohols. Its network of β-O-4, α-O-4, and various C–C bonds, combined with abundant methoxyl, aliphatic hydroxyl, and phenolic hydroxyl groups, imparts both chemical resilience and rich functional diversity.

However, its native high molecular weight, heterogeneous composition, and limited availability of reactive sites hinder direct incorporation into high-performance polymers, typically capping lignin substitution in phenolic and polyurethane systems at 20–30% before mechanical properties deteriorate (DSouza et al., 2024; Mahmood et al., 2016). Depolymerization addresses these limitations by cleaving specific inter-unit linkages, especially the prevalent β-O-4 ethers, yielding lower-molecular-weight oligomers enriched in aliphatic hydroxyls. These fragments exhibit enhanced solubility, reduced steric hindrance, and elevated hydroxyl numbers, rendering them suitable bio-polyol replacements at levels exceeding 50% in phenol–formaldehyde, polyurethane, and epoxy formulations (Mahmood et al., 2016; DSouza et al., 2025; Li et al., 2019). The choice of depolymerization strategies, such as thermochemical (e.g., pyrolysis, hydrothermal liquefaction), catalytic (acid/base or metal-catalyzed), biological (enzymatic or microbial), and emerging technologies (microwave-assisted, ionic liquids, deep eutectic solvents), governs product distribution, yield,

DOI: 10.1201/9781003653189-4

and functionality of the lignin-derived oligomers (Agarwal et al., 2018; Pei et al., 2024; Zhou et al., 2023).

This chapter provides a comprehensive overview of lignin depolymerization methods, from established thermal and catalytic processes to recent innovations in green solvents and microwave activation. We discuss how depolymerization conditions influence fragment size, functional group distribution, and downstream polymer performance. Subsequent sections examine the incorporation of depolymerized lignin into phenol–formaldehyde, polyurethane, and epoxy resin systems, highlighting formulation considerations and property trade-offs. Finally, we address analytical techniques such as 2D HSQC NMR, MALDI-TOF MS, and flow-induced dispersion analysis (Kosyakov et al., 2014; Simonsen et al., 2025) that enable detailed characterization of lignin derivatives. By integrating advances in depolymerization chemistry, material formulation, and characterization, this chapter illustrates pathways to valorize lignin as a sustainable, bio-based precursor for high-performance polymeric materials.

4.2 LIGNIN STRUCTURE, SOURCES, AND VARIABILITY

4.2.1 Composition of Lignin

Lignin is a complex, three-dimensional aromatic polymer and one of the three primary components of lignocellulosic biomass, alongside cellulose and hemicellulose. As the second most abundant renewable polymer after cellulose, lignin holds immense potential for conversion into high-value chemicals and fuels due to its rich bio-aromatic structure (Ren et al., 2019). The structure of lignin is highly branched and amorphous, varying depending on the biomass species and isolation methods. Lignin is formed by the radical polymerization of three monolignols: p-coumaryl, coniferyl, and sinapyl alcohols, resulting in a 3-dimensional, heterogeneous biopolymer with various linkage types, such as β-O-4, α-O-4, 4-O-5 ether bonds, β-1, β-5, β-β, and C-C bonds. These linkages, especially β-aryl ether bonds, account for a significant portion of the structure in softwood and hardwood. The reactivity of lignin is influenced by the presence of functional groups, including methoxyl, aliphatic hydroxyl, phenolic, benzyl alcohol, and carbonyl groups (Chen & Wan, 2017). The distribution of these monomers varies by biomass type; for example, coniferyl alcohol is abundant in softwood lignin, while hardwood lignin comprises both coniferyl and sinapyl alcohols; however, grass lignin contains all three monolignols. (Duval & Lawoko, 2014).

Lignin, despite being the most abundant natural polymer and a promising renewable resource, faces several challenges that hinder its widespread use in producing value-added products like polyurethane and phenolic foams (DSouza et al., 2026). Its high molecular weight, complex and heterogeneous structure, and limited reactive sites, due to steric hindrance and the presence of different types of linkages, make it difficult to directly incorporate into industrial formulations. Additionally, the lower reactivity of native lignin, compounded by its bulky and branched structure, often results in weaker and less durable products when used in large quantities. Currently, lignin is primarily underutilized, often burned as a byproduct in industrial processes, or used as a low-value fuel source, due to its relatively low heating value (DSouza

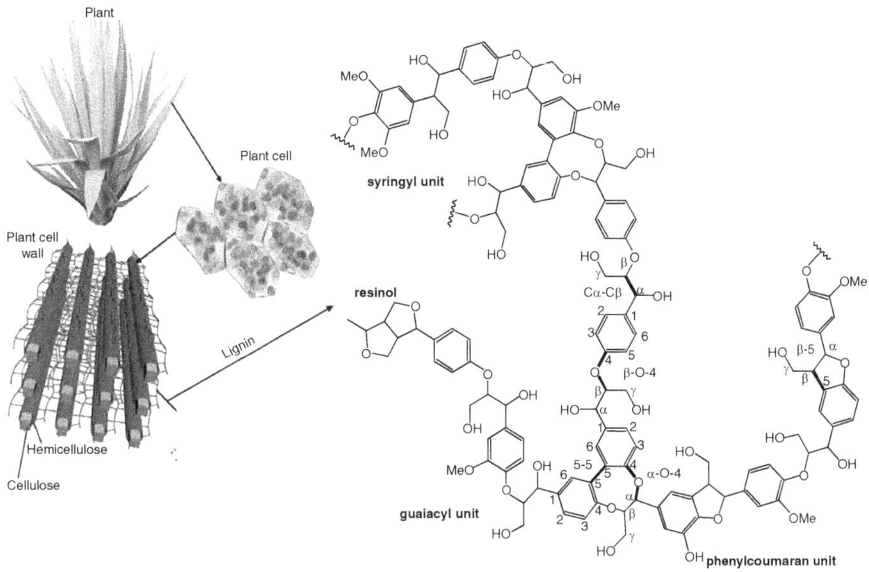

FIGURE 4.1 Structure of lignin in lignocellulosic material. Adapted from Bourbiaux et al. (2021) and Zakzeski et al. (2010)

et al., 2025; Wang et al., 2018). Addressing these issues through depolymerization and modification techniques is crucial to unlocking lignin's full potential as a sustainable bio-based material. Figure 4.1 illustrates the lignin structure in lignocellulosic biomass.

4.2.2 Sources of lignin and influence on depolymerization

Lignin, a major byproduct of pulping and bioethanol production from lignocellulosic biomass, is typically classified based on sulfur content as either sulfur-containing or sulfur-free. Its chemical structure and applications are strongly influenced by both the type of biomass feedstock (e.g., softwood, hardwood, or grasses) and the delignification process employed. Common extraction methods, including kraft, sulfite, soda, and organosolv pulping, significantly modify lignin's native structure by cleaving key interunit linkages such as β-O-4 and β-β bonds. These treatments generally reduce aliphatic hydroxyl content while increasing phenolic, carbonyl, and carboxylic functionalities. Consequently, the resulting lignin fractions vary considerably in terms of solubility, molecular weight, hydrophobicity, and chemical reactivity (Roy et al., 2022; Duval & Lawoko, 2014).

4.2.2.1 Kraft lignin

Kraft lignin (KL) is the predominant technical lignin produced by the kraft pulping process, which employs sodium hydroxide and sodium sulfide at 140–170°C to solubilize wood lignin into black liquor. It is characterized by a high degree of condensation, residual sulfur (3–7%), and ash content (0.5–3%). Extensive cleavage of β-O-4

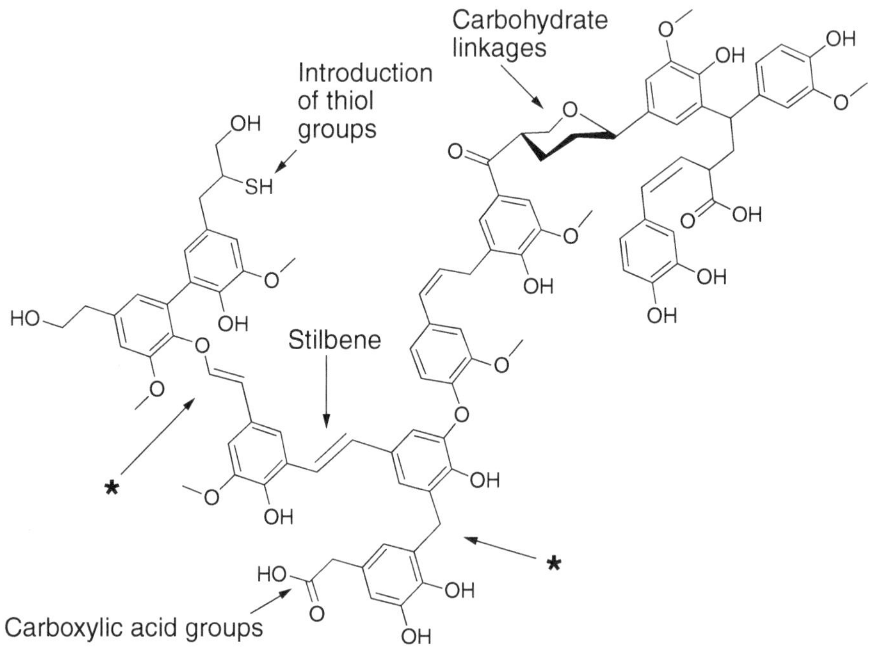

FIGURE 4.2 Structural characteristic of (a) kraft pine lignin and (b) lignosulfonate lignin adapted from Zakzeski et al. (2010)

ether bonds during pulping generates abundant phenolic hydroxyl groups, quinones, and carboxylic acids, yet also promotes new C–C linkages that increase molecular weight and reduce depolymerization efficiency (see Figure 4.2a for KL structure) (Duval & Lawoko, 2014). While KL finds uses in resins, carbon fibers, and specialty chemicals such as vanillin, its condensed structure and sulfur impurities impede catalytic or oxidative depolymerization, often yielding low monomer recoveries and rapid catalyst deactivation (Bajwa et al., 2019).

4.2.2.2 Lignosulfonate lignin

Sulfite pulping converts lignin into water-soluble lignosulfonates by reacting wood chips with sulfur dioxide and bases (Ca^{2+}, Na^+, or NH_4^+) under acidic conditions (see Figure 4.2b for kraft lignosulfonate structure). These lignosulfonates contain 3.5–8% sulfur and 4–8% ash and exhibit high average molecular weights with broad polydispersity (PDI ≈ 6–8) (Vishtal & Kraslawski, 2011; Bajwa et al., 2019). Their sulfonate groups confer excellent dispersant and binding properties, making them valuable in concrete additives, dyes, and agrochemicals. However, residual cations and inorganic impurities can poison depolymerization catalysts, and their high molecular weight limits monomer yields, necessitating careful pretreatment or detoxification for valorization.

4.2.2.3 Soda lignin

Soda lignin, derived from a sulfur-free pulping process using sodium hydroxide (NaOH), is chemically distinct from kraft and lignosulfonate lignin due to the absence

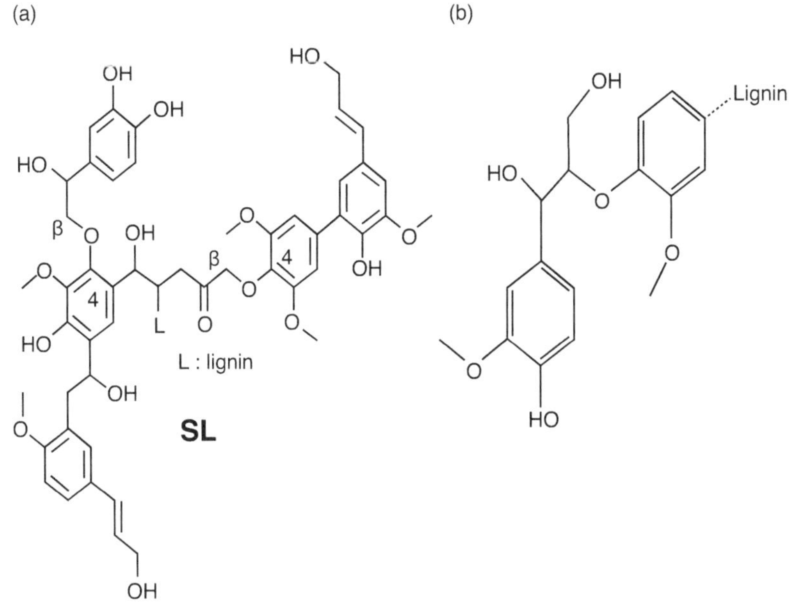

FIGURE 4.3 Structure of (a) Soda lignin, adapted and modified from Kaya et al. (2024), and (b) Organosolv hardwood lignin adapted from Muthuraj et al. (2020)

of sulfur and the presence of vinyl ether linkages (see Figure 4.3a for soda lignin structure). The pulping process typically involves reacting lignocellulosic biomass in a pressurized reactor at 160–170°C with 10–15% w/w NaOH (Zevallos Torres et al., 2020). Modifications, such as the inclusion of anthraquinone as a catalyst, have been shown to improve carbohydrate stabilization and enhance lignin solubilization. SL is used in phenolic resins, dispersants, and polymers and has potential applications in animal nutrition. However, it has a lower pulp yield and is less favored industrially (Gomes et al., 2020).

4.2.2.4 Organosolv lignin

Organosolv lignin (OL), produced through a solvent-based delignification process, is sulfur-free and closely resembles native lignin in structure (see Figure 4.3b for OL structure). It has high purity, low molecular weight, and high chemical reactivity, making it suitable for advanced bioproduct applications. OL is highly soluble in organic solvents but nearly insoluble in water since they are hydrophobic, and their recovery requires precipitation by adjusting pH, temperature, or concentration (Gomes et al., 2020). Additionally, the incorporation of Lewis acids as catalysts has been shown to reduce its molecular mass and hydroxyl content, while greater processing severity enhances syringyl and condensed phenolic structures (Constant et al., 2015).

4.2.2.5 Hydrolysis lignin

Hydrolysis lignin (HL) is a sulfur-free byproduct, amounting to as much as 60% of the original biomass, obtained when fermentable sugars are removed from

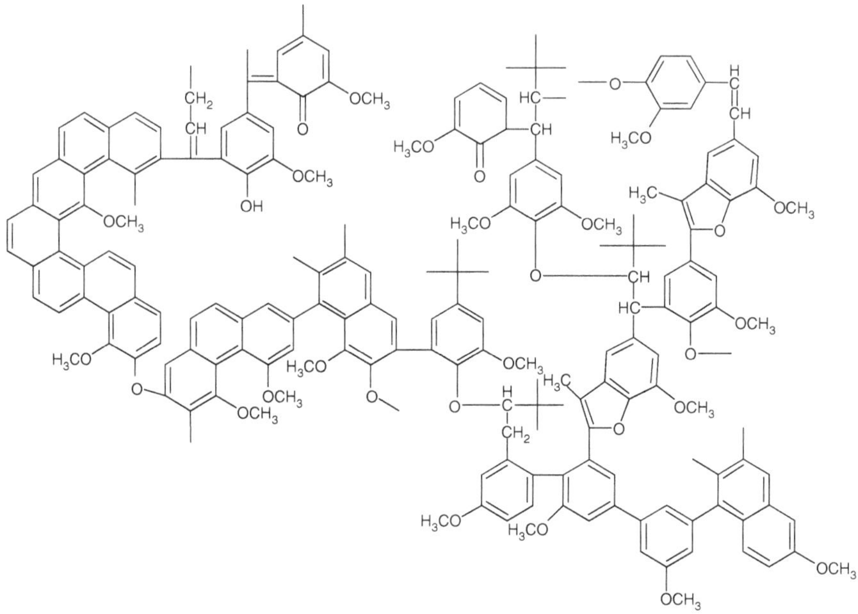

FIGURE 4.4 Structure of a fragment of hydrolysis lignin adapted from Budnyak et al. (2021)

lignocellulose by acid or enzymatic hydrolysis (Evstigneyev et al., 2016; Mahmood et al., 2016). In acid-catalyzed processes, feedstocks such as wheat straw are treated at 50–70°C for 4–12 h with formic acid or dilute H_2SO_4, yielding an insoluble residue lignin (RL) and a soluble lignin (SL) fraction following dioxane–water extraction and precipitation (Zhang et al., 2010). Enzymatic hydrolysis employs cellulase (~60 FPU/g) and β-glucosidase (~64 U/g) at ~50°C and pH 4.7–4.9, often following alkali, dilute-acid, or hot-water pretreatment to depolymerize cellulose and hemicellulose, leaving a lignin-rich residue (Nlewem & Thrash, 2010). This hydrolysis lignin is enriched in condensed aromatic structures and residual polysaccharides, giving it strong sorption capacity but poor solubility and dewaterability. Despite these challenges, HL's high phenolic hydroxyl content enhances its chemical reactivity, making it a promising precursor for biopolymers, graft copolymers, adsorbents, and soil conditioners (see Figure 4.4) (Vishtal & Kraslawski, 2011). However, its rigid β-O-4, 4-O-5, and C–C linkages render depolymerization difficult, necessitating tailored chemical, thermal, or enzymatic treatments to unlock aromatic monomers for sustainable material applications (Roy et al., 2022).

4.3 DEPOLYMERIZATION PATHWAYS

Lignin's complex structure, dominated by ether linkages, with β-O-4 linkages accounting for 50–60% of the bonds, and low chemical reactivity hinder its broader application potential, leading to its predominant use as a fuel for energy generation. As a result, making their selective cleavage a key step in efficient depolymerization

TABLE 4.1
Studies on emerging depolymerization techniques of lignin

Depolymerization Technique	Lignin Type	Process Conditions	Major Product	Yield	Reference
Bubbling fluidized-bed pyrolysis	Wheat-straw organosolv lignin	500°C, Ar fluidization, vapor residence ~1–3 s	Phenolic bio-oil, Char	Oil 48–55 wt%, char 30–40 wt%	Wild et al. (2012)
HTL in subcritical water	Softwood kraft lignin (LignoBoost™)	350°C, 25 MPa, 2 kg/h, 5.5 wt% lignin, ZrO_2 + K_2CO_3/KOH co- catalyst	Bio-oil, water-soluble organics (WSO), Char	73.5 ± 0.3% oil; WSO 12 ± 0.5%; Char 14.6 ± 0.7%	Belkheiri et al. (2018)
Acid-catalyzed β-O-4 ether bond cleavage	Model dimers (phenolic/ non-phenolic)	0.2 M H_2SO_4, 150°C, ≤ 90 min	Phenol/guaiacol, aldehyde and minor by- products	100% phenolic; 40–50% non-phenolic	Sturgeon et al. (2014)
Base-Catalyzed depolymerization	Soda P1000	1 g lignin, 1.8 wt% NaOH, 250°C, 4 h, 40 bar N_2	Lignin oil	29 wt%	Purushothaman et al. (2023)
Catalytic fast pyrolysis lignin depolymerization	Kraft lignin	500–700°C, He, 10°C ms^{-1}, 6 s, 10% Pd/C, 1:5 lignin:catalyst	Creosol, guaiacol, 4- vinylguaiacol	Phenolics 83–90%; aromatics 2–7%	Colina et al. (2025)
Reductive catalytic fractionation	Birch sawdust (hardwood)	240°C, 3 MPa H_2, 4 h, RuN/ ZnO/C on N-doped C	Propyl-syringol (Pr-S), propyl-guaiacol (Pr-G)	46.4 wt%	Wang et al. (2025)
Catalytic depolymerization with metal chloride (Lewis acid)	Alkali lignin	Ethanol, $ZnCl_2$ (0.001 mol), 1 MPa H_2, 300°C, 2 h	Guaiacols, phenols, syringols	Liquid product: 75.8 wt%; volatile fraction: 38.1 wt%	Zhang et al. (2014)
Catalytic oxidation using ionic liquid (metal-free)	Model compounds	[C_8mim][OAc], 1.5 MPa O_2, 100°C, 2 h	Phenol, Benzoic acid, Guaiacol	Phenol 96–98%; benzoic acid 85–86%; guaiacol 98%	Yang et al. (2017)

(Continued)

TABLE 4.1
(Continued)

Depolymerization Technique	Lignin Type	Process Conditions	Major Product	Yield	Reference
Oxidative depolymerization in ionic liquids	Organosolv beech lignin	100°C, 24 h, [Emim][CF_3SO_3], 20 wt% $Mn(NO_3)_2$, air	DMBQ, syringaldehyde, vanillin	11.5 wt% DMBQ; 66.3% conversion	Stärk et al. (2010)
Microwave-assisted selective Cα–Cβ bond cleavage with $Fe_2(SO_4)_3$ catalyst	Organosolv lignin	MeOH/H_2O, $Fe_2(SO_4)_3$, 160°C, 30 min, 150 W	Vanillin, syringaldehyde, methyl derivatives	96.3% Cα–Cβ cleavage	Zhu et al. (2017)
Microwave-assisted liquefaction	Wheat-straw alkali lignin	120°C, 40 min, 10 wt% H_2SO_4, phenol donor, 300 W	Monophenolic compounds	15.77 wt%	Ouyang et al. (2015)
Oxidative redistribution with phenol in ionic liquids	Organosolv/Klason lignin	180°C, 3–9 h, [Emim][ABS] or [Bmim][$MeSO_4$], $CuCl_2 \cdot 2H_2O$/EDTA	Oligomeric lignin fragments	~90 wt%	Nanayakkara et al. (2014)
Anaerobic bacterial biodegradation by Acetoanaerobium sp.	Kraft lignin	31.5°C, pH 6.8, KL:NH_4Cl 6.5:1, 12 d	Low-molecular-weight aromatics and acids	24.9% COD Removal	Xu et al. (2018)
Enzymatic (laccase-catalyzed) depolymerization in DES	Alkali lignin	Room temperature, pH 5–7, ChCl–glycerol or betaine–glycerol (DES)	Oligomers and small aromatics	N/A	Zhou et al. (2023)

OAc (Acetate), Emim (1-Ethyl-3-methylimidazolium), CF_3SO_3 (Triflate anion), ABS (Ammonium bisulfate), Bmim (1-Butyl-3-methylimidazolium), DMBQ (2,6-Dimethoxy-1,4-benzoquinone).

and developing effective methods remains a major challenge in enhancing its value-added utilization (Chio et al., 2019; Roy et al., 2022). Several depolymerization techniques (Table 4.1) have been successfully investigated in various studies, offering advantages in efficiency, yield, and adaptability to different lignin sources.

4.3.1 Thermal depolymerization

Thermal depolymerization is a class of thermochemical processes that break down complex organic materials into simpler molecules using heat. Among the primary techniques are pyrolysis, hydrothermal liquefaction, and hydrothermal carbonization, each operating under distinct conditions and yielding different products. Collectively, these thermal depolymerization methods offer pathways for waste-to-energy and resource recovery. However, challenges remain, including harsh reaction conditions, high coke yields, low selectivity in depolymerization, and difficulties in separating the final products, which hinder their widespread application (Pei et al., 2024).

4.3.1.1 Lignin depolymerization through pyrolysis

Pyrolysis is one of the most extensively researched thermochemical methods for lignin depolymerization, involving the thermal breakdown of lignin at an elevated temperature and in an oxygen-free environment, often with the aid of catalysts (Agarwal et al., 2018). The process typically begins with primary pyrolysis between 200 and 400°C, during which ether bonds, particularly α-O-4 and β-O-4 linkages, are cleaved (see Figure 4.5 for the lignin pyrolysis mechanism). As the temperature exceeds 400°C, secondary pyrolysis takes place, which involves radical-driven reactions that cause significant structural rearrangements and promote C–C bond cleavage, resulting in a mixture of monomeric and oligomeric products. However, secondary reactions may lead to repolymerization, reducing monomer recovery efficiency (Li & Takkellapati, 2018).

The reactivity of lignin during pyrolysis is significantly influenced by its chemical structure, which includes functional groups such as carboxyl, carbonyl, hydroxyl, and methoxyl groups, and the production of biochar is considerably impacted by methyl groups. Lignin with a higher methoxyl concentration generates less char because these groups stabilize molecular fragments and prevent further polymerization. The pyrolysis process is also temperature sensitive; for example, lignin softens below 200°C, whereas depolymerization occurs between 200 and 350°C, yielding aromatic alcohols like coniferyl and sinapyl alcohols. When the temperature rises to 350–500°C, it produces methylphenols and alkylphenols. Methanol and other low-molecular-weight compounds are the byproducts of further breakdown. Different phenolic product distributions are produced by hardwood lignin, which contains a mixture of aliphatic and phenolic groups, and softwood lignin, which is rich in guaiacyl phenolic groups (Hoang et al., 2021).

Lignin pyrolysis produces various volatile compounds (e.g., methanol, acetone), monolignols, phenolics (e.g., phenol, guaiacol, syringol), and gaseous products (e.g., H_2, CO_2, CH_4). Biochar, a stable carbon-rich solid, forms as a byproduct and can be further functionalized for advanced applications (Liu et al., 2015). The efficiency and selectivity of pyrolysis products are highly dependent on factors such as lignin

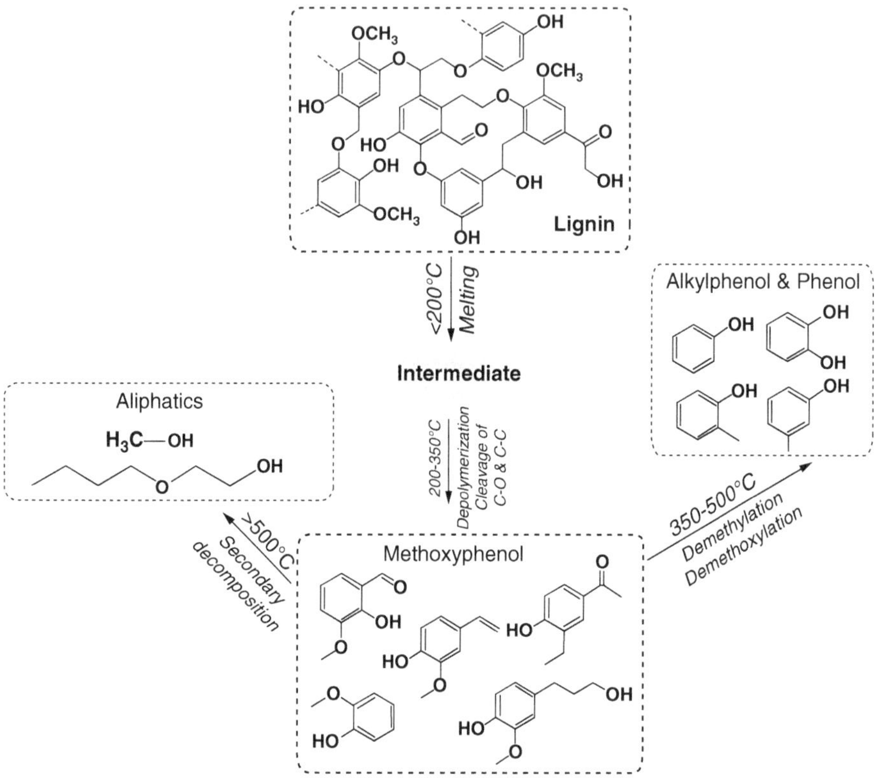

FIGURE 4.5 Mechanism of lignin pyrolysis adapted from Chen et al. (2019)

origin, operational temperature, reactor configuration, heating rate, and feed characteristics (Zhou et al., 2022).

4.3.1.2 Hydrothermal liquefaction of lignin

Hydrothermal liquefaction (HTL) is a thermochemical technique used to convert biomass into liquid fuels by treating it in a high-temperature, high-pressure aqueous environment. This process involves the breakdown of complex biopolymeric structures into predominantly liquid-phase products under typical conditions ranging from 523 to 647 K and pressures between 4 and 22 MPa using hot-compressed or subcritical water (374°C, 22.1 MPa) as the reaction medium, with or without added catalysts. Solvents like water, methanol, ethanol, and formic acid are typically used, often in subcritical or supercritical states to enhance reaction efficiency (Cao et al., 2020). Hydrothermal processing is categorized into three types based on temperature: hydrothermal carbonization (<520 K) produces hydrochar; hydrothermal liquefaction (520–647 K) yields biocrude, a petroleum-like liquid fuel; and hydrothermal gasification (>647 K) generates synthetic fuel gas with high carbon efficiency (Elliott et al., 2015). Among operational parameters, temperature plays a critical role; elevated temperatures enhance lignin depolymerization, improving biocrude yield and reducing reaction time. However, excessive heat also promotes repolymerization of low-molecular-weight intermediates through radical mechanisms, potentially lowering product quality (Ren et al., 2019). Despite its potential, challenges remain in controlling product selectivity, minimizing secondary reactions, and optimizing catalyst systems to enhance monomer yields. Nonetheless, HTL stands out as a viable and sustainable pathway for converting lignin into valuable bio-based chemicals under eco-friendly conditions.

4.3.1.3 Hydrothermal carbonization

Hydrothermal carbonization (HTC) is a thermochemical process that transforms biomass, such as lignin, at elevated temperatures (180–250°C) and pressures (10–25 bar) to the biomass, leading to a series of chemical reactions such as hydrolysis, dehydration, decarboxylation, and polymerization (Durak, 2023). This method enhances the solubility of lignin and promotes interactions with water, leading to the formation of carbon-rich structures. The key chemical reactions involved include hydrolysis, cleavage of C–O–C and C–C bonds, demethoxylation, alkylation, and condensation, with β-O-4 and Cα–Cβ bonds being the most susceptible to breakdown, while the aromatic rings typically remain intact. At lower temperatures and shorter reaction times, HTC favors the production of phenolic monomers and dimers, whereas higher temperatures and prolonged durations promote demethoxylation and alkylation, resulting in alkyl phenol formation.

The resulting char from HTC at elevated temperatures tends to have a rough surface, minimal vesicles, and a high degree of crystallinity, with most functional groups removed except hydroxyl groups. One of the main advantages of HTC over pyrolysis is its ability to process high-moisture feedstocks without the need for energy-intensive drying. However, commercialization of HTC remains limited due to the need for high-pressure systems, complex reactor and separator designs, difficulties in continuous feeding under pressure, and potential wall effects in heat and mass transfer during scale-up. Despite these challenges, HTC holds significant potential for converting lignin into valuable chemicals and functional carbon materials (Liu et al., 2015).

4.3.2 Chemical catalytic lignin depolymerization

Chemical-mediated technologies offer an efficient pathway for the conversion of lignin into high-value products. Among these, catalytic depolymerization stands out as a promising approach due to its ability to selectively break ether and C–C bonds through hydrolysis, oxidation, and non-reductive fractionation. The effectiveness of lignin depolymerization is highly dependent on the type and design of catalysts. Common catalysts include nano-structured precious metals, transition metals, and acid/base bifunctional materials, each with distinct advantages and limitations. Compared to traditional thermal methods, catalytic approaches generally operate under milder conditions and offer improved selectivity. However, concerns remain regarding the environmental impact associated with the use of certain chemical catalysts (Ji et al., 2023).

4.3.2.1 Acid-catalyzed degradation of lignin

Acid-catalyzed depolymerization of lignin primarily involves the hydrolytic cleavage of α- and β-aryl ether linkages, with α-aryl ethers depolymerizing more readily due to their lower activation energy (Agarwal et al., 2018). Various acids, including hydrochloric, sulfuric, formic, and peracetic acid, have been employed as catalysts for the depolymerization of lignin by targeting ether linkages. These acids facilitate hydrolysis by providing hydronium ions (H_2O^+), which promote the cleavage of ether bonds, leading to the formation of valuable phenolic compounds (Mahmood et al., 2015b). The depolymerization of β-O-4 linkages proceeds via two major pathways: path A, the formation of Hibbert's ketones (C3 fragments), and path B, which involves the loss of formaldehyde followed by the formation of ethylbenzene derivatives (C2 fragments), with the former being dominant due to the instability of C2 aldehydes, which often undergo repolymerization (see Figure 4.6).

A study by Fraile et al. (2018) has demonstrated that triflic acid and heterogeneous solid-acid catalysts like Nafion SAC-13 are effective in cleaving β-O-4 ether linkages in lignin. Nafion, a highly fluorinated sulfonic acid polymer, exhibits acidity similar to concentrated sulfuric acid. The acidolysis of α-aryl ether linkages follows an SN1-type mechanism and proceeds faster in phenolic and alcoholic solvents compared to water. Various Lewis acids, including chlorides of zinc, boron, iron, and copper, have also been used for lignin depolymerization. Among them, zinc chloride in ethanol showed the highest efficiency, yielding up to 56 wt% of liquid products (Zhang et al., 2014). The use of Lewis acids in alcoholic media or ionic liquids facilitates bond cleavage by generating Bronsted acidity in situ. Ionic liquids improve lignin solubility and stabilize carbocation intermediates, enhancing the cleavage of β-O-4 bonds. Although acid-catalyzed depolymerization of lignin has been widely explored, it is hindered by significant challenges such as the need for harsh conditions (high temperature, pressure, and corrosive reagents), generation of hazardous waste, and product repolymerization (Forsythe et al., 2013), which reduces the yield

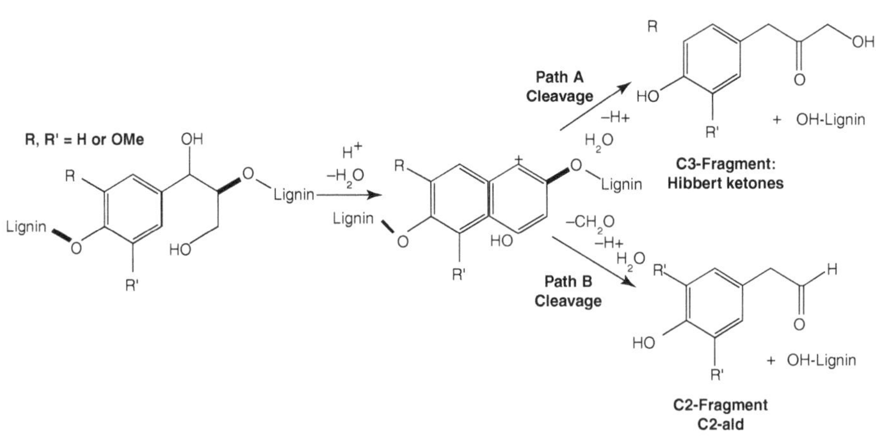

FIGURE 4.6 Major pathways for the acidolysis of lignin β-O-4 linkage, showing C2 and C3 fragments adapted from Agarwal et al. (2018)

of valuable aromatics. Addressing these issues is crucial to enhance the process's efficiency and sustainability.

4.3.2.2 Base-catalyzed degradation of lignin

Base-catalyzed depolymerization (BCD) is one of the most extensively studied methods for lignin valorization. NaOH is the most widely used base catalyst owing to its low cost, high effectiveness, and industrial applicability. The β-O-4 ether bond, which is the most abundant linkage in lignin, begins to cleave at approximately 270°C. Under strong alkaline conditions, cations from bases such as NaOH or KOH form cationic adducts that promote a six-membered transition state, efficiently breaking the β-O-4 bond to yield monomeric phenolics including catechol, guaiacol, and syringol. BCD typically operates under harsh conditions (temperatures > 300°C and pressures > 200 bar), which also cleave aryl-alkyl linkages and methoxy groups. NaOH enhances lignin degradation, giving over fourfold higher product yield than base-free reactions, and strong bases such as NaOH and KOH outperform weaker alternatives like calcium hydroxide ($Ca(OH)_2$) and lithium hydroxide (LiOH) by improving depolymerization efficiency and reducing repolymerization and char formation (Toledano et al., 2012).

In aqueous systems, increasing NaOH concentration correlates with higher monomer yield, reaching 25.8 wt% pyrocatechol at 315°C, while added phenol aids lignin solubilization and suppresses char. Efficiency depends on base type and concentration, reaction temperature and pressure, and the solvent environment (Roberts et al., 2011). For example, 5% NaOH at 260°C for one hour achieves near-complete alkaline lignin degradation with minimal residue, whereas low base concentrations suffer from acid formation, pH drop, and repolymerization (Chio et al., 2019). To overcome these limitations, organic solvents such as ethanol, isopropanol, and tetrahydrofuran (THF) are used to boost monomer yields and suppress side reactions, though they may also condense with lignin fragments. Recently, magnesium oxide (MgO), a solid base catalyst, has emerged as a cost-effective, recyclable alternative; when combined with THF, MgO shows strong catalytic activity, simplifies post-reaction separation, and improves overall monomer yield. Reaction parameters, including base concentration, temperature, solvent choice, and reaction time, critically determine the yield and distribution of lignin-derived aromatic monomers (Roy et al., 2022).

4.3.2.3 Metal-catalyzed depolymerization

The depolymerization of lignin through traditional acid and base catalysis is often hindered by low product selectivity and the requirement for harsh reaction conditions. These limitations have prompted the exploration of more efficient and selective catalytic approaches under milder conditions. Among these, metal-based catalysts have emerged as a promising solution due to their ability to selectively cleave specific bonds within the lignin structure. Metal catalysts for lignin depolymerization are generally divided into three classes: transition metal catalysts, precious metal catalysts, and bimetallic catalysts. Transition metals such as iron (Fe), nickel (Ni), copper (Cu), cobalt (Co), and molybdenum (Mo) are widely studied for their low cost, abundance, and catalytic efficiency; nickel in particular shows strong potential for selectively cleaving ether bonds and hydrolyzing carbon–hydroxyl groups on lignin side chains to yield phenolic compounds and alkanes. Precious metals, including ruthenium (Ru), rhodium (Rh), palladium (Pd), and platinum (Pt), excel at hydrogenation and hydrodeoxygenation reactions, but their high cost and limited availability challenge industrial scalability. The choice of catalyst support, including activated carbon, silica, alumina, or zeolites, has crucially influenced metal dispersion, prevented agglomeration, and enhanced stability and reusability (Pei et al., 2024).

For example, Song et al. (2012) identified Ni/AC (Ni on activated carbon) as the most effective heterogeneous catalyst: under mild conditions (160°C, 3 MPa H_2) in ethylene glycol, Ni/AC cleaves β-O-4 ether bonds in lignosulfonate with up to 91% conversion and 75–95% selectivity toward 4-propylguaiacol and 4-ethylguaiacol. Unlike Pd/AC, Ni/AC preserves aromatic rings, tolerates sulfur by converting sulfonate groups to H_2S (forming NiS), and retains activity after H_2 regeneration. To balance performance and cost, bimetallic catalysts leverage synergistic effects between two metals. Zhang et al. (2014) developed a NiAu bimetallic catalyst that converts organosolv lignin to aromatic monomers at 14 wt% yield in pure water at 170°C, addressing the selective cleavage of β-O-4 linkages at low temperature. However, the use of precious metals in bimetallic systems may still incur high costs and risk excessive hydrogenation, which can diminish yields of valuable phenolic products.

4.3.2.4 Oxidative/reductive lignin depolymerization

Oxidative depolymerization offers a mild, sustainable route to valorize lignin by selectively cleaving ether linkages (β-O-4, β-1) while retaining its aromatic core. Common oxidants include H_2O_2, permanganate, nitrobenzene, copper oxide, and chlorine-based reagents; H_2O_2 is favored for its low cost, eco-friendliness, and efficacy at low concentrations, which can be further enhanced by Cu^{2+}/Fe^{3+} catalysts or acidic media (Roy et al., 2022; Zhang et al., 2017). Microwave-assisted wet peroxide oxidation over Ti-silicalite-1 boosts dicarboxylic acid yields (e.g., succinic, malic), while nitrobenzene oxidation of kraft lignin can produce up to 13 wt% vanillin and peroxodicarbonate up to 6.2 wt%. Advanced systems like $NaVO_2/H_2SO_4$, polyoxometalate ionic liquids, and Co–Schiff base catalysts enable high-selectivity conversion to formic, acetic, maleic, and adipic acids (Yin et al., 2025). Collectively, these methods create versatile pathways to aromatic monomers, acids, and renewable fuel precursors under moderate conditions.

Reductive depolymerization employs catalytic hydrogenolysis over heterogeneous metal catalysts (e.g., Ni, Ru, Pt) to selectively cleave C–O and C–C bonds, particularly the abundant β-O-4 linkages—by supplying hydrogen from pressurized H_2 or in situ donors such as methanol or isopropanol. While conventional approaches demand high H_2 pressures and temperatures above 200°C, transfer hydrogenolysis using alcohols and non-precious metals under milder conditions has proven an effective alternative. Wei et al. (2023) introduced an acid-promoted Ni/C system that operates at just 160°C in methanol, incorporating 0.4 vol% H_2SO_4 and 0.4 mmol/g p-hydroxybenzyl alcohol as a capping agent to suppress condensation. This protocol achieves 20.42 wt% phenolic monomers and 31.70 wt% oligomers while keeping undepolymerized lignin below 10 wt%. By competitively trapping Cα carbocations, p-hydroxybenzyl alcohol minimizes repolymerization and enhances selectivity, offering a scalable, energy-efficient route for lignin valorization and lignin-first biomass fractionation.

4.3.3 Ionic liquid-assisted lignin depolymerization

Ionic liquid (IL)-assisted lignin depolymerization has emerged as a versatile and environmentally benign approach to lignin valorization by operating under mild, controllable conditions and leveraging ILs' negligible vapor pressure, exceptional thermal and chemical stability, and high solubilizing capacity. Lignin's intricate hydrogen-bond network, arising from its abundant hydroxyl functionalities, tends to restrict chain mobility and complicate the cleavage of ether linkages (Lei et al., 2015). By forming hydrogen bonds with terminal hydroxyl groups and engaging in π-stacking interactions with imidazolium cations, ILs effectively disrupt this network, enhancing lignin dissolution and facilitating selective β-aryl ether bond cleavage at moderate temperatures (100–170°C) (Stärk et al., 2010). Imidazolium-based solvents such as 1-ethyl-3-methylimidazolium acetate [C_2mim] [OAc], 1-ethyl-3-methylimidazolium trifluoromethanesulfonate [Emim] [CF_3SO_3], and di-isopropyl-ethyl-ammonium acetate and di-iso-propyl-ethyl-ammonium octanoate have demonstrated significant depolymerization efficiencies and high product selectivity, while acidic ILs (e.g., 1-H-3-methylimidazolium chloride) can simultaneously serve as catalysts to further promote bond scission (Cox et al., 2012). Environmental impact assessments indicate that optimizing IL recovery and minimizing potential ecotoxicity are critical to ensuring a net sustainability benefit over traditional solvents. Nonetheless, the high cost of ILs and the complexity of their separation and reuse remain major obstacles to large-scale adoption. Continued innovation in IL design, coupled with robust recovery and recycling strategies, will be essential to realize the full potential of IL-based lignin depolymerization.

4.3.4 Microwave-assisted depolymerization

Microwave-assisted depolymerization (MAD) has emerged as a powerful, green approach for lignin valorization owing to its rapid, volumetric heating, negligible thermal inertia, high energy efficiency, and enhanced reaction selectivity (Wang et al., 2020). By operating in the 300 MHz–300 GHz frequency range, microwave irradiation couples directly with lignin's polar functional groups (hydroxyl, carbonyl, and methoxyl groups), inducing rapid molecular oscillations that convert electromagnetic energy into localized heat. This unique heating mechanism lowers activation energies, accelerates reaction kinetics, and enables targeted cleavage of β-O-4 ether bonds and C–C linkages, which are critical for effective lignin depolymerization. Microwave heating offers several advantages, including improved control over the heating rate, faster processing times, more uniform heat distribution, and direct interaction with the reaction medium.

Depending on operating conditions, MAD of lignin falls into two main categories: microwave-assisted pyrolysis, carried out at temperatures above 400°C under inert or oxygen-free atmospheres to produce bio-oils and char, and microwave-assisted solvolysis, conducted below 200°C in polar solvents (e.g., alcohols, water, or ionic liquids), often under mild pressure, to yield soluble lignin fragments (Ouyang et al., 2015).

Incorporating heterogeneous catalysts further boosts efficiency and selectivity; for instance, ferric sulfide-catalyzed MAD increased the soluble lignin fraction from 67 to 86% at temperatures under 160°C, outperforming conventional heating (Zhu et al., 2017). Beyond depolymerization, MAD effectively enhances lignin liquefaction in ionic liquids, reducing reaction times and temperatures through unique electromagnetic field effects. The resulting bio-polyols can be directly applied in polyurethane foam synthesis, underscoring MAD's promise as a scalable, sustainable technology for high-value lignin conversion (Chio et al., 2019).

4.3.5 Biological depolymerization

Biological depolymerization of lignin offers a green and sustainable alternative to high-temperature, high-pressure methods, leveraging bacteria, fungi, and enzymes evolved to cleave lignin's complex ether and C–C bonds into valuable aromatic compounds under mild conditions, thereby enhancing selectivity, minimizing side reactions, and reducing byproduct formation such as coke while lowering equipment costs (Pei et al., 2024). Fungal systems, notably white-rot fungi like *Phanerochaete chrysosporium*, secrete a suite of extracellular oxidoreductases—laccases, lignin peroxidases (LiPs), manganese peroxidases (MnPs), and versatile peroxidases (VPs)—that employ oxidative mechanisms to target both phenolic and recalcitrant non-phenolic structures. Laccases catalyze one-electron oxidations of phenolic moieties and, when coupled with low-molecular-weight mediators (e.g., acetosyringone and syringaldehyde), extend their action to non-phenolic linkages by generating reactive radicals that prevent repolymerization (Davinia et al., 2013).

LiPs, with high redox potential, directly oxidize non-phenolic lignin units via long-range electron transfer, while MnPs produce Mn^{3+} chelates that oxidize phenolic substrates through lipid peroxidation pathways. VPs combine catalytic motifs of LiPs and MnPs, enabling broad substrate specificity. Dye-decolorizing peroxidases (DyPs), found in fungi and bacteria such as *Irpex lacteus* and *Rhodococcus jostii*, further expand lignin's degradable repertoire by attacking anthraquinone and other complex aromatic dyes. Accessory enzymes, like aryl-alcohol oxidases, glyoxal oxidases, and multicopper oxidases, generate in situ hydrogen peroxide or act as redox mediators, fueling peroxidase cycles. Anaerobic gut fungi operate under oxygen-free conditions via distinct radical-mediated pathways, while genetically tractable yeasts (*Aspergillus niger*, *Yarrowia lipolytica*, and *Pichia pastoris*) tolerate inhibitors and convert lignin fragments into vanillin, ferulic acid, and polyhydroxyalkanoates for industrial applications (Shrestha et al., 2024). Integrating these enzymatic systems with Fenton chemistry and metabolic engineering underpins emerging strategies for scalable, high-efficiency lignin valorization (Abdelaziz et al., 2016).

4.3.6 Emerging techniques

4.3.6.1 Selective electrochemical lignin depolymerization

Selective electrochemical methods offer an efficient and sustainable route to lignin depolymerization by combining direct electron transfer with redox-mediated pathways. For example, kraft lignin treated on a $Ti/TiO_2NT/PhO_2$ electrode exhibited a 13% decrease in C–O–C linkages and a 44% increase in carbonyl (C=O) groups, reflecting significant structural transformation (Pan et al., 2012). Electrochemical potential crucially dictates product distribution; for instance, Reichert et al. (2012) demonstrated the application of higher potentials on a ruthenium-vanadium-titanium oxide electrode in protic ionic liquids shifted products toward lower-molecular-weight compounds.

In alkaline media, Zhu et al. (2014) showed that a RuO_2–IrO_2/Ti anode cleaved C9 lignin linkages to release over 20 identifiable low-molecular-weight aromatic species, including p-hydroxybenzene, guaiacyl, and syringyl phenylpropane units. The use of redox mediators like laccase-derived compounds (e.g., N-hydroxyphthalimide, 1-hydroxybenzotriazole, violuric acid, 2,2,6,6-tetramethylpiperidine-N-oxyl, and 2,2'-azinobis-(3-ethylbenzothiazoline-6-sulfonate) (ABTS)) has been shown to enhance Cα-Cβ bond cleavage, promoting aldehyde formation. Notably, oxidation of β-O-4 linkages, a key structural motif in lignin, facilitates bond weakening and selective fragmentation. These findings support the development of targeted electrochemical approaches for lignin valorization into valuable aromatic chemicals (Gao et al., 2017).

4.3.6.2 Lignin depolymerization by deep eutectic solvents

Deep eutectic solvents (DESs) have emerged as sustainable, greener alternatives to conventional organic solvents, offering low toxicity, biodegradability, and ease of synthesis. A distinct subclass, amino acid-based DESs (AADESs), incorporates at least one amino acid as either a hydrogen-bond donor (HBD) or acceptor (HBA). They are prepared simply by heating and vacuum evaporation and have found applications in pharmaceuticals, food, cosmetics, and biomass processing. They are classified as ionic or non-ionic based on noncovalent interactions, with hydrogen bonds, primarily involving –COOH and $–NH_2$ groups, providing structural cohesion. Alcohol-based HBDs such as glycerol are often employed for their high diffusivity, which enhances solvent performance (Liang et al., 2020).

Rahman et al. (2022) formulated two novel AADESs using L-(+)-arginine (Arg) and L-glutamic acid (Glu) as HBAs with glycerol (Gly) as the HBD in a 1:3 molar ratio to explore acid-base effects on lignin depolymerization. The eutectic points of Glu-Gly (0.14°C) and Arg-Gly (1.36°C) reflect strong noncovalent interactions. FTIR, ^1H NMR, and mass spectrometry confirmed successful component interactions with minimal ester impurities in Glu-Gly. Thermal stability studies showed Glu-Gly to be more stable than Arg-Gly, exhibiting two degradation stages: 210–271°C (40.47% weight loss from glycerol decomposition and partial Glu breakdown) and 271–300°C (continuous degradation, primarily of Glu). In lignin pretreatment, Arg-Gly isolated a single monomer (4-methyl benzaldehyde), whereas Glu-Gly released three monomeric compounds. The superior performance of Glu-Gly is attributed to its lower viscosity, higher thermal stability, acidic environment, and stronger hydrogen-bonding interactions with glycerol. These attributes highlight Glu-Gly's promise for efficient biomass processing and lignin depolymerization.

4.4 DEPOLYMERIZED LIGNIN CHARACTERIZATION THROUGH CONVENTIONAL AND ADVANCING ANALYTICAL TOOLS

Accurate characterization of lignin depolymerization products is essential for the rational design and optimization of conversion processes. Lignin has a complex structure and yields a range of products from monomeric phenolics to oligomeric fragments and residual high-molecular-weight polymers; a multi-technique analytical approach is required. Two-dimensional heteronuclear single quantum coherence (2D HSQC) NMR spectroscopy provides direct correlations between ^{1}H and ^{13}C nuclei, enabling unambiguous identification and quantification of key interunit linkages (β-O-4, β-β, β-5) and aromatic subunits (guaiacyl, syringyl, p-hydroxyphenyl) (Ali et al., 2024). Despite its structural resolution, conventional HSQC requires highly purified samples and extended acquisition times (6–12 h), impeding high-throughput analysis. Despite its structural resolution, conventional HSQC requires highly purified samples and extended acquisition times (6–12 h), impeding high-throughput analysis. Fourier-transform infrared (FTIR) spectroscopy offers rapid, in situ monitoring of functional group transformations, such as the attenuation of ether stretching vibrations at ~1,100 cm^{-1} and the emergence of carbonyl absorptions near 1,700 cm^{-1} but lacks the specificity to resolve individual linkages.

Gel permeation chromatography (GPC), especially when coupled with multi-angle light scattering (MALS), yields precise molecular-weight distributions; however, reliance on polystyrene calibration can lead to systematic underestimation of lignin's true molar mass. For compound-specific analyses, high-performance liquid chromatography–mass spectrometry (HPLC-MS) and high-resolution mass spectrometry afford accurate monomer and oligomer identification, albeit through offline workflows that necessitate rigorous sample purification (Zhou et al., 2024). To address the need for higher throughput and real-time capability, several innovative techniques have been introduced. Non-uniform sampling in 2D HSQC NMR reduces data acquisition by up to 75%, decreasing experiment duration to approximately 1.4 h while maintaining over 98% of spectral integrity (Hao et al., 2025). In mass spectrometric analysis, optimized MALDI-TOF protocols employ matrices such as α-cyano-4-hydroxycinnamic acid (CHCA) and 2,4,6-trihydroxyacetophenone in sequential analyte-matrix-analyte deposition (10:1–100:1), achieving reproducible ionization of high-mass lignin fragments; ionic-liquid matrices (e.g., IMTBA-CHCA) further enhance signal intensity and spectral fidelity (Kosyakov et al., 2014).

Finally, flow-induced dispersion analysis, a microfluidic adaptation of Taylor dispersion analysis, quantifies the hydrodynamic radii of lignin fragments (0.55–1.45 nm) under laminar flow conditions, correlating closely with DOSY NMR and GPC results, yet requiring minimal sample preparation and no stationary phase (Simonsen et al., 2025). By integrating these classical and emerging techniques, each addressing specific aspects of lignin structure, size distribution, and functional group evolution, researchers can achieve a comprehensive, multi-scale understanding of depolymerization pathways. Such a concerted analytical strategy is essential for closing mass balances, elucidating reaction mechanisms under high-pressure and high-temperature conditions, and ultimately guiding the development of more efficient, selective, and scalable lignin valorization processes.

4.5 CONVERSION OF DEPOLYMERIZED LIGNIN INTO BIOPOLYMER RESINS

Lignin's rigid, aromatic structure offers a renewable platform for polymer synthesis but requires chemical modification to overcome brittleness. Hydroxyl functionalization via hydroxypropylation, glycidylation, or other routes enhances lignin's reactivity, enabling its integration into polyurethanes, polyesters, and phenolic resins (Laurichesse & Avérous, 2014). Depolymerized lignin has successfully replaced up to 50 wt% of phenol in phenol–formaldehyde and epoxy formulations, improving cure kinetics and mechanical properties in adhesives and coatings. Likewise, direct use of lignin-derived polyols with polyisocyanates produces rigid and flexible polyurethane foams at a lower cost. Although full commercialization is still emerging, lignin's low price, abundance, and intrinsic chemical resistance make it an attractive co-monomer for multi-component polymer systems.

4.5.1 Depolymerized lignin-substituted phenol–formaldehyde resin and foam

Phenol–formaldehyde (PF) resins are thermosetting polymers renowned for their thermal stability, flame retardance, chemical resistance, and insulating properties (DSouza et al., 2025). Synthesized under acidic (novolac) or alkaline (resole) conditions, PF resins serve as electrical insulators, circuit boards, adhesives, and composites but are limited by brittleness and reliance on petroleum-derived phenol. PF foams, employed in insulation (closed-cell) and horticultural (open-cell) applications, derive their properties from cell morphology, which is controlled by blowing agents, acid-catalyzed curing, and surfactants such as Tween 40/60/80 and DC-193, with wetting agents (e.g., sodium laureth sulfate and sodium dodecyl benzenesulfonate) enhancing water retention. Several studies have worked on these formulation parameters (Table 4.2), including the formaldehyde-to-phenol ratio, cure temperature, and additives like sodium polyacrylate, which influence foam density, thermal conductivity, resin viscosity, and structural stability (DSouza et al., 2024).

To reduce environmental impact and material cost, lignin has been explored as a bio-phenol substitute. Technical lignin's high molecular weight and low reactivity are overcome by depolymerization via thermochemical routes such as pyrolysis or hydrothermal liquefaction, yielding lower-mass lignin fractions suitable for PF formulations (Li et al., 2019). DSouza et al. (2024) replaced 0–50 wt% of phenol with kraft lignin in open-cell PF foams. Although resin viscosity increased from 3.50 P to 22.20 P, the resulting foams maintained low densities (39–55 kg/m³), high open-cell contents (55–100%), and water absorption capacities of 1,500–2,500%, with uptake rates of 0.24–0.66 cm³/s. KL-substituted foams also demonstrated superior biodegradability, achieving up to 68% weight loss under soil burial, making them well suited for hydroponics and floral support.

Building on these findings, DSouza et al. (2025) produced depolymerized Kraft lignin (DKL) via NaOH treatment (250°C, 2 h), which exhibits lower molecular weight and increased hydroxyl functionality. When used to replace 50% of phenol, DKL-based foams retained low density (~62 kg/m), near-complete open-cell structure, and achieved 2,557% water absorption by weight. They also reached 39% biodegradability under soil burial and 26% in composting tests. These studies confirm that up to 50% phenol substitution with KL or DKL yields PF foams matching or exceeding conventional performance, while significantly enhancing hydrophilicity, biodegradability, and sustainability, paving the way for fully bio-based thermosetting materials (see Figure 4.7).

TABLE 4.2
Studies on depolymerization techniques of lignin in bio-based phenol–formaldehyde resin and foam production

Foam/Resin Type	Lignin Source	Depolymerization Technique	Max. Substitution (%)	Resin and Foam Characteristics				Applications	Reference
				F/P Ratio	Thermal Conductivity W.m−1. K−1	Foam Density (kg/m³)	Resin Viscosity mPa.s		
Hydrophilic phenol–formaldehyde (PF) foam	Alkaline hydrolysis of kraft lignin	Base-catalyzed hydrolytic cleavage	50	N/A	N/A	~ 62	~878	Floral arrangements & hydroponic germination	DSouza et al. (2025)
One-pot solid-acid catalytic phenolation (SACP)-Enzymatic hydrolysis lignin (EHL) phenolic foam	EHL	SACP	50	1.6:1	0.038	38.89 ± 2.29	8900 ±500	Lightweight thermal insulation	Wang et al. (2018)
Bio-based lignin phenol-formaldehyde resin (BPF)	Corncob lignin	Organosolv ultrasound-assisted depolymerization	55	2.5:1	N/A	N/A	500–1000	Plywood adhesive	Ma et al. (2025)
BPF	Hydrolysis lignin	Low-T/P Hydrolysis-lignin depolymerization	50	1:1	0.033–0.040	~108	~194	Load-bearing insulation & fire-resistance	Li et al. (2017)

(Continued)

TABLE 4.2 (Continued)

Foam/Resin Type	Lignin Source	Depolymerization Technique	Max. Substitution (%)	Resin and Foam Characteristics				Applications	Reference
				F/P Ratio	Thermal Conductivity W.m−1. K−1	Foam Density (kg/m³)	Resin Viscosity mPa.s		
BPF	Industrial alkali lignin (wheat straw)	Hydrothermal degradation	40	N/A	N/A	N/A	1000–5000	Closed-cell flame-retardant insulating foam	Zhao et al. (2023)
Nitrogen-phosphorus lignin-based phenolic foam	EHL	Enzymatic hydrolysis of lignin.	30	1:1.7 (mol)	N/A	69.2	N/A	Rigid thermal-insulation & fire-protection foams	Zhou et al. (2024)
LPF resin for wood adhesive	Soda bagasse black liquor	Base-catalyzed hydrolytic cleavage	50	2.2:1	N/A	N/A	430	Wood adhesives for particleboard panels	Younesi-Kordkheili and Pizzi (2021)
Phenolic foam	Sulfite pulping process	Acid-catalyzed sulfonation	N/A	1.7:1	N/A	28.55	4500–6000	Fire-resistance	Hu et al. (2012)

N/A – Not Applicable

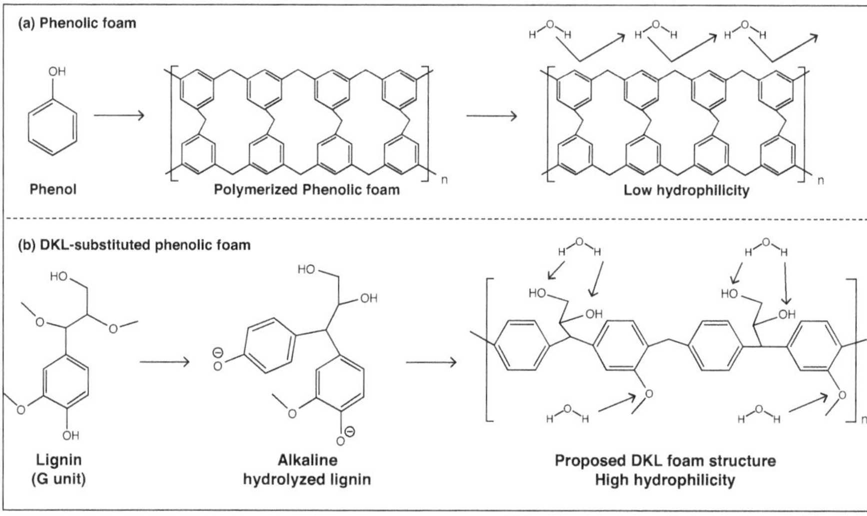

FIGURE 4.7 Structure of (a) phenolic foam and (b) DKL-substituted phenolic foams obtained after biodegradation by *Bacillus sp.* A4. Adapted from DSouza et al. (2025)

In another study, Younesi-Kordkheili and Pizzi (2021) investigated the effects of several lignin modification techniques on the wood adhesive properties of lignin phenol–formaldehyde (LPF) resins. Kraft lignin was chemically altered using glyoxal, phenol, maleic anhydride (MA), and an IL. Each modified lignin replaced 50% of the phenol in the resin composition. The synthesized LPF resins were examined for gel time, viscosity, solids content, and curing behavior before being used to make particleboard panels. Among the investigated alterations, MA-modified lignin resulted in resins with the shortest gel time, maximum viscosity and solids content, and lowest curing temperature, indicating increased reactivity. Particleboards glued with LPF resins treated with MA or IL demonstrated greater mechanical strength and much lower formaldehyde emissions. All modification treatments resulted in a lower curing temperature than unmodified lignin, showing the usefulness of lignin functionalization in enhancing LPF resin characteristics. (Younesi-Kordkheili & Pizzi, 2021).

Hu et al. (2012) synthesized a phenolic foam by partially substituting phenol with phenolated lignosulfonate (PLS). The resin was developed under alkaline conditions using phenol, formaldehyde, and PLS, then foamed with n-pentane as the blowing agent, 50% sulfuric acid as the catalyst, and Tween-80 as the surfactant (Hu et al., 2012). The resulting lignosulfonate-based phenolic foam was tested for TGA, SEM, and mechanical properties. When compared to conventional phenolic foams, the modified foam had lower density, increased toughness, and provided excellent thermal insulation. SEM research indicated a homogeneous and stable cellular structure, whereas TGA showed excellent thermal stability. The study revealed that PLS can serve as an effective, bio-based partial phenol substitution in phenolic foams, and the final product meets performance requirements for practical thermal insulation applications (Hu et al., 2012).

4.5.2 Depolymerized lignin-substituted polyurethane foams

Polyurethane (PU) foams, particularly the rigid grades used in construction and insulation, are traditionally produced by reacting petroleum-derived polyols with isocyanates (commonly polymeric MDI) in the presence of catalysts, blowing agents, surfactants, and crosslinkers to form a closed-cell urethane network (Mahmood et al., 2016; see Table 4.3). However, sustainability concerns have driven efforts to replace fossil polyols with lignin-based alternatives. Native lignin's high molecular weight and low solubility limit its direct use to below 30 wt% replacement (Verdolotti et al., 2013), but depolymerization techniques such as hydrothermal liquefaction (250°C, 2–3 MPa, ethanol–water) or alkaline hydrolysis (250°C, NaOH under N_2) produce oligomeric phenolics rich in aliphatic hydroxyls. These lignin-derived polyols exhibit hydroxyl values of 70–80 mg KOH/g and viscosities of 80–100 cP, closely matching commercial polyols while reducing steric hindrance. For example, Li et al. (2017) substituted 30 wt% of petroleum polyols with wheat-straw pyrolysis oil—characterized by a 49% phenolic content, 77.8 mg KOH/g hydroxyl value, and 87 cP viscosity—in flexible PU foams. By optimizing the bio-polyol/petroleum-polyol ratio, polyol-to-isocyanate ratio (2.5), and isocyanate index (0.66–0.72) via response surface methodology, they achieved 37% resilience and thermal stability to 200°C, meeting automotive cushioning standards despite a modest drop in tensile strength.

Similarly, Mahmood et al. (2013) incorporated hydrothermally liquefied lignin (70% recovery) at 30 wt% and 50 wt% levels in rigid PU foams. At 30 wt%, density fell from 58.0 to 55.6 kg/m, glass transition temperature rose from 138 to 146°C, and thermal conductivity decreased from 0.045 to 0.038 W/m/K, while compressive strength remained near 521 kPa at 10% deformation. At 50 wt%, density increased to 99.0 kg/m³, and water absorption doubled (0.12 → 0.20 wt%), reflecting altered cell morphology. In sucrose-polyol foams, 30–50 wt% lignin similarly enhanced density, stiffness, and glass transition temperature (178 → 189°C) while reducing moisture uptake (0.99 → 0.29 wt%) and thermal conductivity (0.044 → 0.035–0.037 W/m/K). Collectively, these studies demonstrate that depolymerized lignin derivatives can replace 30–50 wt% of petroleum polyols in both flexible and rigid PU foams, yielding materials with mechanical, thermal, and moisture-resistant properties comparable to or exceeding conventional formulations, thus advancing bio-derived PU technologies. A different investigation by Mahmood et al. focused on developing bio-based rigid polyurethane foams by incorporating hydrolytically DKL as a partial substitute for conventional polyols at a 50 wt% loading. Two approaches were examined: direct replacement of either PPG400 or sucrose-derived polyols with DKL, and the use of oxypropylated DKL as a modified polyol to enhance its reactivity and miscibility with isocyanates. Among the formulations tested, the foam containing oxypropylated DKL exhibited superior performance, achieving the highest compressive modulus (10,986 kPa) and the lowest thermal conductivity (0.029 W/m·K), indicating its strong potential for thermal insulation applications. Additionally, all the resulting foams demonstrated thermal stability up to approximately 200°C, reinforcing the viability of using depolymerized lignin as a sustainable and functional polyol alternative in rigid PU systems (Mahmood et al., 2015a). Vieira et al. (2023) developed fully bio-based rigid polyurethane foams (RPUFs) using 100% crude lignin-based polyol (LBP), synthesized via oxyalkylation of LignoBoost kraft lignin with propylene

TABLE 4.3
Studies on depolymerization techniques of lignin in bio-based polyurethane resin and foam production

Foam/Resin Type	Lignin Source	Depolymerization Technique	Max. Substitution (%)	Resin and Foam Characteristics				Applications	Reference
				Isocyanate index	Thermal Conductivity W/m/K	Foam Density (kg/m³)	Resin Viscosity mPa.s		
RPU foam	Corncob lignin	Base-catalyzed hydrolytic cleavage	N/A	1.0–1.2	0.028–0.032	30–50	800–1200	Designed for building/ appliance insulation	Xue et al. (2017)
RPU foam	*Eucalyptus globulus* lignin	Oxidative depolymerization	100	1.1	0.00372–0.0490	44.7–112.2	4 800–456500	Thermal insulation panels and appliance insulation	Pinto et al. (2021)
RPU foam	Pinus Wood	Hydrogenolysis lignin depolymerization	50	1.3	N/A	67–56	N/A	Insulation (construction and refrigeration), water treatment systems, flame retardants, optical sensors.	Quinsaat et al. (2021)
Bio-based rigid PU foam (BRPU)	Hydrolysis lignin	Hydrolytic liquefaction of lignin	70	1:1	0.030 ± 0.001	~64	390	Building insulation	Mahmood et al. (2016)
RPU foam	Pyrolytic lignin	Fast pyrolysis of lignin	20	N/A	N/A	48.94 ± 1.35	8900	Insulation, packaging, energy-absorption	Garcia et al. (2024)
Non- isocyanate-BRPU	Kraft lignin	Cyclocarbonation of lignin with glycerol carbonate	50.6	N/A	N/A	241 ± 34	N/A	Insulation, structural panels	Sternberg and Pilla (2023)
Rigid PU foam	Kraft lignin	Hydrolytic depolymerization	50	1.1	0.029–0.038	54–104	610	Insulation materials	Mahmood et al. (2015a)
Rigid PU foam	Kraft lignin	N/A	100	N/A	0.029	33.2	N/A	Thermal insulation	Vieira et al. (2023)

N/A – Not Applicable

carbonate. Using a design of experiments (DoE) approach, the formulation was optimized to achieve lightweight insulation foams with low thermal conductivity and minimal catalyst usage. The optimized bio-based RPUF exhibited a thermal conductivity of 0.0289 W/m·K and a low density of 33.2 kg/m³, meeting industry requirements for thermal insulation panels. Although the foams demonstrated slightly lower thermo-oxidative stability and mechanical strength compared to conventional polyol-based RPUFs, they showed improved fire performance—reducing the average heat release rate by 18.5% and increasing burn time by 25%, in compliance with the EN45545 standard. The study pointed out the importance of unpurified lignin-based polyols as sustainable alternatives in rigid PU foam production (Vieira et al., 2023).

4.5.3 Depolymerized lignin-substituted epoxy resins

Epoxy resins, most notably diglycidyl ether of bisphenol A (DGEBA), are prized for their superior thermal stability and mechanical strength, yet their reliance on bisphenol A raises environmental and health concerns. Lignin, a renewable aromatic biopolymer rich in phenolic hydroxyl groups, offers a sustainable alternative feedstock. However, unmodified lignin typically exhibits poor solubility, low reactivity, and high molecular weight, which limit its incorporation to below 50 wt% before phase separation, reduced tensile strength, and compromised thermal stability. To address these limitations, researchers have employed chemical functionalization strategies such as glycidylation, amination, and esterification to enhance lignin's compatibility with conventional epoxy systems (see Table 4.4). Glycidylation with epichlorohydrin (ECH) converts lignin's phenolic –OH groups into epoxy functionalities, facilitating co-curing with commercial resins. In particular, glycidylated depolymerized lignin with molecular weights in the 1,330–1,600 g/mol range has demonstrated effective co-curing behavior with commercial epoxy resins like EP828, yielding composites with improved thermal stability, mechanical performance, and crosslink density relative to neat DGEBA (Lu & Gu, 2023).

Ferdosian et al. (2014) illustrated that the organosolv lignin was reductively depolymerized over a Ru/C catalyst in supercritical acetone at 350°C and 10 MPa H_2, producing low-molecular-weight oligomers with enhanced hydroxyl functionality. These oligomers were then epoxidized using ECH in the presence of NaOH and a phase-transfer catalyst. Through central composite design optimization, the ideal conditions—55°C, 8 h reaction time, and an NaOH to DOL molar ratio of 6.3—yielded an epoxy prepolymer with a 99% yield and an epoxy equivalent weight of approximately 125 g eq^{-1}. Spectroscopic (^{1}H-NMR, FTIR) and GPC analyses confirmed successful depolymerization and epoxidation, showing a substantial increase in reactive epoxy sites and a suitable molecular-weight distribution for network formation. Lignin-based epoxy foams have been produced by integrating blowing agents into functionalized lignin resins. In particular, syntactic epoxy foams reinforced with hollow microspheres exhibit reduced density and enhanced compressive strength, making them suitable for demanding applications (Biswal et al., 2023). Moreover, strategic lignin depolymerization, such as reductive fractionation followed by glycidylation, yields bio-based epoxy resins and foams with thermal, mechanical, and flame-retardant properties comparable to or exceeding those of petroleum-derived bisphenol systems.

TABLE 4.4
Studies on depolymerization techniques of lignin in production of epoxy resin

Resin Type	Lignin Source	Depolymerization Technique	Resin Characteristics: Max. Substitution	Tg °C	(EEW) (g/eq)	Crosslink Density (mol/cm)	Flexural Strength (MPa)	Applications	Reference
Demethylated lignin epoxy resin (DMLE)	Alkali lignin	One-pot demethylation	20% DMLE in BADGE	140.6	240.1	6.64	62.3	Thermosetting epoxy networks for coatings, adhesives, composites	Zhen et al. (2021)
DKL epoxy resin	Kraft lignin	Low-pressure alkaline hydrolysis	50% DKL in BADGE	~101	768	3.5×10^{-3}	163	Glass-fiber-reinforced composites, structural adhesives	Feghali et al. (2018)
Lignin Hydrogenolysis Oil Glycidyl Ether (LHOGE) epoxy polymer	Native softwood lignin (LHOG 4–5)	Catalytic hydrogenolysis	67% LHOGE in BADGE replaced	LHOGE 4: 77 LHOGE 5: 82	LHOGE4: 422 LHOGE5: 451	N/A	LHOGE4: 152 LHOGE5: 163	High-performance coatings, adhesives, and composite matrices	Feghali et al. (2020)
Lignin Hydrogenolysis Oil Epoxy Prepolymer (LHOEP)	Native Pinus radiata lignin	Mild hydrogenolysis	50% LHOEP in BADGE	80	359	N/A	129	Coatings, adhesives, composites	Pas and Torr (2017)
Depolymerized hydrolysis lignin (DHL)-Epoxy/DDM	Hydrolysis lignin	Low-pressure alkaline depolymerization	100% DHL-Epoxy in matrix	N/A	N/A	N/A	47 ± 2	Polymer matrix in glass-fiber-reinforced composites	Ferdosian et al. (2016)

N/A - Not Applicable; Tg - Glass transition temperature; EEW - Epoxy Equivalent Weight; BADGE - Bisphenol A Diglycidyl Ether; DDM -4,4′-diaminodiphenyl methane

4.6 FACTORS INFLUENCING THE QUALITY OF DEPOLYMERIZED LIGNIN-BASED BIOPOLYMERS

The effective valorization of lignin into high-performance biopolymers depends on several interrelated factors, including the lignin's origin and purity, its functional group makeup, reaction conditions, and the use of additives or catalysts. Each of these influences both the efficiency of depolymerization and the performance of downstream polymeric materials.

4.6.1 Lignin source and purity

The structural characteristics of technical lignin are set by its botanical origin and isolation process. Kraft lignin, the most abundant technical stream, is produced under strongly alkaline, sulfide-rich conditions that cleave ether bonds but also promote condensation, yielding a high-molecular-weight, sulfur-containing material with relatively few β-O-4 linkages (Vásquez-Garay et al., 2021). In contrast, organosolv lignin extracted in organic solvents under milder acid or alkaline conditions retains more native β-O-4 and other labile linkages, translating into lower molecular weight and higher reactivity toward depolymerization. Soda lignin (alkali only) and lignosulfonates (sulfite pulping) occupy intermediate positions in terms of condensation and functional group distribution. Furthermore, the monomeric composition ratios of p-hydroxyphenyl (H), guaiacyl (G), and syringyl (S) units vary by feedstock; for example, softwoods are G-rich, hardwoods contain both G and S, and grasses exhibit all three units but low H content. Since β-O-4 ether bonds (bond dissociation 221–295 kJ/mol) are the most cleavable, lignin richer in these linkages (e.g., organosolv, hardwood) yields higher monomer fractions upon depolymerization, making them preferable for synthesizing high-performance polymers (Obrzut et al., 2023).

4.6.2 Functional group distribution

Lignin contains a complex mixture of functional groups such as aliphatic and phenolic hydroxyls, methoxyls, benzyl alcohols, and carboxyls, each imparting distinct reactivity. In polyurethane formulations, for example, hydroxyl groups react with diisocyanates via nucleophilic attack. Aliphatic hydroxyls (primary and secondary) readily form urethane linkages, whereas phenolic hydroxyls display lower nucleophilicity and often require activation or derivatization. Moreover, phenolic –OH subtypes (syringyl, non-condensed or condensed guaiacyl, and p-hydroxyphenyl) differ in steric accessibility and acidity. Uneven distribution can lead to phase separation, incomplete crosslinking, and heterogeneous mechanical properties. To address this, chemical modifications such as hydroxypropylation (introducing aliphatic chains), acetylation, or esterification to convert phenolic –OH into more reactive aliphatic esters, and homogenizing reactivity (Antonino et al., 2021). Alternatively, direct functionalization with diisocyanates produces lignin-based prepolymers rich in urethane linkages, although the selectivity toward specific hydroxyl subtypes remains an active research area.

4.6.3 pH of the Reaction Environment

The effect of pH affects both depolymerization pathways and byproduct formation. In BCD, strong alkaline media (pH ~12 with 2 M NaOH) solubilize lignin and cleave β-O-4 bonds efficiently, achieving up to ~25 wt% monomer yield while suppressing char to 2 wt% (Purushothaman et al., 2023). However, as lignin fragments ionize and phenolic acids accumulate, pH can drop, promoting repolymerization and char. Thus, maintaining high and stable pH is crucial. Under subcritical water conditions, mild alkalinity (pH ~8) minimizes coke formation (~13 wt%) and solids, whereas more extreme pH values either acid-catalyze condensation or foster excessive side reactions (Belkheiri et al., 2016). In acid-catalyzed or enzymatic depolymerization, mildly acidic conditions (pH 2.5–5.0) facilitate protonation of ether oxygen, yielding Hibbert's ketones and other intermediates. LiPs, for instance, show peak activity near pH 2.6, boosting β-O-4 cleavage from ~38 to ~92% conversion on model dimers (Pham et al., 2021). Yet overly low pH (<2) encourages condensation, lowering solubility. According to Yang et al. (2020), maintaining a pH of around 4.0 during hydrothermal pretreatment has been shown to best preserve β-O-4 linkages and stabilize lignin–carbohydrate complexes. Neutral or weakly acidic water–ethanol mixtures (no added acid) can balance reactivity and solubility, achieving ~70 wt% yield of depolymerized hydrolysis lignin with lower molecular weight, likely by reducing condensation and avoiding acid-catalyzed char (Mahmood et al., 2015b).

4.6.4 Additives and Modifiers

Preventing undesirable radical-induced repolymerization and char formation is vital for achieving high monomer yields and functional integrity in lignin valorization. Capping agents such as boric acid and phenol selectively mask phenolic –OH sites. Boric acid forms borate esters, and phenol yields phenolic adducts, thereby suppressing condensation and increasing monomer recovery to over 80 wt% in alkaline systems. Phenol additions (5–10 wt%) in ethanol–water mixtures similarly reduce repolymerization by 50–70%, although they require careful dosing to avoid excess volatiles. Hydrogenolytic catalysts (Ru, Ni, Pt, Pd) under H_2 effectively quench radical intermediates and cleave β-O-4 and C–C bonds. For instance, Ni/activated carbon at 160°C and 3 MPa H_2 converts lignosulfonate to 4-propylguaiacol with 91% conversion and 75–95% selectivity, even tolerating sulfur via in situ sulfide formation (Song et al., 2012). When paired with hydrogen-donor solvents (ethanol or isopropanol), these catalysts stabilize lignin fragments and minimize char, yielding clean, high-functionality lignin oils well suited for advanced biopolymer applications.

4.7 CHALLENGES, ADVANCEMENTS, AND FUTURE PERSPECTIVES

4.7.1 Technical Challenges in Lignin Depolymerization

Lignin depolymerization faces several technical and scientific challenges that hinder its effective utilization. One major obstacle is the inherently complex, three-dimensional, and amorphous structure of lignin, which makes it highly resistant to biological and chemical degradation. This necessitates the development of novel and efficient pretreatment strategies to weaken interunit linkages and facilitate breakdown. Additionally, the high molecular weight of lignin and its low solubility result in reduced yields of target products and increased side reactions during catalytic conversion, highlighting the need for more selective and robust catalytic systems (Ji et al., 2023). Catalyst design further complicates the process due to the structural variability of lignin depending on its plant source, which necessitates tailored catalytic approaches. Achieving selective cleavage of C–O and C–C bonds without disrupting aromatic structures remains a significant hurdle, as does catalyst deactivation caused by fouling from impurities or byproducts.

Noble-metal hydrogenation catalysts (e.g., Pd, Ru) can cleave C–O and C–C bonds, but they are expensive and prone to deactivation by coke formation or oxidation of active sites (Gale et al., 2020; Wang et al., 2023). Heterogeneous catalysts face additional challenges because bulky lignin polymers must adsorb or diffuse to active sites on the support. New catalyst designs (e.g., bimetallic particles on high-surface-area supports, polyoxometalates, and base catalysts) are under investigation, but many show low selectivity or rapid loss of activity. Catalyst recovery and recycling also complicate process scale-up (Gale et al., 2020; Wang et al., 2023). Therefore, the development of efficient, selective, and stable catalysts that operate under mild conditions is critical for improving the sustainability and viability of lignin-based chemical production. Additionally, the depolymerization of lignin often results in a complex mixture of monomers with similar molecular structures, making separation and purification processes challenging. This complexity leads to increased costs and hinders the economic viability of lignin-based chemical production compared to fossil-based routes (Ziwei et al., 2023).

Variety in lignin feedstocks is another inherent limitation that affects the efficiency and reproducibility of depolymerization processes. Lignin is a highly heterogeneous polymer whose structural features depend significantly on the biomass source and extraction process. For instance, hardwood lignins, rich in syringyl units, exhibit distinct reactivity compared to softwood lignins, which are predominantly guaiacyl-based, or grass lignins containing p-coumaryl units. Furthermore, technical lignins, including kraft, organosolv, and lignosulfonates, differ in key elements such as sulfur concentration, molecular weight, and residual carbohydrate levels. This structural and compositional diversity produces inconsistent depolymerization performance and product profiles, limiting cross-study comparisons and process standardization. (Jeffri et al., 2024; Wang et al., 2023). Accurately determining laccase activity on complex, industrially relevant lignin substrates is a significant challenge in the field of enzymatic lignin valorization. Most available assays use model compounds like ABTS or 2,6-dimethylphenol, which do not replicate the structural complexity and variability of native lignin. As a result, the enzymatic activity and product distribution measured with these surrogates frequently do not correspond to actual catalytic performance. Traditional solution-based spectrophotometric assays are notably inadequate for determining laccase reactivity with non-soluble lignin polymers. To overcome this constraint, advanced analytical techniques like 2D NMR, NIMS, EPR spectroscopy, and ITC have been developed, providing more precise and thorough information about bond cleavage patterns, radical generation rates, and thermodynamic parameters.

Nonetheless, the development and standardization of effective, high-throughput assays for measuring laccase activity on real lignin substrates remains a major challenge to advancing scaled lignin biorefinery applications (Zhou et al., 2023). Product separation and purification are additional challenges in the lignin valorization process, as the depolymerization output typically includes a highly heterogeneous mixture of compounds. These include water-soluble phenolics, organic-phase oils, solid chars, and catalyst residues, all of which must be effectively separated to extract high-value products. The requirement for multi-step separations, such as solvent extraction, distillation, filtration, and chromatographic purification, not only increases process complexity and operational costs but also leads to possible material losses (Jeffri et al., 2024; Wang et al., 2023). Furthermore, residual contaminants from upstream processing can pollute product streams and foul equipment, making proper separation an important bottleneck in the conversion strategy.

Scaling to industrial production remains a significant challenge in lignin depolymerization, as most technologies are only verified at the lab scale. Scaling up presents challenges such as inefficient heat and mass transfer, management of viscous lignin slurries, and difficulties with continuous operation. High-temperature processes such as hydrogenolysis necessitate pressure-resistant reactors and hydrogen infrastructure. Furthermore, contaminants in technical lignin, such as sulfur and chlorine, might result in corrosion and catalyst deactivation. These factors, combined with high capital and energy costs, hinder commercialization. Bridging this gap will require pilot-scale demonstrations, integration with existing biorefineries, and thorough techno-economic assessments (Wang et al., 2023). Key challenges in lignin depolymerization include: (i) variability in lignin composition across biomass sources, complicating reaction optimization; (ii) absence of a universal solvent that can fully dissolve lignin yet remain insoluble in organic phases; (iii) limited availability of genetically engineered microorganisms capable of tolerating solvent conditions for biocatalytic depolymerization; (iv) lack of effective post-depolymerization separation strategies due to the complexity of product mixtures; (v) high cost and limited availability of efficient catalysts, with noble metals being expensive and transition metals requiring further development; (vi) difficulty in fine-tuning process parameters to enhance efficiency and reduce costs; and (vii) challenges in scaling up to industrial applications, particularly due to catalyst deactivation and the need for long-lasting, selective catalysts (Ji et al., 2024).

4.7.2 Knowledge Gaps and Research Needs

- The wide range of lignin sources, including native and genetically modified types, demands tailored recovery and valorization strategies. Scaling these methods for industrial use remains a key challenge.
- Knowledge of how to modify lignin's native structure during pretreatment for easier recovery is limited. Future research should focus on designing lignin structures suited to specific end uses such as fuels, chemicals, or materials.
- Lignin valorization must be integrated with processes that retain the value of cellulose and hemicellulose, enabling efficient and sustainable use of the entire lignocellulosic biomass.

- There is a significant gap in advanced computational models that predict lignin behavior. Developing simulation tools that integrate experimental data is critical for guiding lignin redesign (Ragauskas et al., 2014).
- Many catalysts currently employed for lignin conversion are adapted from petroleum refining and are not ideally suited for biomass due to differences in feedstock properties. New catalysts specifically engineered for biomass conversion are necessary.
- Due to lignin's variability, a single depolymerization method is not feasible. Research must focus on developing lignin-type-specific catalysts and testing them on real samples to understand structure-performance relationships (Jeffri et al., 2024).
- Most catalytic studies rely on model lignin compounds instead of real biomass-derived lignin. Consequently, data on catalyst performance amid real impurities like proteins and inorganic salts are limited, hindering assessment of their industrial viability (Zakzeski et al., 2010).
- Current lignocellulose fractionation methods often damage lignin's native β-O-4 linkages, reducing depolymerization efficiency. Improved methods that preserve these structures via chemical stabilization or gentler conditions need further development and wider application.
- The gap between lab-scale innovations and industrial implementation persists. Technologies must be adapted to accommodate existing industrial lignin streams, especially kraft lignin, to ensure practical and scalable solutions.
- The predominant industrial processes, such as kraft pulping, yield lignins with significant structural degradation. These lignins are less reactive and more difficult to valorize into chemicals. New methods that can selectively cleave C–C bonds in degraded lignins are needed, but they often require harsh conditions that reduce yield and selectivity (Schutyser et al., 2018).
- Thorough characterization is essential to match lignin sources with suitable depolymerization or material pathways, due to the structural complexity and variability from different feedstocks and isolation methods.
- Current analytical techniques, such as [1H;13C]-HSQC NMR, are not able to detect many of the interunit linkages in technical lignins. This suggests the presence of extensive tertiary or quaternary carbon-carbon bonds, highlighting the need for alternative approaches to fully elucidate lignin's structure.
- Despite multiple analytical methods, lignin's full structure remains partially understood. Properties like ash and sulfur content, functional groups, and molar mass provide insights but are insufficient for fully informed feedstock selection (Constant et al., 2016).

4.8 CONCLUSION

Lignin's complex, three-dimensional polyphenolic network has long limited its direct use in high-performance materials. This chapter has reviewed a comprehensive suite of depolymerization strategies, from thermal (pyrolysis, hydrothermal liquefaction, and carbonization) and catalytic (acid, base, and metal-catalyzed) to emerging electrochemical, ionic-liquid, microwave-assisted, and enzymatic methods, that selectively cleave key interunit linkages. Advances in analytical techniques, including accelerated 2D HSQC NMR, optimized MALDI-TOF, and microfluidic dispersion analysis, now permit real-time insight into bond-breaking pathways and molecular-weight distributions. Depolymerized lignin derivatives have been successfully formulated into phenol–formaldehyde resins (up to 50 wt% substitution), polyurethane foams (30–50 wt% replacement of petroleum polyols), and epoxy networks (up to 50 wt% of bisphenol A) without sacrificing thermal stability or mechanical integrity. These achievements emphasize lignin's viability as a renewable co-monomer in adhesives, coatings, insulation, and structural composites. Looking forward, the integration of lignin-first fractionation, tailored catalyst design, scalable biocatalytic processes, and rigorous techno-economic and life-cycle assessments will be essential. By aligning process efficiency with sustainability metrics, future research can bridge laboratory innovations and industrial implementation, advancing lignin valorization as a cornerstone of the circular bioeconomy.

REFERENCES

Abdelaziz, O. Y., D. P. Brink, J. Prothmann, et al. 2016. Biological valorization of low molecular weight lignin. *Biotechnol. Adv.* 34: 1318–46. https://doi.org/10.1016/j.biotechadv.2016.10.001

Agarwal, A., M. Rana, and J. H. Park. 2018. Advancement in technologies for the depolymerization of lignin. *Fuel Process. Technol.* 181: 115–32. https://doi.org/10.1016/j.fuproc.2018.09.017

Ali, S., A. Rani, M. Dar, et al. 2024. Recent advances in characterization and valorization of lignin and its value-added products: Challenges and future perspectives. *Biomass* 4(3): 947–77. https://doi.org/10.3390/biomass4030053

Antonino, L. D., J. R. Gouveia, R. R. Sousa Júnior, et al. 2021. Reactivity of aliphatic and phenolic hydroxyl groups in Kraft lignin towards 4,4′ MDI. *Molecules* 26(8): 2131. https://doi.org/10.3390/molecules26082131

Bajwa, D. S., G. Pourhashem, A. H. Ullah, and S. G. Bajwa. 2019. A concise review of current lignin production, applications, products and their environment impact. *Ind. Crops Prod.* 139: 111526. https://doi.org/10.1016/j.indcrop.2019.111526

Belkheiri, T., S. I. Andersson, C. Mattsson, L. Olausson, H. Theliander, and L. Vamling. 2018. Hydrothermal liquefaction of Kraft lignin in sub-critical water: The influence of the sodium and potassium fraction. *Biomass Conv. Bioref.* 8(3): 585–95. https://doi.org/10.1007/s13399-018-0307-9

Belkheiri, T., C. Mattsson, S. I. Andersson, et al. 2016. Effect of pH on Kraft lignin depolymerisation in subcritical water. *Energy Fuels* 30(6): 4916–24. https://doi.org/10.1021/acs.energyfuels.6b00462

Biswal, T., T. R. Sethy, and P. K. Sahoo. 2023. Epoxy foams. *ACS Symp. Ser.* 1439: 161–81. https://doi.org/10.1021/bk-2023-1439.ch008

Bourbiaux, D., J. Pu, F. Rataboul, L. Djakovitch, C. Geantet, and D. Laurenti. 2021. Reductive or oxidative catalytic lignin depolymerization: An overview of recent advances. *Catal. Today* 373: 24–37. https://doi.org/10.1016/j.cattod.2021.03.027

Budnyak, T. M., J. Onwumere, I. V. Pylypchuk, et al. 2021. LignoPhot: Conversion of hydrolysis lignin into the photoactive hybrid lignin/Bi4O5Br2/BiOBr composite for simultaneous dyes oxidation and Co_2+ and Ni_2+ recycling. *Chemosphere* 279: 130538. https://doi.org/10.1021/acs.iecr.0c01617

Cao, Y., C. Zhang, D. C. W. Tsang, J. Fan, J. H. Clark, and S. Zhang. 2020. Hydrothermal liquefaction of lignin to aromatic chemicals: Impact of lignin structure. *Ind. Eng. Chem. Res.* 59: 16957–69. https://doi.org/10.1016/j.fuproc.2019.106180

Chen, X., Q. Che, S. Li, et al. 2019. Recent developments in lignocellulosic biomass catalytic fast pyrolysis: Strategies for the optimization of bio-oil quality and yield. *Fuel Process. Technol.* 191: 106180. https://doi.org/10.1016/j.fuproc.2019.106180

Chen, Z., and C. Wan. 2017. Biological valorization strategies for converting lignin into fuels and chemicals. *Renew. Sust. Energy Rev.* 73: 610–23. https://doi.org/10.1016/j.rser.2017.01.166

Chio, C., M. Sain, and W. Qin. 2019. Lignin utilization: A review of lignin depolymerization from various aspects. *Renew. Sust. Energy Rev.* 107: 232–49. https://doi.org/10.1016/j.rser.2019.03.008

Colina, J. R., M. Ortega, J. Norambuena-Contreras, S. Ghysels, F. Ronsse, and L. E. Arteaga-Pérez. 2025. Thermo-catalytic depolymerization of lignin over Pd-based catalysts: Role of catalyst support on monoaromatics selectivity. *Biomass Bioenergy* 193: 107547. https://doi.org/10.1016/j.biombioe.2024.107547

Constant, S., C. Basset, C. Dumas, F. Di Renzo, M. Robitzer, A. Barakat, and F. Quignard. 2015. Reactive organosolv lignin extraction from wheat straw: Influence of Lewis acid catalysts on structural and chemical properties of lignins. *Ind. Crops Prod.* 65: 180–89. https://doi.org/10.1016/j.indcrop.2014.12.009

Constant, S., H. L. J. Wienk, A. E. Frissen, et al. 2016. New insights into the structure and composition of technical lignins: A comparative characterisation study. *Green Chem.* 18: 2651–65. https://doi.org/10.1039/C5GC03043A

Cox, B. J., and J. G. Ekerdt. 2012. Depolymerization of oak wood lignin under mild conditions using the acidic ionic liquid 1-H-3-methylimidazolium chloride as both solvent and catalyst. *Bioresour. Technol.* 118: 584–88. https://doi.org/10.1016/j.biortech.2012.05.012

Davinia, S., A. Prieto, Á. T. Martínez, and M. J. Martínez. 2013. Characterization of a novel dye-decolorizing peroxidase (DyP)-type enzyme from Irpex lacteus and its application in enzymatic hydrolysis of wheat straw. *Appl. Environ. Microbiol.* 79: 4316–24. https://doi.org/10.1128/AEM.00699-13

DSouza, G. C., C. Chio, M. B. Ray, A. Prakash, W. Qin, and C. Xu. 2024. Synthesis and characterization of biodegradable Kraft lignin-based hydrophilic phenol formaldehyde foams. *Sustain. Mater. Technol.* 41: e01064. https://doi.org/10.1016/j.susmat.2024.e01064

DSouza, G. C., C. Chio, A. Venkatesh, et al. 2025. Microbially degradable phenolic foams based on depolymerized Kraft lignin for hydrophilic applications. *Bioresour. Technol.* 419: 132082. https://doi.org/10.1016/j.biortech.2025.132082

DSouza, G. C., F. Dodangeh, G. B. Venkata, M. B. Ray, A. Prakash, and C. Xu. 2025. A comprehensive review of biobased polyurethane and phenol formaldehyde hydrophilic foams for environmental remediation, floral, and hydroponics applications. *Biomass Bioenergy* 192: 107493. https://doi.org/10.1016/j.biombioe.2024.107493

DSouza, G. C., H. Li, Z. Yuan, C. C. Xu, M. B. Ray, and A. Prakash. 2024. Investigating the hydrophilicity of phenol formaldehyde foams: Effects of synthesis parameters. *J. Appl. Polym. Sci.* 141: e54877. https://doi.org/10.1002/app.54877

DSouza, G. C., M. Ray, A. Prakash, and C. C. Xu. 2026. Synthesis, performance, and applications of biobased phenol formaldehyde resins. In Sandhya P.K., Sreekala Meyyarappallil Sadasivan and Sabu Thomas (Eds.), *Phenolic Resins* (pp. 587–612). Elsevier. https://doi.org/10.1016/B978-0-323-95660-4.00025-8

Durak, H. 2023. Comprehensive assessment of thermochemical processes for sustainable waste management and resource recovery. *Processes* 11: 2092. https://doi.org/10.3390/pr11072092

Duval, A., and M. Lawoko. 2014. A review on lignin-based polymeric, micro- and nano- structured materials. *React. Funct. Polym.* 85: 78–96. https://doi.org/10.1016/j.reactfunctpolym.2014.09.017

Elliott, D. C., P. Biller, A. B. Ross, A. J. Schmidt, and S. B. Jones. 2015. Hydrothermal liquefaction of biomass: Developments from batch to continuous process. *Bioresour. Technol.* 178: 147–56. https://doi.org/10.1016/j.biortech.2014.09.132

Evstigneyev, E. I., O. S. Yuzikhin, A. A. Gurinov, et al. 2016. Study of structure of industrial acid hydrolysis lignin, oxidized in the H_2O_2–H_2SO_4 system. *J. Wood Chem. Technol.* 36: 259–69. https://doi.org/10.1080/02773813.2015.1137945

Feghali, E., D. J. Van De Pas, and K. M. Torr. 2020. Toward bio-based epoxy thermoset polymers from depolymerized native lignins produced at the pilot scale. *Biomacromolecules* 21: 1548–59. https://doi.org/10.1007/s41061-018-0211-6

Feghali, E., K. M. Torr, D. J. van de Pas, et al. 2018. Thermosetting polymers from lignin model compounds and depolymerized lignins. *Top. Curr. Chem.* 376: 21. https://doi.org/10.1007/s41061-018-0211-6

Ferdosian, F., Z. Yuan, M. Anderson, and C. Xu. 2014. Synthesis of lignin-based epoxy resins: Optimization of reaction parameters using response surface methodology. *RSC Adv.* 4: 31745–53. https://doi.org/10.1039/c4ra03978e

Ferdosian, F., Z. Yuan, M. Anderson, and C. Xu. 2016. Synthesis and characterization of hydrolysis lignin-based epoxy resins. *Ind. Crops Prod.* 91: 295–301. https://doi.org/10.1016/j.indcrop.2016.07.020

Forsythe, W. G., M. D. Garrett, C. Hardacre, M. Nieuwenhuyzen, and G. N. Sheldrake. 2013. An efficient and flexible synthesis of model lignin oligomers. *Green Chem.* 15: 3031–38. https://doi.org/10.1039/C3GC41110A

Fraile, J. M., J. I. García, Z. Hormigón, J. A. Mayoral, C. J. Saavedra, and L. Salvatella. 2018. Role of substituents in the solid acid-catalyzed cleavage of the β-O-4 linkage in lignin models. *ACS Sustain. Chem. Eng.* 6: 1837–47. https://doi.org/10.1021/acssuschemeng.7b03218

Gale, M., C. Cai, and K. Gilliard-Abdul-Aziz. 2020. Heterogeneous catalyst design principles for the conversion of lignin into high-value commodity fuels and chemicals. *ChemSusChem* 13: e202000822. https://doi.org/10.1002/cssc.202000822

Gao, W. J., C. M. Lam, B. G. Sun, R. D. Little, and C. C. Zeng. 2017. Selective electrochemical C–O bond cleavage of β-O-4 lignin model compounds mediated by iodide ion. *Tetrahedron* 73(17): 2447–54. https://doi.org/10.1016/j.tet.2017.03.027

Garcia, G. E. S., A. Potthast, I. Sumerskii, et al. 2024. Pyrolytic lignin fractionation for rigid polyurethane foams: Relationship between the pyrolysate's chemical structure and foam properties. *Sustain. Chem. Environ.* 5: 100065. https://doi.org/10.1016/j.scenv.2024.100065

Gomes, J. B., F. Eloy de Souza, E. O. Brito, and R. C. C. Lelis. 2020. A review on lignin sources and uses. *J. Appl. Biotechnol. Bioeng.* 7: 100–5. https://doi.org/10.15406/jabb.2020.07.00222

Hao, J. H., H. F. Huang, J. P. Zhang, et al. 2025. A novel approach for characterization of lignin structure in tobacco via fast 2D HSQC NMR spectroscopy with non-uniform sampling. *Ind. Crops Prod.* 223: 120236. https://doi.org/10.1016/j.indcrop.2024.120236

Hoang, A. T., H. C. Ong, I. M. R. Fattah, et al. 2021. Progress on the lignocellulosic biomass pyrolysis for biofuel production toward environmental sustainability. *Fuel Process. Technol.* 212: 106997. https://doi.org/10.1016/j.fuproc.2021.106997

Hu, L., Y. Zhou, M. Zhang, and R. Liu. 2012. Characterization and properties of a lignosulfonate-based phenolic foam. *BioResources* 7: 554–64. https://doi.org/10.15376/biores.7.1.554-564

Jeffri, N., N. Rawi, M. Kassim, and C. Abdullah. 2024. Unlocking the potential: Evolving role of technical lignin in diverse applications and overcoming challenges. *Int. J. Biol. Macromol.* 274: 133506. https://doi.org/10.1016/j.ijbiomac.2024.133506

Ji, D., Y. Wang, J. Peng, et al. 2024. Recent advances in depolymerization and high-value utilization of lignin: A review. *Ind. Eng. Chem. Res.* 63: e4c02617. https://doi.org/10.1021/acs.iecr.4c02617

Ji, Q., X. Yu, L. Chen, et al. 2023. Comprehensive depolymerization of lignin from lignocellulosic biomass: A review. *Crit. Rev. Environ. Sci. Technol.* 53: 2190314. https://doi.org/10.1080/10643389.2023.2190314

Kariim, I., R. Bakari, M. W. Syed, J. Y. Park, I. G. Lee, and T. Kivevele. 2025. Biomass depolymerization into liquid biofuel: A review of process conditions and an insight into the technoeconomic evaluation. *ACS Symp. Ser.* 1501: 105–35. https://doi.org/10.1021/bk-2025-1501.ch006

Kaya, K., A. Atsay, H. Gunduz, A. Slabon, and Y. Yagci. 2024. Nature-inspired depolymerization of soda lignin: Light-induced free radical promoted cleavage of β- O-4 bonds. *Adv. Sustain. Syst.* 8(3): 2300366. https://doi.org/10.1002/adsu.202300366

Kosyakov, D. S., N. V. Ul'yanovskii, E. A. Sorokina, and N. S. Gorbova. 2014. Optimization of sample preparation conditions. *J. Anal. Chem.* 69: 1345.

Kumar, A., A. J. Kumar, and T. Bhaskar. 2020. Utilization of lignin: A sustainable and eco-friendly approach. *J. Energy Inst.* 93(1): 235–71. https://doi.org/10.1016/J.JOEI.2019.03.005

Laurichesse, S., and L. Avérous. 2014. Chemical modification of lignins: Towards biobased polymers. *Prog. Polym. Sci.* 39: 1266–90. https://doi.org/10.1016/j.progpolymsci.2013.11.004

Lei, Z. P., Z. Q. Hu, H. F. Shui, S. B. Ren, Z. C. Wang, S. G. Kang, and C. X. Pan. 2015. Pyrolysis of lignin following ionic liquid pretreatment at low temperature. *Fuel Process. Technol.* 138: 612–5. https://doi.org/10.1016/j.fuproc.2015.06.049

Li, B., Y. Wang, N. Mahmood, Z. Yuan, J. Schmidt, and C. Xu. 2017. Preparation of bio-based phenol formaldehyde foams using depolymerized hydrolysis lignin. *Ind. Crops Prod.* 97: 409–16. https://doi.org/10.1016/j.indcrop.2016.12.063

Li, B., Z. Yuan, J. Schmidt, and C. Xu. 2019. New foaming formulations for production of bio-phenol formaldehyde foams using raw Kraft lignin. *Eur. Polym. J.* 111: 1–10. https://doi.org/10.1016/j.eurpolymj.2018.12.011

Li, H., N. Mahmood, Z. Ma, et al. 2017. Preparation and characterization of bio-polyol and bio-based flexible polyurethane foams from fast pyrolysis of wheat straw. *Ind. Crops Prod.* 103: 64–72. https://doi.org/10.1016/j.indcrop.2017.03.042

Li, T., and S. Takkellapati. 2018. The current and emerging sources of technical lignins and their applications. *Biofuels Bioprod. Biorefin.* 12(5): 756–87. https://doi.org/10.1002/bbb.1913

Liang, Y., W. Duan, X. An, Y. Qiao, Y. Tian, and H. Zhou. 2020. Novel betaine-amino acid based natural deep eutectic solvents for enhancing the enzymatic hydrolysis of corncob. *Bioresour. Technol.* 310: 123389. https://doi.org/10.1016/j.biortech.2020.123389

Liu, W. J., H. Jiang, and H. Q. Yu. 2015. Thermochemical conversion of lignin to functional materials: A review and future directions. *Green Chem.* 17: 4888–906. https://doi.org/10.1039/c5gc01054c

Lu, X., and X. Gu. 2023. A review on lignin-based epoxy resins: Lignin effects on their synthesis and properties. *Int. J. Biol. Macromol.* 230: 144–59. https://doi.org/10.1016/j.ijbiomac.2022.12.322

Ma, L., X. Shi, J. Liu, et al. 2025. Optimization of ultrasound-assisted lignin depolymerization using response surface methodology for low-formaldehyde emission wood adhesives. *Ind. Crops Prod.* 227: 120760. https://doi.org/10.1016/j.indcrop.2025.120760

Mahmood, N., Z. Yuan, J. Schmidt, M. Tymchyshyn, and C. C. Xu. 2016. Hydrolytic liquefaction of hydrolysis lignin for the preparation of bio-based rigid polyurethane foam. *Green Chem.* 18(8): 2385–98. https://doi.org/10.1039/C5GC02876K

Mahmood, N., Z. Yuan, J. Schmidt, and C. C. Xu. 2013. Valorization of hydrolysis lignin for polyols/polyurethane foam. *J-FOR J. Sci. Technol. Forest Prod. Process.* 139: 13–20. https://www.researchgate.net/publication/264655151

Mahmood, N., Z. Yuan, J. Schmidt, and C. Xu. 2015a. Preparation of bio-based rigid polyurethane foam using hydrolytically depolymerized Kraft lignin via direct replacement or oxypropylation. *Eur. Polym. J.* 68: 1–9. https://doi.org/10.1016/J.EURPOLYMJ.2015.04.030

Mahmood, N., Z. Yuan, J. Schmidt, and C. C. Xu. 2015b. Hydrolytic depolymerization of hydrolysis lignin: Effects of catalysts and solvents. *Bioresour. Technol.* 190: 416–19. https://doi.org/10.1016/j.biortech.2015.04.074

Mahmood, N., Z. Yuan, J. Schmidt, and C. Xu. 2016. Depolymerization of lignins and their applications for the preparation of polyols and rigid polyurethane foams: A review. *Renew. Sustain. Energy Rev.* 60: 317–29. https://doi.org/10.1016/j.rser.2016.01.037

Muthuraj, R., A. R. Horrocks, and B. K. Kandola. 2020. Hydroxypropyl-modified and organosolv lignin/bio-based polyamide blend filaments as carbon fibre precursors. *J. Mater. Sci.* 55(16): 7066–83. https://doi.org/10.1007/s10853-020-04486-w

Nanayakkara, S., A. F. Patti, and K. Saito. 2014. Lignin depolymerization with phenol via redistribution mechanism in ionic liquids. *ACS Sustain. Chem. Eng.* 2(9): 2159–64. https://doi.org/10.1021/sc5003424

Nlewem, K. C., and M. E. Thrash. 2010. Comparison of different pretreatment methods based on residual lignin effect on the enzymatic hydrolysis of switchgrass. *Bioresour. Technol.* 101(14): 5426–30. https://doi.org/10.1016/j.biortech.2010.02.031

Obrzut, N., R. Hickmott, L. Shure, and K. A. Gray. 2023. The effects of lignin source and extraction on the composition and properties of biorefined depolymerization products. *RSC Sustain.* 1(9): 2328–40. https://doi.org/10.1039/d3su00262d

Ouyang, X., G. Zhu, X. Huang, and X. Qiu. 2015. Microwave assisted liquefaction of wheat straw alkali lignin for the production of monophenolic compounds. *J. Energy Chem.* 24(1): 72–6. https://doi.org/10.1016/S2095-4956(15)60286-8

Pan, K., M. Tian, Z. Jiang, B. Kjartanson, and A. Chen. 2012. Electrochemical oxidation of lignin at lead dioxide nanoparticles photoelectrodeposited on TiO_2 nanotube arrays. *Electrochim. Acta.* 60: 147–53. https://doi.org/10.1016/j.electacta.2011.11.025

Pei, Z., X. Liu, J. Chen, H. Wang, and H. Li. 2024. Research progress on lignin depolymerization strategies: A review. *Polymers* 16 (17): 2388. https://doi.org/10.3390/polym16172388.

Pham, L. T. M., K. Deng, T. R. Northen, et al. 2021. Experimental and theoretical insights into the effects of pH on catalysis of bond-cleavage by the lignin peroxidase isozyme H8 from *Phanerochaete chrysosporium. Biotechnol. Biofuels.* 14(1). https://doi.org/10.1186/s13068-021-01953-7

Pinto, J. A., I. P. Fernandes, V. D. Pinto, et al. 2021. Valorization of lignin side-streams into polyols and rigid polyurethane foams—a contribution to the pulp and paper industry biorefinery. *Energies* 14(13). https://doi.org/10.3390/en14133825

Purushothaman, R. K. P., R. Kumar, G. van Erven, et al. 2023. New insights into the base catalyzed depolymerization of technical lignins: A systematic comparison. *RSC Adv.* 13(8): 4898–909. https://doi.org/10.3390/polym16172388

Quinsaat, J. E. Q., E. Feghali, D. J. Van de Pas, R. Vendamme, and K. M. Torr. 2021. Preparation of mechanically robust bio-based polyurethane foams using depolymerized native lignin. *ACS Appl. Polym. Mater.* 3(11): 5845–56. https://doi.org/10.1021/acsapm.1c01081

Ragauskas, A. J., G. T. Beckham, M. J. Biddy, et al. 2014. Lignin valorization: Improving lignin processing in the biorefinery. *Science* 344(6185). https://doi.org/10.1126/science.1246843

Rahman, M. S., R. Roy, C. Montoya, M. A. Halim, and D. E. Raynie. 2022. Acidic and basic amino acid-based novel deep eutectic solvents and their role in depolymerization of lignin. *J. Mol. Liq.* 362: 119751. https://doi.org/10.1016/j.molliq.2022.119751

Reichert, E., R. Wintringer, D. A. Volmer, and R. Hempelmann. 2012. Electro-catalytic oxidative cleavage of lignin in a protic ionic liquid. *Phys. Chem. Chem. Phys.* 14(15): 5214–21. https://doi.org/10.1039/C2CP23596J

Ren, T., W. Qi, R. Su, and Z. He. 2019. Promising techniques for depolymerization of lignin into value-added chemicals. *ChemCatChem* 11(2). https://doi.org/10.1002/cctc.201801428

Roberts, V. M., V. Stein, T. Reiner, A. Lemonidou, X. Li, and J. A. Lercher. 2011. Towards quantitative catalytic lignin depolymerization. *Chem. Eur. J.* 17(21): 5939–48. https://doi.org/10.1002/chem.201002438

Roy, R., M. S. Rahman, T. A. Amit, and B. Jadhav. 2022. Recent advances in lignin depolymerization techniques: A comparative overview of traditional and greener approaches. *Biomass* 2(3). https://doi.org/10.3390/biomass2030009

Schutyser, W., T. Renders, S. Van den Bosch, S. F. Koelewijn, G. T. Beckham, and B. F. Sels. 2018. Chemicals from lignin: An interplay of lignocellulose fractionation, depolymerisation, and upgrading. *Chem. Soc. Rev.* 47: 852–908. https://doi.org/10.1039/c7cs00566k

Seidi, F., Y. Liu, Y. Huang, H. Xiao, and D. Crespy. 2025. Chemistry of lignin and condensed tannins as aromatic biopolymers. *Chem. Soc. Rev.* 54: 3140–232. https://doi.org/10.1039/d4cs00440j

Shrestha, S., S. Goswami, D. Banerjee, et al. 2024. Perspective on lignin conversion strategies that enable next generation biorefineries. *ChemSusChem* 17: e202301460. https://doi.org/10.1002/cssc.202301460

Simonsen, T. I., D. T. Djajadi, A. Ponzecchi, et al. 2025. Lignin molar mass estimation by dispersion analysis. *Macromol. Rapid Commun.* https://doi.org/10.1002/marc.202400751

Song, Q., F. Wang, and J. Xu. 2012. Hydrogenolysis of lignosulfonate into phenols over heterogeneous nickel catalysts. *Chem. Commun.* 48: 7019–21. https://doi.org/10.1039/c2cc31414b

Stärk, K., N. Taccardi, A. Bösmann, and P. Wasserscheid. 2010. Oxidative depolymerization of lignin in ionic liquids. *ChemSusChem* 3: 719–23. https://doi.org/10.1002/cssc.200900242

Sternberg, J., and S. Pilla. 2023. Chemical recycling of a lignin-based non-isocyanate polyurethane foam. *Nat. Sustain.* 6: 316–24. https://doi.org/10.1038/s41893-022-01022-3

Sturgeon, M. R., S. Kim, K. Lawrence, et al. 2014. A mechanistic investigation of acid-catalyzed cleavage of aryl-ether linkages: Implications for lignin depolymerization in acidic environments. *ACS Sustain. Chem. Eng.* 2: 472–85. https://doi.org/10.1021/sc400384w

Toledano, A., L. Serrano, and J. Labidi. 2012. Organosolv lignin depolymerization with different base catalysts. *J. Chem. Technol. Biotechnol.* 87: 1593–99. https://doi.org/10.1002/jctb.3799

Van de Pas, D. J., and K. M. Torr. 2017. Biobased epoxy resins from deconstructed native softwood lignin. *Biomacromolecules* 18(8): 2640–48. https://doi.org/10.1039/d2ra06998a

Vásquez-Garay, F., I. Carrillo-Varela, C. Vidal, P. Reyes-Contreras, M. Faccini, and R. T. Mendonça. 2021. A review on the lignin biopolymer and its integration in the elaboration of sustainable materials. *Sustainability* 13. https://doi.org/10.3390/su13052697

Verdolotti, L., M. Lavorgna, E. Di Maio, and S. Iannace. 2013. Hydration-induced reinforcement of rigid polyurethane–cement foams: The effect of the co-continuous morphology on the thermal-oxidative stability. *Polym. Degrad. Stab.* 98: 64–72. https://doi.org/10.1016/j.polymdegradstab.2012.10.027

Vieira, F. R., N. V. Gama, D. V. Evtuguin, et al. 2023. Bio-based polyurethane foams from Kraft lignin with improved fire resistance. *Polymers* 15: 1074. https://doi.org/10.3390/polym15051074

Vishtal, A., and A. Kraslawski. 2011. Challenges of lignins. *BioResources* 6 (3): 3547–68.

Wang, G., X. Liu, J. Zhang, W. Sui, J. Jang, and C. Si. 2018. One-pot lignin depolymerization and activation by solid acid catalytic phenolation for lightweight phenolic foam preparation. *Ind. Crops Prod.* 124: 216–25. https://doi.org/10.1016/j.indcrop.2018.07.080

Wang, S., X. Li, R. Ma, and G. Song. 2025. Catalytic hydrogenolysis of lignin into serviceable products. *Acc. Chem. Res.* https://doi.org/10.1021/acs.accounts.4c00644

Wang, W., Z. Ma, X. Zhao, S. Liu, L. Cai, S. Q. Shi, and Y. Ni. 2020. Effect of various microwave absorbents on the microwave-assisted lignin depolymerization process. *ACS Sustain. Chem. Eng.* 8: 16086–90. https://doi.org/10.1021/acssuschemeng.0c04658

Wang, Y., L. Wei, Q. Hou, Z. Mo, X. Liu, and W. Li. 2023. A review on catalytic depolymerization of lignin towards high-value chemicals: Solvent and catalyst. *Fermentation* 9: 386. https://doi.org/10.3390/fermentation9040386

Wei, N., S. Qi, G. Wang, et al. 2023. Acid-promoted lignin reductive depolymerization under mild conditions via a condensation-minimizing approach: From organosolv lignin to woody biomass. *Fuel* 338: 127311. https://doi.org/10.1016/j.fuel.2022.127311

De Wild, P. J., W. J. J. Huijgen, and H. J. Heeres. 2012. Pyrolysis of wheat straw-derived organosolv lignin. *J. Anal. Appl. Pyrolysis.* 93: 95–103.

Xu, R., K. Zhang, P. Liu, et al. 2018. Lignin depolymerization and utilization by bacteria. *Bioresour. Technol.* https://doi.org/10.1016/j.biortech.2018.08.118

Xue, B. L., P. L. Huang, Y. C. Sun, X. P. Li, and R. C. Sun. 2017. Hydrolytic depolymerization of corncob lignin in the view of a bio-based rigid polyurethane foam synthesis. *RSC Adv.* 7: 6123–30. https://doi.org/10.1039/c6ra26318f

Yang, Z., L. Cao, Y. Li, M. Zhang, F. Zeng, and S. Yao. 2020. Effect of pH on hemicellulose extraction and physicochemical characteristics of solids during hydrothermal pretreatment of eucalyptus. *BioResources* 15: 6627–35. https://doi.org/10.15376/biores.15.3.6627-6635

Yang, Y., H. Fan, Q. Meng, Z. Zhang, G. Yang, and B. Han. 2017. Ionic liquid [OMIm][OAc] directly inducing oxidation cleavage of the β-O-4 bond of lignin model compounds. *Chem. Commun.* 53: 8850–53. https://doi.org/10.1039/c7cc04209d

Yin, Y., C. Suo, C. Ma, and S. Liu. 2025. Recent progress towards fuels and value-added chemicals through lignin depolymerization. *J. Environ. Chem. Eng.* 13: 116321. https://doi.org/10.1016/j.jece.2025.116321

Younesi-Kordkheili, H., and A. Pizzi. 2021. A comparison among lignin modification methods on the properties of lignin–phenol–formaldehyde resin as wood adhesive. *Polymers* 13: 3502. https://doi.org/10.3390/polym13203502

Zakzeski, J., P. C. A. Bruijnincx, A. L. Jongerius, and B. M. Weckhuysen. 2010. The catalytic valorization of lignin for the production of renewable chemicals. *Chem. Rev.* 110: 3552–99. https://doi.org/10.1021/cr900354u

Zevallos Torres, L. A., A. L. Woiciechowski, V. O. de A. Tanobe, et al. 2020. Lignin as a potential source of high-added value compounds: A review. *J. Clean. Prod.* https://doi.org/10.1016/j.jclepro.2020.121499

Zhang, C., H. Li, J. Lu, et al. 2017. Promoting lignin depolymerization and restraining the condensation via an oxidation–hydrogenation strategy. *ACS Catal.* 7: 3419–29. https://doi.org/10.1021/acscatal.7b00148

Zhang, J., H. Asakura, J. van Rijn, et al. 2014. Highly efficient, NiAu-catalyzed hydrogenolysis of lignin into phenolic chemicals. *Green Chem.* 16: 2432–37. https://doi.org/10.1039/C3GC42589D

Zhang, J., H. Deng, Y. Sun, C. Pan, and S. Liu. 2010. Isolation and characterization of wheat straw lignin with a formic acid process. *Bioresour. Technol.* 101: 2311–16.

Zhang, X., Q. Zhang, J. Long, et al. 2014. Phenolics production through catalytic depolymerization of alkali lignin with metal chlorides. *BioResources.* https://doi.org/10.15376/biores.9.2.3347–3360

Zhao, S., X. Chen, Z. Fan, et al. 2023. Using lignin degraded to synthesize phenolic foams with excellent flame-retardant property. *Colloids Surf. A* 666: 131373. https://doi.org/10.1016/j.biortech.2009.11.037

Zhen, X., H. Li, Z. Xu, et al. 2021. Demethylation, phenolation, and depolymerization of lignin for the synthesis of lignin-based epoxy resin via a one-pot strategy. *Ind. Crops Prod.* 173: 114135. https://doi.org/10.1016/j.indcrop.2021.114135

Zhou, M., O. A. Fakayode, M. Ren, et al. 2023. Laccase-catalyzed lignin depolymerization in deep eutectic solvents: Challenges and prospects. *Bioresour. Bioproc.* https://doi.org/10.1186/s40643-023-00640-9

Zhou, M., L. Zhong, L. Hu, Y. Zhou, and X. Yang. 2024. Synthesis of a reactive lignin-based flame retardant and its application in phenolic foam. *Environ. Technol.* (U. K.) 45: 2506–18. https://doi.org/10.1080/09593330.2023.2176792

Zhou, N., W. P. D. W. Thilakarathna, Q. S. He, and H. P. V. Rupasinghe. 2022. A review: Depolymerization of lignin to generate high-value bio-products: Opportunities, challenges, and prospects. *Front. Energy Res.* https://doi.org/10.3389/fenrg.2021.758744

Zhou, Z., C. Cui, L. Zhu, et al. 2024. Online mass spectrometric characterization of oligomeric products in high-pressure liquid-phase lignin depolymerization reactions. *ACS Meas. Sci. Au.* https://doi.org/10.1021/acsmeasuresciau.4c00067

Zhu, G., D. Jin, L. Zhao, X. Ouyang, C. Chen, and X. Qiu. 2017. Microwave-assisted selective cleavage of Cα–Cβ bond for lignin depolymerization. *Fuel Process. Technol.* 161: 155–61. https://doi.org/10.1016/j.fuproc.2017.03.020

Zhu, H., L. Wang, Y. Chen, et al. 2014. Electrochemical depolymerization of lignin into renewable aromatic compounds in a non-diaphragm electrolytic cell. *RSC Adv.* 4: 29917–24. https://doi.org/10.1039/C4RA03793F

Ziwei, W., S. Hao, C. Yizhen, et al. 2023. Thermal, photonic, and electrocatalysis in lignin depolymerization research. *RSC Adv.* https://doi.org/10.1039/d3ra06880c

5 Catalytic valorization of lignin

From conventional to advanced catalytic technologies

Vivek Raj, Santosh Kumar Karn, Virat Vaibhav, Moni Kumari, Anne bhambri, Farheen Firdaus, Aman Prakash, Alok Prasad Das, and Vishnu D. Rajput

5.1 INTRODUCTION

Recently, the heavy reliance on fossil fuels has caused significant environmental damage—contributing to climate change, pollution, and the depletion of natural resources. These growing concerns have sparked worldwide efforts to find cleaner, renewable, and more sustainable alternatives to traditional fossil-based materials (Al-Naji et al., 2021). Lignocellulosic biomass (LCB) has become among the most promising and abundant feedstocks to generate fuels, chemicals, and value-added products in a sustainable fashion among the numerous alternatives available. LCB is the total amount of organic matter formed by plants during the process of photosynthesis, which is a natural process by which carbon dioxide present in the atmosphere is mixed with water using solar energy, whereby the end product is the formation of carbohydrates and oxygen. Since the carbon in LCB is obtained directly by capturing atmosphere, it is referred to as carbon-neutral, or even carbon-negative, and aids in reducing the net emission of greenhouse gases and climate effects (Al-Naji et al., 2021; Isikgor & Becer, 2015). Worldwide, LCB is the largest renewable source of biomass on the planet and has been estimated to be up to 550 gigatonnes in volume (Yang & Antonietti, 2020). This makes it the second-largest carbon reservoir after humic substances found in soils. LCB has been broadly accessible in the form of agro-waste, forestry residues, and rapidly growing perennial grasses, which are often produced in vast quantities (Kudakasseril Kurian et al., 2013). Notably, more than 4 gigatonnes of LCB are processed industrially each year, with 1 gigatonne being consumed by the pulp and paper industry alone (Martin & Haggith, 2018).

 DOI: 10.1201/9781003653189-5

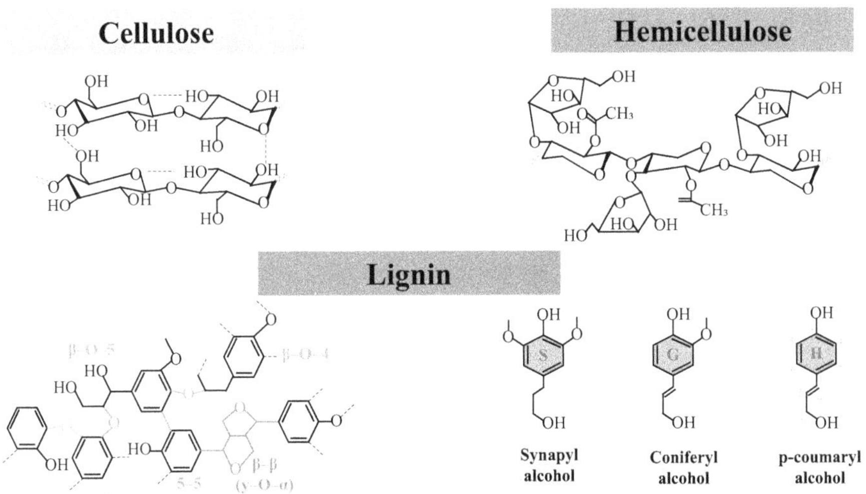

FIGURE 5.1 Representation of the major components of the LCB (cellulose, hemicellulose, and lignin) and the monolignols (synapyl alcohol, coniferyl alcohol, and p-coumaryl alcohol)

This clearly demonstrates that LCB is already integrated into industrial supply chains, both as a primary raw material and as a byproduct, which supports its feasibility for large-scale applications in biorefineries (Isikgor & Becer, 2015). From a chemical perspective, LCB is primarily composed of three key biopolymers: cellulose (35–50%), hemicellulose (20–30%), and lignin (10–25%), as shown in Figure 5.1.

These major components are supported by small amounts of proteins, lipids, and minerals, which together make up about 1–5% of the total biomass. The proportion and structural arrangement of these components vary depending on age, plant species, and tissue type. Based on these features, LCB is typically classified into four main categories: hardwoods, softwoods, grasses, and agricultural residues such as straw. The three biopolymers in LCB are not freely floating or easily separable; they are bound together in a complex, non-uniform three-dimensional structure through a variety of heterogeneous and supramolecular interactions (Isikgor & Becer, 2015; Martin & Haggith, 2018). These characteristics evolved over millions of years to give plants structural integrity and resistance to environmental stresses, microbial degradation, and mechanical damage. However, while beneficial for plant survival, these same features make LCB highly recalcitrant, or resistant to chemical and biological breakdown, posing challenges in its processing and conversion (Bosch et al., 2015).

Cellulose, the chief abundant component of LCB, is a long-chain polysaccharide composed of glucose molecules linked to β (1→4) glycosidic bonds (Brandi et al., 2021). These chains form tightly packed, crystalline fibrils through a dense network of hydrogen bonds, providing exceptional tensile strength and making cellulose difficult to dissolve or degrade. Depending on the plant source, each cellulose chain can consist of a few hundred or more than ten thousand glucose units (Al-Naji et al., 2021; Isikgor & Becer, 2015). In contrast, hemicellulose is a more heterogeneous, lightly connected heteropolysaccharide. It is a combination of C5 and C6 sugar units that include arabinose, mannose, xylose, and glucose (Al-Naji et al., 2021; Alonso et al., 2010). Hemicellulose, unlike cellulose, is branched and amorphous and therefore can be easily hydrolyzed, although it is less mechanically rigorous. Hemicellulose surrounds the cellulose fibers and combines with lignin to make up a protective structure of the plant cell wall (Alonso et al., 2010). Cellulose and hemicellulose have become subjects of significant interest in modern catalytic conversion technology, as they can potentially be hydrolyzed to simple sugars, which are precursors in the manufacturing process of fuel and chemicals. In fact, 12 high-value chemicals have been formulated by the U.S. Department of Energy (DOE), which include ethanol, glycerol, lactic acid, succinic acid, and furfural that can be generated using biomass sugars. Their value in industry can be seen in chemicals that can be used as precursors by the polymer industry, drugs and food industries, and cosmetics as a more sustainable alternative when compared to fossil-based products (Al-Naji et al., 2021; Alonso et al., 2010). Various chemical transformation processes, including hydrogenation, oxidation, dehydration, hydrolysis, and isomerization, have been successfully applied to convert these sugar-derived molecules into valuable intermediates (Al-Naji et al., 2021; Mesa et al., 2020). These processes can be conducted in batch or continuous-flow systems, making them adaptable for industrial-scale production. However, this chapter focuses specifically on the valorization of lignin, and thus, details of cellulose and hemicellulose conversion technologies will not be discussed in depth. Unlike cellulose and hemicellulose, lignin isn't sugar-based. It is a more complex and irregular aromatic polymer made from methoxylated phenylpropanoid units (Sun et al., 2018; Zakzeski et al., 2010). Lignin is identified as the world's most abundant renewable source of aromatic compounds, and it contributes significantly to the energy density of biomass because of its high carbon content and low oxygen concentration. Despite its potential, lignin's heterogeneous, cross-linked structure and chemical resistance have earned it the label of "supramolecular chaos," making it one of the most challenging biopolymers to process (Zakzeski et al., 2010; Achyuthan et al., 2010). Lignin is naturally synthesized in plants through the oxidative integration of three key monomer units: sinapyl (S), coniferyl (G), and p-coumaryl (H) alcohols (Rinaldi et al., 2016; Schutyser et al., 2018). These units form several ether and carbon–carbon linkages, with the β-O-4 ether bond being prominently common and also the most easily cleavable. Relative abundance of these linkages and proportion of S, G, and H units vary widely among plant types (Zakzeski et al., 2010; Rinaldi et al., 2016). As an example, hardwoods are more likely to have more S-units and ether bonds and are easier to depolymerize, and softwoods are more likely to have more G-units, forming more C-C bonds and are harder to break (Rinaldi et al., 2016). Grasses have a blend of all three compounds, i.e., monomers. Additionally, biomass has varying levels of lignin, with herbaceous plants containing 15–24%, hardwoods 18–25%, and softwoods 21–29% (Schutyser et al., 2018). These compositional and structural differences significantly influence the reactivity and molecular mass, along with prospective valorization of lignin into biofuels and value-added chemicals. Due to its availability, its aromaticity, and its energy content, lignin has immense potential as a renewable source of feedstock to displace fossil-based aromatic chemicals (Schutyser et al., 2018). The chapter seeks to offer a general account of the structure, reactivity, and valorization of lignin, along with current progress in addressing the challenges of fully realizing its potential with regard to a sustainable bioeconomy.

5.2 CLASSICAL APPROACHES FOR LIGNIN VALORIZATION

Lignin contains high levels of aromatic compounds and, because of its polyaromatic structure, is a promising source of renewable high-value aromatic compounds. Lignin as a component of LCB can be hydrolyzed into smaller, lower-molecular-weight lignin components that could be used as alternatives to petrochemical-based aromatics. Depending on the extraction and depolymerization approach, as well as the particular catalysts employed, the resulting type of aromatic products obtained includes monomers and dimers to low-molecular-weight fragments. Drawing insights from traditional petroleum refining, researchers have developed four classical strategies for lignin valorization: (1) thermal treatment (especially fast pyrolysis), (2) solvolytic processes, (3) acid-catalyzed methods, and (4) base-catalyzed processes. These methods primarily target ether bonds in lignin structure, and under certain conditions like pyrolysis, even carbon–carbon bonds can be broken. The following sections describe these processes in more detail.

5.2.1 Thermal valorization: fast pyrolysis

Thermal treatment of lignin through rapid pyrolysis has been broadly studied in the field of biomass conversion. However, researchers have yet to reach a consensus on the exact degradation mechanisms (Li et al., 2015). One ongoing debate revolves around whether bond cleavage, particularly C–O and C–C bonds, happens through homolytic scission, forming free radicals, or through a concerted mechanism such as retro-ene fragmentation (Choi et al., 2016). A key example is the breaking of the β-O-4 bond, planned to follow both pathways, though definitive evidence is still lacking. Recent computational advances have allowed modeling of lignin structures and thermal behavior, pointing to fresh promise in deciphering these counterintuitive reactions (Yanez et al., 2018). Remarkably, other studies have demonstrated that the pyrolysis of unprocessed biomass may produce alternative products compared to isolated lignin. In a study, the synthesis of synapyl and coniferyl alcohol was observed upon taking whole LCB through a process called pyrolysis, but not on isolated lignin (Evans et al., 1986). Additional optimization of conditions has been promising to enhance yields of beneficial phenolic compounds. A research study showed that replacing purified lignin with lignocellulosic materials like cedar, cypress, and red pine results in a substantial rise in monomer yields of the order of 6–8% to 18–23% (Kotake et al., 2015). The main products included guaiacol, methylguaiacol, vinylguaiacol, eugenol, and isoeugenol. However, using whole biomass introduces complexity due to the presence of cellulose and hemicellulose, especially when no catalysts are involved, resulting in more complicated pyrolysis oils.

5.2.2 Solvolytic valorization

Solvolytic processes involve breaking down lignin using solvents under heat, without adding oxidants, reductants, acids, or bases. This approach is economically attractive due to its lower operational cost and simpler product recovery, since it does not require expensive solid catalysts. Typically performed at temperatures between 523 K and 723 K, solvents like water, organic liquids, or alcohol-water mixtures are used (Güvenatam et al., 2016; Onwudili & Williams, 2014). The nature of the products depends heavily on the reaction temperature. Higher temperatures tend to produce methoxyphenol units with alkyl or unsubstituted side chains, while lower temperatures lead to a broader mix, including oxygenated and unsaturated derivatives. However, compared to catalyzed processes, monomer yields are generally lower, averaging around ~10 wt%. Still, certain conditions have yielded higher outputs, such as ~22 wt% from softwood lignin in water at 537 K and up to ~24 wt% using tetrahydrofuran (THF) at 573 K with organosolv lignin. Recently, continuous-flow solvothermal processing of sodium lignosulfonate in alcohol-water systems (e.g., methanol-water) has shown promising results. At 523 K, this method produced low-molecular-weight lignin fragments with a dispersity of 5.3 and a minimum molecular weight around 2,994 g/mol (Brandi et al., 2021).

5.2.3 Acid-Catalyzed Depolymerization

Lignin's ether bonds can be efficiently broken under acidic conditions through hydrolysis. However, acid treatment often changes the structure of lignin, as both fragmentation and repolymerization can occur simultaneously. The dominant pathway in this type of depolymerization involves cleavage of β-O-4 linkages (Figure 5.2). The process initiates with acidolysis, where a hydroxyl group at the α-position is removed, forming a benzylic carbocation. This reactive intermediate can then rearrange into enol ether structures, sometimes breaking the Cβ–Cγ bond and forming formaldehyde (Mu et al., 2013; Sturgeon et al., 2014). The cleavage pattern is influenced by the acid used. For example, H_2SO_4 tends to promote Cβ–Cγ bond cleavage, while HCl does not. Hydrolysis of the resulting enol ethers produces various phenolics, such as Hibbert's ketones and aldehyde-substituted aromatics (Deuss et al., 2015). These compounds can participate in further repolymerization reactions, forming complex, condensed lignin networks (Sannigrahi et al., 2010). The extent of repolymerization

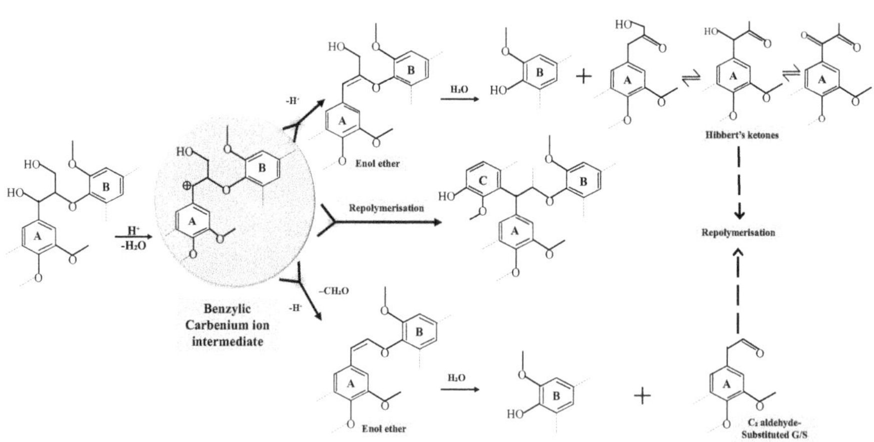

FIGURE 5.2 An outlined mechanism for lignin fragmentation under acidic circumstances

greatly affects the final yield and type of products, and it is influenced by reaction parameters like temperature, pH, catalyst type, and duration. Yields can range from as low as 2 wt% (in cases of significant repolymerization) to as high as 61 wt% when repolymerization is minimized (Schutyser et al., 2018). Common products include monomers and oligomers with unsaturated or carboxylic side groups. Both homogeneous (Brønsted or Lewis acids) and heterogeneous catalysts have been explored at temperatures up to 525 K (Schutyser et al., 2018).

5.2.4 Base-Catalyzed Depolymerization

In alkaline solutions, lignin becomes more soluble because its phenolic hydroxyl groups are deprotonated. This property is widely used in processes like kraft pulping to separate lignin from biomass (Brandi et al., 2021; Ragnar et al., 2000). Alkaline conditions help break lignin–carbohydrate bonds and also initiate the fragmentation and partial repolymerization of lignin itself. One key focus is β-O-4 linkage, whose reactivity relies on whether the unit is phenolic or non-phenolic (Figure 5.3). In non-phenolic units, cleavage proceeds through an epoxide intermediate, which is slower but eventually yields phenolic units. These can then form quinone methide intermediates if suitable leaving groups (like –OH or –OR) are present at the α-position (Gierer, 1985). Quinone methides are highly reactive and prone to nucleophilic attack, leading to depolymerization or, in some cases, repolymerization. In kraft processes, sulfide ions ($H\overline{S}$) act as strong nucleophiles, forming sulfur-containing intermediates that undergo further transformations. Another common pathway involves retro-aldol cleavage, where the terminal γ-CH_2OH group is removed, forming alkali-stable enol ethers and formaldehyde (Lapierre, 2010). As with acid catalysis, a range of homogeneous and heterogeneous bases can be used. Typical phenolic monomers produced include methoxyphenols, aldehydes,

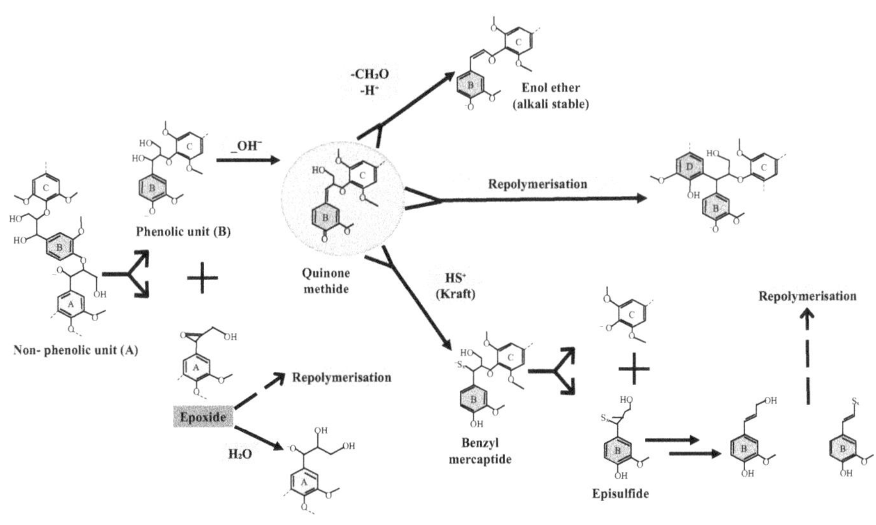

FIGURE 5.3 An outlined mechanism for lignin fragmentation under alkaline circumstances

acetophenones, catechols, and their alkylated derivatives. Yields generally range between 10 and 20 wt% (Schutyser et al., 2018).

5.3 THREE MODERN CATALYTIC TECHNOLOGIES FOR VALORIZING LIGNIN: A LIGNIN-FIRST BIOREFINERY PERSPECTIVE

Traditionally, the paper and pulp manufacturing units have focused mainly on extracting and processing the cellulose portion of LCB, while the lignin fraction, which is a complex and irregular polymer, has mostly been burned as fuel to generate energy. Only a small fraction of this lignin, which typically has low ether bond content and high molecular weight, has been utilized to produce specialty chemicals like vanillin (Questell-Santiago et al., 2020; Renders et al., 2017). Recently, emphasis on biomass valorization has shifted toward a more comprehensive approach where all the prominent components of LCB, including lignin, cellulose, and hemicellulose, are effectively utilized in a sustainable biorefinery model. This new model is known as the lignin-first biorefinery, and it represents a major promotion in the sector of biomass conversion (Renders et al., 2017). The core idea of the lignin-first strategy is to actively protect and stabilize the lignin fraction during the extraction process, rather than treating it as a waste byproduct or dealing with it only after cellulose processing. This method prevents unwanted side reactions such as recondensation, which make lignin more difficult to process. Importantly, this strategy not only maximizes the yield of valuable aromatic compounds from lignin but also preserves carbohydrate (cellulose and hemicellulose) fractions for further downstream processing (Abu-Omar et al., 2021). The lignin-first approach generally involves three interconnected steps: (a) Catalytic extraction of lignin from biomass using solvents. (b) Stabilization of reactive intermediates to prevent repolymerization. (c) Depolymerization into valuable aromatic monomers. Currently, three major catalytic technologies are being applied under the lignin-first framework to carry out these processes efficiently:

A. **Reductive Catalytic Fractionation (RCF)**
   RCF is among the most studied and promising lignin-first technologies. In this procedure, lignin is extracted from biomass in a solvent, typically an alcohol like methanol or ethanol, under a hydrogen atmosphere in the presence of a metal-based hydrogenation catalyst (e.g., Ru, Pd, or Ni) (Abu-Omar et al., 2021). This approach not only breaks down the lignin into smaller phenolic molecules but also stabilizes the reactive intermediates by hydrogenating them. This prevents lignin fragments from recombining into higher molecular weight polymers. RCF is particularly effective at producing high yields of monomeric phenols while preserving the cellulose pulp, which can then be processed separately into biofuels, bioplastics, or paper materials. Although RCF has mainly been performed in batch reactors, recent developments are exploring continuous-flow systems to improve process efficiency and scalability (Liao et al., 2020; Van den Bosch, 2017).

B. **Oxidative Catalytic Fractionation (OCF)**
OCF is an alternative lignin-first approach where the stabilization step involves oxidation instead of hydrogenation. In this method, lignin is depolymerized in the presence of oxidative catalysts, such as copper or vanadium-based catalysts, under a controlled oxygen or air atmosphere. This causes cleavage of ether bonds of lignin structure to form useful aromatic aldehydes and acids: syringaldehyde and vanillin (Galkin & Samec, 2016). The OCF often works under less extreme temperatures than the RCF but maintains a more strict oxidative control condition to preclude product over-oxidation and degradation of the product molecules. Similar to RCF, the carbohydrate fraction is left essentially unchanged and can be valorized independently. The approach is also in its initial phases of incorporation into continuous-flow systems (Anderson et al., 2016).

C. **Catalytic Protection Group Chemistry**
One more recently developed method of lignin valorization is the application of the protection group chemistry to lignin extraction. In this approach, the reactive sites of lignin fragments are chemically treated, typically with acetal or ester, upon bond cleavage. These protective groupings keep reactive intermediaries separate and do not allow recombination to produce unwanted, condensed structures (Renders et al., 2017). The method is particularly applicable when applied with either metal catalysts or acid environments, leading to controlled fragmentation of lignin. This method involves the use of protective agents during the extractive process and can be used to leave all of the L- and C- fractions chemically intact and separable. This approach was demonstrated when applied to laboratory-scale biorefinery models and exhibited notable outcomes. Although still in research stages, the flexible platform can enable integration into either oxidative or reductive pathways.

5.3.1 Lignin-first via reductive catalytic fractionation

Over the last few years, RCF has arisen as among the broadest researched lignin-first approaches that facilitate effective extraction and valorization of lignin within LCB. In RCF, the solvent, in reductive conditions, solubilizes the lignin, and frequently this is mediated by transition metal catalysis with the presence of solvent molecules of deposited hydrogen (H_2) or sacrificial hydrogen donors (Anderson et al., 2016; Galkin & Samec, 2016). Interestingly, despite the seeming modernity of the RCF, its roots stretch back to the early 20th century when it was first formulated to gain a deeper insight into lignin structures and the discovery of alternative pulping procedures. Nevertheless, the process never got to commercial use fast enough and was lost to memory until its rediscovery in recent years as a potential pathway to biomass valorization (Anderson et al., 2016). During the RCF process, lignin undergoes solvolytic extraction from the biomass matrix, leading to its partial breakdown (Galkin & Samec, 2016). (Figure 5.4). The fragmented lignin intermediates, which are unstable, are then immediately stabilized by a metal catalyst through reduction reactions. Simultaneously, the catalyst promotes further cleavage of lignin's internal

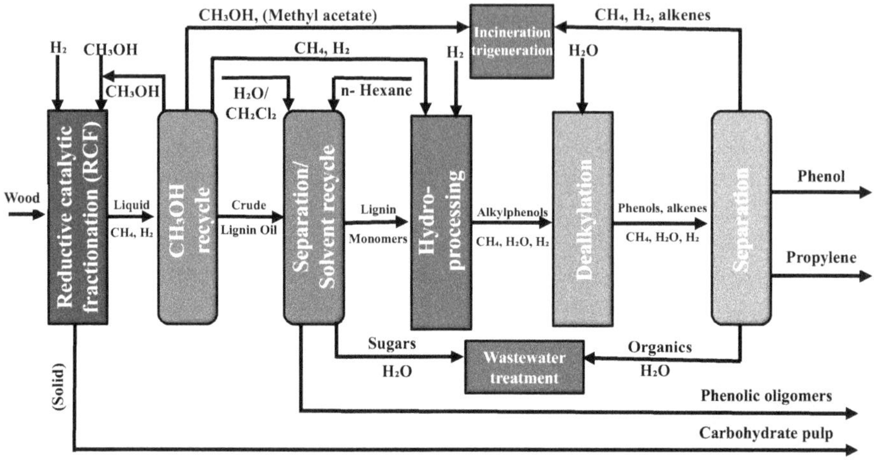

FIGURE 5.4 A flow diagram depicting the modern process of valorization of lignocellulosic biomass

ether bonds, especially β-O-4 linkages, through hydrogenolysis, producing a lignin oil rich in phenolic monomers, dimers, and oligomers. This method also preserves the carbohydrate-rich pulp, which remains available for additional valorization (Renders et al., 2017; Abu-Omar et al., 2021). The low-molecular-weight lignin oil, which is enhanced with phenolic chemicals and carbohydrate pulp RCF, produces typical advantages. RCF selectively generates para-substituted methoxyphenols from lignin. However, biomass type substantially affects product composition. Hardwoods, softwoods, and herbaceous crops differ in S, G, and H units, affecting product type and distribution (Anderson et al., 2016; Galkin & Samec, 2016). Reaction temperature, solvent type, catalyst type, and hydrogen source affect phenolic monomer yield and selectivity. Sels' group produced 51 wt% phenolic monomer and 98% delignification from birch wood utilizing Ru/C as a catalyst and methanol (MeOH) as a solvent (Renders et al., 2017; Cooreman et al., 2020). Similarly, Abu-Omar's group reported a 54 wt% monomer yield from poplar wood by employing a bimetallic Zn-Pd/C catalyst (Abu-Omar et al., 2021). A study demonstrated that using Ni@ZIF-8 in MeOH at 493 K yielded 44 wt% of phenolic monomers with 95% delignification (Parsell et al., 2015).

The choice of solvent is vital for RCF efficiency. Polar solvents like MeOH or EtOH are preferred as they facilitate high delignification; however, if the solvent polarity is too high, it can lead to the undesired loss of carbohydrate fractions (Renders et al., 2017; Cooreman et al., 2020). Different solvents also affect monomer yields, reporting 50 wt% in MeOH, 48 wt% in EtOH, and only 27 wt% in isopropanol (iPrOH) using a Ni/C catalyst (Xu et al., 2014). Solvent mixtures, especially alcohol-water blends, offer further tunability. For example, during a study, it was observed that a 70% water content optimized carbohydrate extraction, while a 30% water blend favored selective lignin extraction (Renders et al., 2017). Similarly, another study achieved a 49% monomer yield and 82% selectivity toward 4-propylsyringol using a Pt catalyst with a 1:1 MeOH-water solvent at 230°C (Ouyang et al., 2022). The role of additives such as co-catalysts (e.g., H_3PO_4 or NaOH) has also been explored. Acidic and basic conditions enhance delignification (up to 95% and 80%, respectively) compared to neutral environments but can differently impact monomer yields and risk repolymerization (Anderson et al., 2016; Huang et al., 2017). Another significant innovation has been conducting RCF without external hydrogen pressure. A study reported successful lignin depolymerization from pine and birch wood using Pd/C in an EtOH-water system, yielding 4-propylguaiacol and 4-propylsyringol at 23% and 49%, respectively (Galkin & Samec, 2016). Similarly, a study obtained up to 24 wt% monomers using Ru/C in acidified ethylene glycol at atmospheric pressure (Ren et al., 2021). The type of feedstock significantly influences RCF outcomes. Studies have explored various sources, including low-value residues like black locust bark (yielding 35 wt% lignin oil with Pd/C), hemp (38% phenolic monomers), vanilla seed (C-lignin), and agricultural remains such as stover of corn, straw of wheat, bagasse from sugarcane, and energy crops like miscanthus and switchgrass (yielding up to 45 wt% phenolic monomers) (Stone et al., 2018). Moving toward industrial application, reactor design and process optimization are key. Traditionally, RCF has been conducted in batch systems, but there is growing interest in continuous-flow systems, which can separate solvolytic extraction and catalytic depolymerization into two reactors, offering better control, reduced side reactions, and enhanced scalability (Galkin & Samec, 2016). Continuous-flow RCF has demonstrated promising results. For instance, the Samec group achieved a 21 wt% monomer yield using a Pd/C catalyst with MeOH-water solvent under flow conditions (Galkin & Samec, 2016). A researcher designed a dual-bed continuous reactor system that yielded 15 wt% monomers from poplar wood (Dixon et al., 2024). Furthermore, studies on corn stover lignin solvolysis in continuous-flow setups showed higher monomer yields (10 wt%) compared to batch systems (4 wt%) (Brandi et al., 2021). Hybrid systems combining batch solvolysis with flow catalytic depolymerization have also been explored, showing improved selectivity and product quality. For example, flow systems using Ni/C catalysts with MeOH-water solvents at elevated temperatures demonstrated better separation of catalytic and extraction steps, preserving native lignin structures (Brandi et al., 2021; Stone et al., 2018). Despite these advances, continuous-flow RCF still faces challenges such as lower monomer yields compared to batch processes and limited biomass processing scales. Further fundamental research, economic assessments, and life-cycle analyses are necessary to assess its commercial viability fully (Anderson et al., 2016; Galkin & Samec, 2016). In parallel, alternative lignin-first methods like diol-assisted fractionation (DAF) are emerging, aiming for active stabilization of lignin during extraction using protection chemistry techniques (Renders et al., 2017).

5.3.2 Oxidative Catalytic Fractionation (OCF) as a Lignin-First Approach

Modern lignin-first OCF is similar to reductive catalytic fractionation (RCF) but has a different reaction mechanism. OCF breaks down lignin into oxygen-rich monomers and low-molecular-weight compounds using oxidizing agents like oxygen (O_2), hydrogen peroxide (H_2O_2), nitrobenzene, or metal oxides (Rinaldi et al.,

2016; Schutyser et al., 2018). OCF breaks strong carbon–carbon (C–C) bonds in lignin, unlike RCF, which emphasizes ether connections. Selected C–C bond cleavage in lignin's aliphatic side chains produces useful methoxy-phenolic aldehydes, ketones, and carboxylic acids. Some oxidative circumstances may also cleave the aromatic ring, which forms tiny aliphatic (di)carboxylic acids, which is unfavorable (Schutyser et al., 2018; Galkin & Samec, 2016). Vanillin (4-hydroxy-3-methoxybenzaldehyde) is the most economically significant OCF product. It is mostly formed by selective lignin side chain oxidation and C–C bond cleavage (Fache et al., 2016). Vanillin production from lignin had once dominated globally. Lignin-based vanillin production dropped as cheaper fossil-derived technologies emerged. Norwegian business Borregaard is the only large vanillin producer that employs lignin. Molecular oxygen from air oxidizes lignin in Borregaard's industrial procedure utilizing a concentrated aqueous NaOH solution (Fache et al., 2016). Other OCF studies have experimented with different bases like potassium hydroxide (KOH) (Kagawa & Rokugawa, 1971) and even acidic conditions using agents such as acetic acid (Rahimi et al., 2014), sulfuric acid, or heteropoly acids like phosphomolybdic acid ($H_3PMo_{12}O_{40}$) (Lange et al., 2013). Some processes also involve organic solvents like methanol to assist the reaction (Werhan et al., 2013). High pH levels enable lignin phenolic groups to ionize rapidly, making alkaline media optimal for OCF. This ionization produces phenolate ions that are more easily oxidized into phenoxy radicals, an intermediate that cleaves C–C bonds under OCF conditions (Galkin & Samec, 2016; Schutyser et al., 2018). OCF reactions typically occur at 393–463 K and 2–15 bar oxygen pressures. Under these conditions, p-hydroxybenzaldehyde, vanillin, syringaldehyde, and their acids and acetyl derivatives are the primary products. OCF is challenging since these monomers tend to undergo secondary reactions such as recondensation, reducing the ultimate yield of functional monomers to 10% (Huang et al., 2017; Cao et al., 2018). Recent results reveal that OCF in a flow-through reactor with short residence durations at high temperatures can reduce secondary degradation and enhance phenolic monomer yields (Song et al., 2013).

5.4 CHALLENGES IN CATALYTIC LIGNIN FRACTIONATION FOR CHEMICAL PRODUCTION FROM LIGNOCELLULOSIC BIOMASS

5.4.1 Source of lignocellulosic biomass

Lignin valorization efficiency and consistency depend on LCB supply. LCB is inherently heterogeneous, and this variability arises from factors such as the species, age, and anatomical parts of the plant, including the presence of bark, heartwood, and sapwood. In the case of woody biomass, lignin yield and chemical composition are also independent of the harvesting time. When the residues are agricultural products, such as stalks, leaves, and corn cobs, the external processes, such as the nature of fertilizers and herbicides being used in cultivation, further make the characterization of the biomass difficult (Stone et al., 2018). One major complication is that it is hard to take these various variables into consideration because it is not always easy to learn exactly where the biomass originates. Therefore, a sustainable biorefinery strategy needs to be flexible to various feedstocks and is expected to produce consistent and high-quality products irrespective of the source of biomass. Furthermore, non-edible LCB should be utilized, ideally by-products or waste material, in the ethical and ecological sense so as to avoid competing with food production and make the process sustainable (Valentini et al., 2024).

5.4.2 Composition analysis

The LCB is very complex and heterogeneous, and it is really impossible to predict the products yielded in the catalytic fractionations. Due to increased complexity, randomization makes determining reaction conversion rates, yield, and selectivity challenging. Thus, optimal lignin valorization requires careful measurement and composition analysis of each biomass fraction. The U.S. Department of Energy and ASTM International created LCB specification protocols (Jenkins et al., 2019). Nevertheless, lignin quantification is complex, particularly the soluble lignin proportion. The most widespread technique is the Klason lignin method, only giving the insoluble lignin content, which is less applicable when high levels of bark and leaves are present in the biomass since they too have proteins and suberins (Galkin & Samec, 2016). To increase precision in determining lignin, it is vital to quantify various lignin units, including S, G, and H. These units perform a vital function in the study of the depolymerization process and anticipating derived products (Schutyser et al., 2018). The S/G/H ratio must, however, be quantified by first producing a complex aromatic mixture by reacting these with cellulose and then removing the mix of cellulose-derived products. Gas chromatography analyses using 1D and 2D NMR (principally, HSQC) are required to measure and detect the different lignin components (Huang et al., 2017). The comprehensive analysis is not only essential to the comprehension of the composition of the feedstock; it is also a valuable tool when comparing different lignin valorization strategies present in the literature (Galkin & Samec, 2016).

5.4.3 Active and stable catalysts

Another major problem in lignin valorization is using active and stable catalysts. Most developments involve powdered precious metal catalysts on traditional materials. The high metal loading of these catalysts normally limits them to low-volume laboratory testing because of the low operational lifetimes. These catalysts may rapidly lose their activity under the austere hydrothermal conditions relevant to lignin valorization because of the problems of metal leaching, agglomeration, or sintering (Schutyser et al., 2018). Moreover, the solid catalysts are not easily restored in the biomass slurry and thus complicate the process. Recycling such catalysts is also not very efficient because, when recycled, a series of functions can lead to the wearing down of the catalysts, both in terms of performance and destruction of the support material. Scaling up the valorization of lignin requires the development of more robust and readily recuperable catalysts (Schutyser et al., 2018).

5.4.4 Reactor Design (Batch vs. Continuous-Flow Reactors)

Reactor design is crucial to the scale and efficiency of lignin valorization. Research has focused on batch reactors, which mix biomass and catalyst effectively with a small amount of solvent, which is hard to recycle. Nevertheless, batch reactors do not apply to industrially scaled lignin valorization, especially beyond the 20 kilotons yearly levels. Low heating rate and inability to separate the catalyst and the biomass slurry render batch reactors poorly suited to large-scale processes (Rahimi et al., 2014). Moreover, diffusion usually becomes a limitation when using a catalyst basket in batch reactors, which makes the process less efficient. By contrast, continuous-flow reactors provide a number of benefits, among them the opportunity to physically separate feedstock and catalyst beds. Such fractioning enables more control over the conditions in each of the process steps and leads to higher effectiveness of both the reductive and oxidative catalytic fractionation (OCF) steps (Galkin & Samec, 2016). Nevertheless, continuous-flow systems have their own set of problems. As an example, channeling may occur in the case of non-ideal solvent flow and thus diminish the performance of the lignin extraction and solvolysis. In addition, a low solvent flow rate to reduce the amount of solvent used in the process may result in low monomer yields as lignin monomers are fragmented and will repolymerize before they are subjected to hydrogenation (Stone et al., 2018). To address these challenges, reactor design optimization should further be achieved to enhance the overall process efficiency of continuous-flow lignin valorization processes.

5.5 CONCLUSION

To shift toward a more sustainable biorefinery model based on lignocellulose biomass, it is important that all products, consisting of cellulosic and lignin fractions, are optimized with minimal production of waste and generation of CO_2. Developing biorefinery systems requires the effective valorization of lignin to produce commercially valuable drop-in chemicals and bio-based products. Lignin-first techniques using reductive or OCF seem promising since they can transform lignin into useful compounds based on the intended products. These technologies have a technological readiness level (TRL) of 1 to 4, suggesting that they are not, however, ready for commercial application. Another challenge is that these advanced processes are typically carried out at a batch scale, which is a limitation compared to the continuous-flow operations commonly used in traditional petrochemical refineries. Continuous-flow processes are more efficient for large-scale production, and without this capability, the new technologies cannot yet compete with the established petrochemical industry. The success of these modern lignin valorization techniques hinges on their ability to reduce the carbon footprint, especially CO_2 emissions, compared to current petrochemical processes. Thus, life-cycle assessment and techno-economic studies need to be completed to understand the commercial feasibility of these technologies in terms of environmental and economic costs. These studies are the only way to find out whether these new methods can provide a viable, sustainable alternative to the traditional petrochemical processes.

REFERENCES

Abu-Omar, M. M., K. Barta, G. T. Beckham, et al. 2021. Guidelines for Performing Lignin-First Biorefining. *Energy & Environmental Science* 14 (1): 262–92.

Achyuthan, K. E., A. M. Achyuthan, P. D. Adams, et al. 2010. Supramolecular Self-Assembled Chaos: Polyphenolic Lignin's Barrier to Cost-Effective Lignocellulosic Biofuels. *Molecules* 15 (12): 8641–88.

Al-Naji, M., H. Schlaad, and M. Antonietti. 2021. New (and Old) Monomers from Biorefineries to Make Polymer Chemistry More Sustainable. *Macromolecular Rapid Communications* 42 (3): 2000485. https://doi.org/10.1002/marc.202000485

Alonso, D. M., J. Q. Bond, and J. A. Dumesic. 2010. Catalytic Conversion of Biomass to Biofuels. *Green Chemistry* 12 (9): 1493–513. https://doi.org/10.1039/C004654J

Anderson, E. M., R. Katahira, M. Reed, et al. 2016. Reductive Catalytic Fractionation of Corn Stover Lignin. *ACS Sustainable Chemistry & Engineering* 4 (12): 6940–50.

Bosch, S. V., W. Schutyser, R. Vanholme, et al. 2015. Reductive Lignocellulose Fractionation into Soluble Lignin-Derived Phenolic Monomers and Dimers and Processable Carbohydrate Pulps, June. https://doi.org/10.1039/C5EE00204D

Brandi, F., M. Antonietti, and M. Al-Naji. 2021. Controlled Lignosulfonate Depolymerization via Solvothermal Fragmentation Coupled with Catalytic Hydrogenolysis/Hydrogenation in a Continuous Flow Reactor. December. https://doi.org/10.1039/D1GC01714D

Cao, Z., M. Dierks, M. T. Clough, I. B. D. De Castro, and R. Rinaldi. 2018. A Convergent Approach for a Deep Converting Lignin-First Biorefinery Rendering High-Energy-Density Drop-in Fuels. *Joule* 2 (6): 1118–33.

Choi, Y. S., R. Singh, J. Zhang, et al. 2016. Pyrolysis Reaction Networks for Lignin Model Compounds: Unraveling Thermal Deconstruction of β-O-4 and α-O-4 Compounds. *Green Chemistry* 18 (6): 1762–73.

Cooreman, E., T. Vangeel, K. V. Aelst, et al. 2020. Perspective on Overcoming Scale-Up Hurdles for the Reductive Catalytic Fractionation of Lignocellulose Biomass. *Industrial & Engineering Chemistry Research* 59 (39): 17035–45. https://doi.org/10.1021/acs.iecr.0c02294

Deuss, P. J., M. Scott, F. Tran, N. J. Westwood, J. G. de Vries, and K. Barta. 2015. Aromatic Monomers by in Situ Conversion of Reactive Intermediates in the Acid-Catalyzed Depolymerization of Lignin. *Journal of the American Chemical Society* 137 (23): 7456–67. https://doi.org/10.1021/jacs.5b03693

Dixon, R. A., A. Puente-Urbina, G. T. Beckham, and Y. Román-Leshkov. 2024. Enabling Lignin Valorization through Integrated Advances in Plant Biology and Biorefining. *Annual Review of Plant Biology* 75 (1): 239–63.

Evans, R. J., T. A. Milne, and M. N. Soltys. 1986. Direct Mass-Spectrometric Studies of the Pyrolysis of Carbonaceous Fuels: III. *Primary Pyrolysis of Lignin. Journal of Analytical and Applied Pyrolysis* 9 (3): 207–36. https://doi.org/10.1016/0165-2370(86)80012-2

Fache, M., B. Boutevin, and S. Caillol. 2016. Vanillin Production from Lignin and Its Use as a Renewable Chemical. *ACS Sustainable Chemistry & Engineering* 4 (1): 35–46.

Galkin, M. V., and J. S. M. Samec. 2016. Lignin Valorization through Catalytic Lignocellulose Fractionation: A Fundamental Platform for the Future Biorefinery. *Chem Sus Chem* 9 (13): 1544–58.

Gierer, J. 1985. Chemistry of Delignification: Part 1: General Concept and Reactions during Pulping. *Wood Science and Technology* 19: 289–312.

Güvenatam, B., E. H. J. Heeres, E. A. Pidko, and E. J. M. Hensen. 2016. Lewis-Acid Catalyzed Depolymerization of Protobind Lignin in Supercritical Water and Ethanol. *Catalysis Today, Catalytic Science and Technology in Sustainable Energy* 259: 460–66. https://doi.org/10.1016/j.cattod.2015.03.041

Huang, X., O. M. M. Gonzalez, J. Zhu, T. I. Korányi, M. D. Boot, and E. J. M. Hensen. 2017. Reductive Fractionation of Woody Biomass into Lignin Monomers and Cellulose by Tandem Metal Triflate and Pd/C Catalysis. *Green Chemistry* 19 (1): 175–87.

Isikgor, H. F., and C. R. Becer. 2015. Lignocellulosic Biomass: A Sustainable Platform for the Production of Bio-Based Chemicals and Polymers. June. https://doi.org/10.1039/C5PY00263J

Jenkins, B. M., L. L. Baxter, and J. Koppejan. 2019. Biomass Combustion. *Thermochemical Processing of Biomass: Conversion into Fuels, Chemicals and Power* 49–83.

Kagawa, S., and M. Rokugawa. 1971. Studies on the Utilization of Alkali Lignin (Part 3). *Japan Tappi Journal* 25 (10): 506–11. https://doi.org/10.2524/jtappij.25.10_506

Kotake, T., H. Kawamoto, and S. Saka. 2015. Pyrolytic Formation of Monomers from Hardwood Lignin as Studied from the Reactivities of the Primary Products. *Journal of Analytical and Applied Pyrolysis* 113 (May): 57–64. https://doi.org/10.1016/j.jaap.2014.09.029

Kudakasseril Kurian, J., G. R. Nair, A. Hussain, and G. S. V. Raghavan. 2013. Feedstocks, Logistics and Pre-Treatment Processes for Sustainable Lignocellulosic Biorefineries: A Comprehensive Review. *Renewable and Sustainable Energy Reviews* 25: 205–19. https://doi.org/10.1016/j.rser.2013.04.019

Lange, H., S. Decina, and C. Crestini. 2013. Oxidative Upgrade of Lignin–Recent Routes Reviewed. *European Polymer Journal* 49 (6): 1151–73.

Lapierre, C. 2010. Determining Lignin Structure by Chemical Degradations. *Lignin and Lignans: Advances in Chemistry* 11–48.

Li, C., X. Zhao, A. Wang, G. W. Huber, and T. Zhang. 2015. Catalytic Transformation of Lignin for the Production of Chemicals and Fuels. *Chemical Reviews* 115 (21): 11559–624.

Liao, Y., S.-F. Koelewijn, G. Van den Bossche, et al. 2020. A Sustainable Wood Biorefinery for Low–Carbon Footprint Chemicals Production. *Science* 367 (6484): 1385–90.

Martin, J., and M. Haggith. 2018. The State of the Global Paper Industry. *Environmental Paper Network* 5.

Mesa, J. A. M., F. Brandi, I. Shekova, M. Antonietti, and M. Al-Naji. 2020. P-Xylene from 2,5–Dimethylfuran and Acrylic Acid Using Zeolite in a Continuous Flow System. November. https://doi.org/10.1039/D0GC01517B

Mu, W., H. Ben, A. Ragauskas, and Y. Deng. 2013. Lignin Pyrolysis Components and Upgrading—Technology Review. *Bioenergy Research* 6: 1183–204.

Onwudili, J. A., and P. T. Williams. 2014. Catalytic Depolymerization of Alkali Lignin in Subcritical Water: Influence of Formic Acid and Pd/C Catalyst on the Yields of Liquid Monomeric Aromatic Products. *Green Chemistry* 16 (11): 4740–48.

Ouyang, B., F. Liu, C. Liang, et al. 2022. Toxicity and Activity Inhibition of Metal-Organic Framework MOF-199 to Nitrogen-Fixing Bacterium Azotobacter Vinelandii. *Science of The Total Environment* 813 (March): 151912. https://doi.org/10.1016/j.scitotenv.2021.151912

Parsell, T., S. Yohe, J. Degenstein, et al. 2015. A Synergistic Biorefinery Based on Catalytic Conversion of Lignin Prior to Cellulose Starting from Lignocellulosic Biomass. *Green Chemistry* 17 (3): 1492–99.

Questell-Santiago, Y. M., M. V. Galkin, K. Barta, and J. S. Luterbacher. 2020. Stabilization Strategies in Biomass Depolymerization Using Chemical Functionalization. *Nature Reviews Chemistry* 4 (6): 311–30.

Ragnar, M., G. Henriksson, M. E. Lindström, M. Wimby, J. Blechschmidt, and S. Heinemann. 2000. Pulp. *Ullmann's Encyclopedia of Industrial Chemistry* 1–92.

Rahimi, A., A. Ulbrich, J. J. Coon, and S. S. Stahl. 2014. Formic-Acid-Induced Depolymerization of Oxidized Lignin to Aromatics. *Nature* 515 (7526): 249–52.

Ren, T., S. You, Z. Zhang, et al. 2021. Highly Selective Reductive Catalytic Fractionation at Atmospheric Pressure without Hydrogen. *Green Chemistry* 23 (4): 1648–57.

Renders, T., S. V. Bosch, S. F. Koelewijn, W. Schutyser, and B. F. Sels. 2017. Lignin-First Biomass Fractionation: The Advent of Active Stabilisation Strategies. *Energy & Environmental Science* 10 (7): 1551–57.

Rinaldi, R., R. Jastrzebski, M. T. Clough, et al. 2016. Paving the Way for Lignin Valorisation: Recent Advances in Bioengineering, Biorefining and Catalysis. *Angewandte Chemie International Edition* 55 (29): 8164–215.

Sannigrahi, P., A. J. Ragauskas, and S. J. Miller. 2010. Lignin Structural Modifications Resulting from Ethanol Organosolv Treatment of Loblolly Pine. *Energy & Fuels* 24 (1): 683–89.

Schutyser, W., T. Renders, S. V. Bosch, S. F. Koelewijn, G. T. Beckham, and B. F. Sels. 2018. Chemicals from Lignin: An Interplay of Lignocellulose Fractionation, Depolymerisation, and Upgrading. *Chemical Society Reviews* 47 (3): 852–908.

Song, Q., F. Wang, J. Cai, et al. 2013. Lignin Depolymerization (LDP) in Alcohol over Nickel-Based Catalysts via a Fragmentation–Hydrogenolysis Process. *Energy & Environmental Science* 6 (3): 994–1007.

Stone, M. L., E. M. Anderson, K. M. Meek, et al. 2018. Reductive Catalytic Fractionation of C-Lignin. *ACS Sustainable Chemistry & Engineering* 6 (9): 11211–18.

Sturgeon, M. R., S. Kim, K. Lawrence, et al. 2014. A Mechanistic Investigation of Acid-Catalyzed Cleavage of Aryl-Ether Linkages: Implications for Lignin Depolymerization in Acidic Environments. *ACS Sustainable Chemistry & Engineering* 2 (3): 472–85. https://doi.org/10.1021/sc400384w

Sun, Z., B. Fridrich, A. de Santi, S. Elangovan, and K. Barta. 2018. Bright Side of Lignin Depolymerization: Toward New Platform Chemicals. *Chemical Reviews* 118 (2): 614–78. https://doi.org/10.1021/acs.chemrev.7b00588

Valentini, F., S. Chen, G. Brufani, Y. Gu, and L. Vaccaro. 2024. Ni/PiNe Heterogeneous Catalyst from Biomass Waste: Low-Loading, Ligand-Free Suzuki-Miyaura Cross-Coupling. *Chem Sus Chem* 18 (7): e202402011.

Van den Bosch, S. 2017. Reductive Catalytic Fractionation in the Lignocellulosic Biorefinery.

Werhan, H., N. Assmann, and P. R. von Rohr. 2013. Lignin Oxidation Studies in a Continuous Two-Phase Flow Microreactor. *Chemical Engineering and Processing: Process Intensification* 73: 29–37.

Xu, C., R. A. D. Arancon, J. Labidi, and R. Luque. 2014. Lignin Depolymerisation Strategies: Towards Valuable Chemicals and Fuels. *Chemical Society Reviews* 43 (22): 7485–500.

Yanez, A. J., P. Natarajan, W, Li, R. Mabon, and L. J. Broadbelt. 2018. Coupled Structural and Kinetic Model of Lignin Fast Pyrolysis. *Energy & Fuels* 32 (2): 1822–30.

Yang, F., and M. Antonietti. 2020. Artificial Humic Acids: Sustainable Materials against Climate Change. *Advanced Science* 7 (5): 1902992. https://doi.org/10.1002/advs.201902992

Zakzeski, J., P. C. A. Bruijnincx, A. L. Jongerius, and B. M. Weckhuysen. 2010. The Catalytic Valorization of Lignin for the Production of Renewable Chemicals. *Chemical Reviews* 110 (6): 3552–99.

6 Biological degradation of lignin

Aubrey Dickson Chigwada, Henry Joseph Oduor Ogola, and Memory Tekere

6.1 INTRODUCTION

Lignin, a structurally complex aromatic heteropolymer, ranks as the second most abundant terrestrial biopolymer after cellulose, constituting approximately 15–30% by weight and up to 40% by energy content of lignocellulosic biomass (Abdelaziz & Hulteberg, 2020). Its intricate architecture arises from the oxidative combinatorial polymerization of three primary monolignols, *p*-coumaryl, coniferyl, and sinapyl alcohols, resulting in a heterogeneous polymer network characterized by diverse ether and carbon-carbon linkages, with the β-O-4 aryl ether bond being the most prevalent and critical for polymer integrity (Picart et al., 2015; Pollegioni et al., 2015). This structural heterogeneity and robustness confer significant recalcitrance to both microbial and chemical degradation, posing a substantial barrier to efficient bioconversion processes. Nonetheless, lignin's abundance and rich aromatic content position it as a vast, yet underexploited, renewable resource for the sustainable production of value-added aromatic chemicals and materials (Ma et al., 2016). The global imperative to transition from a fossil fuel-dependent economy to a sustainable, circular bioeconomy necessitates the development of innovative lignin valorization strategies that are economically viable and environmentally sustainable (Park et al., 2022). Biological degradation, leveraging the metabolic versatility and enzymatic machinery of diverse microorganisms, offers a promising, environmentally benign, and potentially cost-effective approach to depolymerize lignin into valuable platform chemicals and biomaterials, integrating lignin valorization into biorefinery frameworks and contributing to carbon-neutral technologies (Park et al., 2022). Recent breakthroughs have expanded our understanding of ligninolytic enzymology, revealing novel oxidative enzymes and multienzyme complexes, such as synergistic enzyme cocktails in *Pseudomonas putida* that enhance lignin depolymerization while preventing repolymerization, highlighting auxiliary enzymes previously unrecognized in classical pathways (Beckham et al., 2016; Liang et al., 2025). These advances mark a transition from descriptive enzymology to a predictive, design-oriented paradigm, further accelerated by protein engineering and artificial intelligence-driven

 DOI: 10.1201/9781003653189-6

enzyme design, enabling the creation of robust biocatalysts tailored for industrial applications (Arnold, 2018; Zhang et al., 2019). Among ligninolytic microorganisms, fungal taxa, particularly white-rot fungi, are widely recognized as the most efficient lignin degraders, capable of complete mineralization through the secretion of a diverse array of oxidative enzymes. Brown-rot fungi primarily modify lignin structure without full degradation, while soft-rot fungi contribute to lignin decay to a lesser extent, often under specific environmental conditions (Weng et al., 2021). Bacterial lignin degraders, distributed across phyla such as *Actinomycetes, Firmicutes, α-proteobacteria*, and *γ-proteobacteria*, include genera like *Bacillus, Rhodococcus, Streptomyces,* and *Pseudomonas*, which have been isolated from diverse ecological niches including soil, decaying wood, plant matter, and animal gastrointestinal tracts (Grgas et al., 2023; Zhu et al., 2017). Recent systems biology approaches have elucidated detailed bacterial catabolic pathways funneling lignin-derived aromatics into central metabolism, and engineered microbial consortia combining fungi and bacteria have demonstrated enhanced lignin degradation efficiencies, underscoring the importance of community-level interactions in natural and engineered ecosystems (Beckham et al., 2016; Hu et al., 2025; Lin, 2022). Grgas et al. (2023) demonstrate that these bacteria have been isolated from a variety of environments, including soil, wood, plant matter, and even the stomachs of animals. Notably, genetically engineered *Corynebacterium glutamicum* has demonstrated the capacity to biotransform lignin-derived monomers like vanillin and vanillate into cis, cis-muconate, a key precursor for industrial polymers such as bio-based PET and nylon, highlighting the translational potential of microbial lignin valorization (Weiland et al., 2023). Central to microbial lignin degradation are extracellular oxidative enzymes that catalyze the cleavage of complex lignin linkages. White-rot fungi such as *Phanerochaete chrysosporium* and *Pleurotus ostreatus* secrete a repertoire of ligninolytic enzymes, including lignin peroxidase (LiP), manganese peroxidase (LnP), laccase (Lac), versatile peroxidase (VP), and dye-decolorizing peroxidase (DyP), which collectively oxidize phenolic and non-phenolic lignin structures (Bautista-Zamudio et al., 2023; Nakamura et al., 1999; Sharma & Kuhad, 2010). Bacterial lignin degradation involves enzymes such as etherases, mixed-function oxygenases, dioxygenases, and multicopper oxidases, which catalyze the depolymerization of high-molecular-weight lignin polymers into low-molecular-weight aromatic compounds, as observed in species like *Pseudomonas putida, Rhodococcus jostii,* and *Amycolatopsis* (Grgas et al., 2023). This chapter aims to elucidate the mechanistic underpinnings of lignin biodegradation and their implications for biotechnological innovations targeting the conversion of lignocellulosic biomass into advanced biomaterials, biofuels, and biochemicals, thereby addressing pressing global energy and environmental challenges. Despite extensive investigation yielding a substantial body of literature on microbial systems, enzymatic pathways, and valorization strategies, the inherent complexity of lignin and the diversity of degradative organisms have engendered divergent perspectives regarding optimal valorization approaches. This chapter systematically synthesizes current knowledge, identifies convergent and divergent findings, and highlights critical research gaps. Key topics include the three principal classes of ligninolytic enzymes—laccases, auxiliary enzymes, and peroxidases—their catalytic mechanisms, synergistic interactions, and prospects for enzyme engineering and application. Additionally, strategies to enhance microbial ligninolytic capacity through metabolic engineering and genetic modification are critically evaluated. The chapter also addresses industrial valorization pathways, encompassing pretreatment technologies, bioconversion processes to platform chemicals and materials, and bioremediation applications. Finally, persistent challenges are discussed, and future research directions are proposed to advance the scalability and economic viability of lignin bioconversion technologies.

6.1.1 Significance, Industrial Potential, and Structural Challenges of Lignin as a Renewable Aromatic Biopolymer

Lignin's prominence as a renewable aromatic biopolymer is underscored by its vast natural abundance and unique chemical architecture, which collectively confer substantial potential as a sustainable feedstock for biorefineries (Ma et al., 2016). Lignin is the second most abundant terrestrial biopolymer after cellulose, representing a major reservoir of renewable aromatic carbon fixed annually through terrestrial photosynthesis (Abdelaziz & Hulteberg, 2020). In plants, lignin confers mechanical strength, rigidity, and hydrophobicity to cell walls while protecting against microbial attack and environmental stresses (Penín et al., 2020). From an industrial perspective, lignin is generated in enormous quantities, exceeding 1.2 million metric tons annually, primarily as a byproduct of the pulp and paper industry, where technical lignins such as kraft lignin and lignosulfonates are produced (Grand View Research, 2023; Reid, 1995). Despite this volume, most industrial lignin is currently combusted for onsite energy, underutilizing its rich chemical potential (Sethupathy et al., 2022). However, the global lignin market is expanding rapidly, valued at approximately USD 1.08 billion in 2023 and projected to reach USD 1.47 billion by 2030, driven by growing demand for animal feed additives, biofuels, specialty chemicals, and biorefinery catalysts (Grand View Research, 2023). This growth trajectory reflects the expanding recognition of lignin as a key renewable feedstock for sustainable chemical production. Chemically, lignin is composed of phenylpropanoid units, p-coumaryl, coniferyl, and sinapyl alcohols, which polymerize into p-hydroxyphenyl (H), guaiacyl (G), and syringyl (S) monomers, respectively. These monolignols serve as versatile precursors for a broad spectrum of value-added chemicals, including vanillin, syringaldehyde, phenolic resins, and high-performance carbon fibers with extensive industrial applications (Sharma & Kuhad, 2010; Weiland et al., 2023). As shown in Figure 6.1, several bacteria and their enzymes target the monolignols to produce their wide range of byproducts.

Particularly, lignin-derived carbon fibers are gaining traction due to their superior environmental profile and performance compared to petroleum-based alternatives (Jardim et al., 2020; Reid, 1995). Additionally, lignin's intrinsic properties—such as biodegradability, antioxidant activity, and ultraviolet absorption—enable the development of advanced materials like bioplastics, coatings, adhesives, and pharmaceuticals, broadening its industrial relevance (Gao et al., 2024). Despite its promise, lignin's structural complexity and heterogeneity present formidable challenges for efficient depolymerization and selective functionalization. Lignin is an

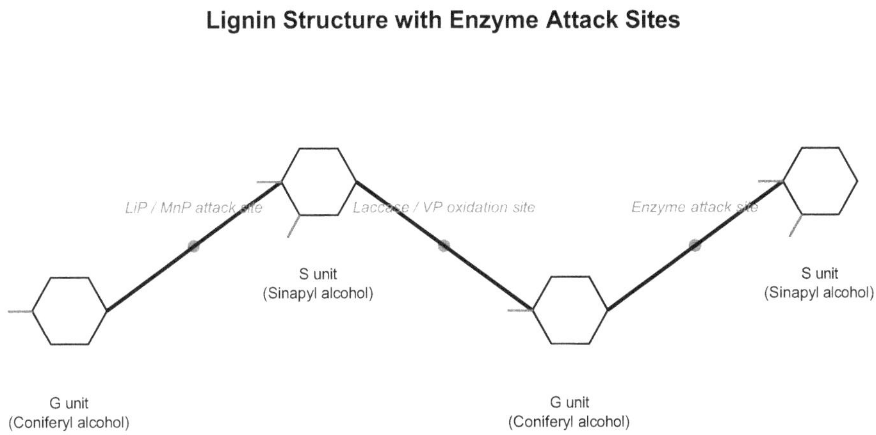

FIGURE 6.1 Coniferyl and sinapyl are targets that many enzymes can transform into useful final products

amorphous, three-dimensional aromatic polymer formed through oxidative radical polymerization of three main phenylpropanoid monomers: p-coumaryl, coniferyl, and sinapyl alcohols, which give rise to p-hydroxyphenyl (H), guaiacyl (G), and syringyl (S) units, respectively (Weiland et al., 2023; Wynands et al., 2023). The relative abundance of these units varies by plant type; softwoods are rich in G units, hardwoods contain both G and S units, while grasses comprise all three (H, G, and S) and are further influenced by factors such as plant age and environment (Pollegioni et al., 2015). These monolignols are interconnected by a diverse array of covalent bonds, predominantly ether linkages (e.g., β-O-4, α-O-4, 4-O-5) and carbon–carbon bonds (e.g., β-β, β-5, 5-5, β-1), with the β-O-4 aryl ether linkage being the most abundant, accounting for approximately 45–60% of all linkages (Picart et al., 2015). Unlike the regular, hydrolyzable polysaccharide backbones, lignin's irregular, cross-linked matrix impedes enzymatic and chemical attack (Zhao et al., 2022). Its hydrophobicity and high molecular weight further restrict enzyme accessibility (Ayuso-Fernández et al., 2018). Moreover, the chemical stability imparted by strong carbon–carbon and ether bonds demands substantial energy input for cleavage, complicating depolymerization strategies (Wang et al., 2018). The variability in lignin composition, affected by plant species, tissue type, age, and environment, necessitates tailored pretreatment and valorization approaches. Softwoods are rich in guaiacyl units, hardwoods contain both guaiacyl and syringyl, while grasses incorporate all three monomers (Pollegioni et al., 2015). This diversity complicates the development of universal bioconversion technologies but also offers opportunities for producing a wide array of lignin-derived products through selective catalytic and biological processes. In summary, lignin's abundant availability, coupled with its rich aromatic structure, underpins its industrial potential as a renewable feedstock for biorefineries. Overcoming its structural recalcitrance through innovative depolymerization and functionalization technologies remains a critical challenge and opportunity for sustainable chemical production.

6.1.2 Overcoming lignin recalcitrance and structural complexity for industrial use

The structural complexity and heterogeneity of lignin, resulting from its variable composition of p-coumaryl (H), coniferyl (G), and sinapyl (S) phenylpropanoid units connected by strong C-O bonds (such as β-O-4) and C-C bonds, present major obstacles to depolymerization and functionalization. These obstacles include low selectivity, repolymerization, and source-dependent variability, which make product separation difficult (Jeffri et al., 2024; Jiju et al., 2025; Pei et al., 2024). Integrated strategies are utilized to solve these issues. According to Pei et al. (2024), pretreatment techniques, such as physical-chemical procedures like milling treatment, acid/alkali, microwave-assisted, or biological-physical hybrids, decrease recalcitrance and fractionate lignin for homogeneity. Selective bond cleavage is enhanced by chemo-catalytic depolymerization, which includes acid/base catalysis (e.g., H_2SO_4 or NaOH, achieving up to 93% recovery), oxidative cleavage (e.g., H_2O_4 yielding 77% aromatic monomers like vanillin), and reductive hydrogenolysis (e.g., using Ni/Pd catalysts to prevent repolymerization) (Jeffri et al., 2024; Jiju et al., 2025; Li et al., 2024). Green solvents, such as deep eutectic solvents and ionic liquids ([Emim]OAc), enhance solubility and selectivity; they are frequently used in conjunction with enzymes or microwave support (Pei et al., 2024; Cai et al., 2024). Although char generation needs to be optimized, thermochemical methods such as microwave pyrolysis (up to 85% bio-oil production) and slow/fast pyrolysis (150–800°C) provide quick breakdown (Jiju et al., 2025; Li et al., 2024). For mild, environmentally friendly depolymerization, biological methods use bacteria (e.g., *Bacillus cereus*, which achieves 89% recovery) and fungi (e.g., *Phanerochaete chrysosporium* with laccases). These methods are enhanced by genetic engineering to tolerate heterogeneous substrates (Jeffri et al., 2024; Gu et al., 2024). Modern methods such as electrocatalysis (Ni/Pb electrodes with ionic liquids) and photocatalysis (using TiO_2 or $g\text{-}C_3N_4$) allow for selective, low-energy bond cleavage (Pei et al., 2024; Cai et al., 2024). Through acetylation, amination, or hydrodeoxygenation, functionalization alters the hydroxyl and methoxy groups in lignin to create polymers, adhesives, and biofuels that find use in thermoplastic polyurethanes and epoxy resins (Jiju et al., 2025; Li et al., 2024; Zhang et al., 2024). Recent research demonstrates scalable solutions, with hybrid enzyme-catalyst systems and stabilization mechanisms further addressing lignin's robustness for sustainable biorefineries. Examples of these include Gu et al.'s 89% bacterial yield and Oruganti et al.'s 93.4% recovery via alkaline hydrolysis (Jeffri et al., 2024; Gu et al., 2024). This exceptional degradative capacity derives from a highly evolved extracellular enzymatic system, primarily composed of ligninolytic heme peroxidases and multicopper oxidases. Among these, LiPs, first characterized in *Phanerochaete chrysosporium*, catalyze hydrogen peroxide (H_2O_2)-dependent one-electron oxidation of non-phenolic lignin units, which constitute the bulk of lignin's complex polymeric structure. This oxidation generates reactive aryl cation radicals that cleave β-O-4 ether linkages, disrupt Cα-Cβ bonds, and open aromatic rings, thereby facilitating lignin depolymerization (Sharma & Kuhad, 2010; Jeyakumar et al., 2013; Wu et al., 2023). Complementing LiPs are manganese peroxidases (MnPs) that oxidize manganese II (Mn^{2+}) to manganese III (Mn^{3+}), a diffusible

TABLE 6.1
Comparative illustration of fungal types in lignin biodegradation

Fungal Type	Lignin Degradation Efficiency Comparison	Key Mechanisms and Extent	References
White-rot	High (complete mineralization to CO_2). Most efficient; higher gene diversity than brown/soft-rot	Oxidative enzymes (such as peroxidases and laccases) with multiple gene copies can degrade all wood components	Floudas et al. (2012); Riley et al. (2014)
Brown-rot	Low (modification, not mineralization). Least efficient for lignin; faster overall wood decay	Limited enzymes; focuses on cellulose/hemicellulose; leaves lignin residue	Arantes et al. (2018); Goodell et al. (2017); Zhang et al. (2016)
Soft-rot	Moderate (partial degradation). Less efficient than white-rot; similar limitations to brown-rot	Attacks in moist, low-lignin woods; creates cell wall cavities; cannot fully mineralize to CO_2	Daniel (2014); Riley et al. (2014)

oxidant stabilized by organic acid chelation (e.g., oxalate), which penetrates the lignocellulosic matrix to oxidize phenolic lignin substructures and indirectly affect non-phenolic moieties via lipid peroxidation pathways (Wu et al., 2023). MnP are another group of heme peroxidases that need (Mn^2+) as a substrate. They convert Mn^{2+} to Mn^{3+}, which functions as a diffusible oxidant that can target phenolic lignin structures and, indirectly, non-phenolic structures through lipid peroxidation pathways. Mn^{3+} is frequently chelated by organic acids such as oxalate. Enzymes cannot pass through the lignocellulose matrix, where Mn^{3+} can because of its tiny size (Wu et al., 2023). It has been demonstrated that bacterial species like *Rhodococcus jostii* use cytochrome P450 monooxygenase, manganese superoxide dismutase, DyP, and laccase to demethylate and hydroxylate lignin. When hydrogen peroxide is not available, lignin can be broken down by the β-ketoadipate pathway (β-KAP) (Grgas et al., 2023). VPs combine the catalytic properties of LiPs and MnPs, expanding substrate specificity to phenolic and high redox-potential aromatic compounds such as veratryl alcohol, thereby enhancing ligninolytic efficiency (Hofrichter et al., 2010; Wang et al., 2018).

6.2 BACTERIAL LIGNIN BIODEGRADATION

In addition to fungal degradation, bacterial lignin biodegradation represents a complementary and increasingly vital avenue for lignin valorization in biorefineries. Diverse bacterial taxa, including *Actinobacteria* (e.g., *Streptomyces, Rhodococcus,* and *Nocardia*), *Proteobacteria* (e.g., *Pseudomonas, Sphingomonas,* and *Comamonas*), and *Firmicutes* (e.g., *Bacillus*), have demonstrated significant capacities to depolymerize lignin and metabolize lignin-derived aromatic compounds (Falade et al., 2017; Grgas et al., 2023). Compared to fungi, bacteria offer advantages for industrial applications such as faster growth rates, amenability to genetic engineering, and greater tolerance to diverse environmental conditions (Pollegioni et al., 2015; Hämäläinen et al., 2018). These attributes position bacteria as promising biocatalysts for tailored lignin bioconversion processes aimed at producing fine chemicals and value-added products.

Bacterial lignin degradation typically initiates with extracellular oxidative enzymes, including laccases, peroxidases, and oxidoreductases, which depolymerize the complex lignin polymer into smaller aromatic fragments (Hämäläinen et al., 2018; Nakamura et al., 1999). A key feature of many lignin-degrading bacteria is their ability to utilize the resulting low-molecular-weight aromatic compounds as carbon and energy sources, funneling them into central metabolic pathways. Intracellular catabolism involves the conversion of diverse aromatic monomers into central intermediates such as protocatechuate and catechol, which are further processed via ring cleavage pathways (ortho- or meta-cleavage) to generate metabolites entering the tricarboxylic acid (TCA) cycle (Weng et al., 2021). These are subsequently processed via ring cleavage pathways such as ortho- or meta-cleavage. New pathways for targeted lignin depolymerization have been made possible by the discovery of novel bacterial enzymes, such as β-etherases (Picart et al., 2015).

A breakthrough in bacterial lignin valorization emerged with the discovery of enzymes like β-etherases, exemplified by *Sphingomonas paucimobilis* LigE, LigF, and LigG, which catalyze the reductive cleavage of the abundant β-O-4 aryl ether bonds in lignin (Picart et al., 2015). This enzymatic specificity enables precise depolymerization, facilitating the production of aromatic monomers that serve as precursors for high-value chemicals. For example, *Rhodococcus jostii* RHA1, a model lignin-degrading bacterium, expresses multiple dye-decolorizing peroxidases (DyPs), laccases, and glyoxal oxidases, enabling it to oxidize phenolic lignin units and convert lignin into valuable compounds such as vanillin and syringaldehyde and other lignin titers, which are widely used as flavoring agents and chemical intermediates (Alruwaili et al., 2023; Rashid et al., 2024).

The metabolic versatility of bacteria extends to the "upper pathways," where enzymes such as O-demethylases, hydroxylases, and decarboxylases modify lignin-derived aromatics by removing methoxyl groups and introducing hydroxyl functionalities, thereby generating protocatechuate and catechol (Pimviriyakul et al., 2020). These intermediates are then cleaved by dioxygenases in the "lower pathways," producing aliphatic compounds that feed into central metabolism. Notably, these catabolic pathways are tightly regulated by inducible operons responsive to lignin-derived substrates, as demonstrated in *Sphingomonas paucimobilis*, where gene clusters encoding β-etherases and related enzymes are upregulated in the presence of lignin fragments (Jha et al., 2021; Kumar et al., 2018). Understanding and manipulating these regulatory networks is critical for engineering bacterial strains with enhanced lignin bioconversion efficiency.

While aerobic bacteria dominate lignin degradation research, emerging studies highlight the potential of anaerobic bacteria inhabiting sediments, wet soils, and animal guts to partially depolymerize lignin under oxygen-limited conditions (DeAngelis et al., 2013). Anaerobic lignin biodegradation often involves reductive or hydrolytic enzymatic pathways distinct from aerobic oxidative mechanisms, enabling cleavage and modification of lignin linkages without molecular oxygen

(Khan & Ahring, 2019). For instance, syntrophic consortia from termite guts and biogas digesters have been shown to demethylate and cleave lignin-derived aromatics, producing intermediates amenable to methane generation or further bioconversion (Rabus et al., 2016; Sijinamanoj et al., 2021). Although complete anaerobic lignin mineralization remains challenging, these processes hold promise for integrated biorefinery designs that couple lignin modification with bioenergy production.

6.2.1 Bacterial taxa contributions to lignin decay in ecological niches

In the breakdown of lignin, bacterial taxa complement fungi by frequently metabolizing lignin-derived aromatics using intracellular funneling pathways that converge on central metabolism, such as acetyl-CoA or benzoyl-CoA, and extracellular enzymes, such as laccases (Bugg et al., 2011; Kamimura et al., 2019). Gram-negative bacteria exhibit significant rates of degradation, and important phyla include Firmicutes, Actinobacteria, and Proteobacteria, which are dominated by the γ-, β-, and α-classes (Bugg et al., 2011; Tian et al., 2014; Kamimura et al., 2019). Soils, symbiotic systems such as termite guts for aromatic breakdown, and man-made habitats such as wastewater for residual lignin are only a few of the several niches in which they contribute (Shi et al., 2013; Kamimura et al., 2019). According to de Boer et al. (2005), microbial interactions frequently promote degradation more than individual taxa. Table 6.1 summarizes the bacterial taxa identified in lignin biodegradation.

Bacterial lignin biodegradation offers a versatile and engineerable platform for lignin valorization in biorefineries. By combining extracellular depolymerization with sophisticated intracellular catabolic pathways, bacteria can convert heterogeneous lignin polymers into defined aromatic monomers and downstream fine chemicals such as vanillin, syringaldehyde, and other phenolics. Advances in enzyme discovery, metabolic engineering, and regulatory network elucidation are accelerating the development of bacterial biocatalysts tailored for efficient lignin conversion, thereby expanding the portfolio of sustainable lignin-derived products and enhancing the economic feasibility of lignocellulosic biorefineries.

6.2.2 Systems biology and omics in lignin degradation

In order to engineer microbial pathways, systems biology integrates omics data from genomics, transcriptomics, proteomics, and metabolomics. These data target enzymes like laccases, manganese peroxidases, and dye-decolorizing peroxidases, which effectively cleave the aromatic structure of lignin in mild conditions, achieving up to 89% degradation in bacteria such as *Bacillus cereus* (Gu et al., 2024). According to studies like Hu et al. (2025), engineered consortia that combine bacteria (e.g., *Pseudomonas* and *Rhodococcus*) and fungi (e.g., *Basidiomycetes*) use synthetic biology to achieve synergistic degradation through complementary enzymes, division of labor, and quorum sensing, increasing efficiency by 20–50% and enabling consolidated bioprocessing that lowers biorefinery costs by more than 30%. By employing engineered cutinases and metabolic funneling to improve biodegradability and recyclability, these developments support a circular bioeconomy by facilitating the valorization of lignin into monomers such as terephthalic acid and adipic acid for bio-based PET and nylon, respectively (Wei et al., 2024; Zhang et al., 2024). Even if issues like enzyme stability still exist, scalable, carbon-neutral production is being fueled by AI-driven modeling and consortium design. Carbios' PET recycling is one example of a process that has commercial promise (López et al., 2025).

6.3 ENZYMATIC SYSTEMS IN LIGNIN BIODEGRADATION

The enzymatic biodegradation of lignin represents a pivotal step in biological lignin valorization, relying on specialized microbial enzymes capable of cleaving the complex and recalcitrant polymeric structure of lignin (Andlar et al., 2018). This section examines the classification, catalytic mechanisms, and synergistic interactions of the principal enzyme systems involved in lignin depolymerization, emphasizing their roles in enhancing degradative efficiency. Key ligninolytic enzymes naturally secreted by fungi and bacteria include heme peroxidases, such as lignin peroxidase, manganese peroxidase, and VP, and multicopper oxidases like laccases. These enzymes operate through oxidative mechanisms that generate reactive radical species, enabling the cleavage of diverse lignin linkages (Ayuso-Fernández et al., 2018b; Wang et al., 2018). Complementing these primary enzymes are auxiliary oxidoreductases and accessory proteins that facilitate the generation of necessary cofactors (e.g., hydrogen peroxide) and enhance substrate accessibility, thereby orchestrating a coordinated enzymatic assault on lignin (Hämäläinen et al., 2018). Furthermore, synthetic biology and protein engineering approaches are being employed to tailor ligninolytic enzymes with enhanced catalytic activity, stability under industrial conditions, and substrate specificity, enabling the design of bespoke biocatalysts optimized for particular lignin feedstocks and process requirements (Weng et al., 2021). These developments underscore the critical importance of understanding both individual enzyme functions and their integrated actions in advancing lignin biorefineries. The ability to engineer and deploy customized enzymatic systems promises to overcome current limitations in lignin valorization, facilitating the sustainable production of biofuels, chemicals, and advanced materials from lignocellulosic biomass.

6.3.1 Key enzymes in lignin biodegradation and their synergistic interactions

Microorganisms degrade lignin primarily through a diverse group of extracellular oxidoreductases collectively known as ligninolytic enzymes (Table 6.2). These enzymes are classified based on their cofactors, active site architectures, and catalytic mechanisms. The principal enzyme classes involved in lignin biodegradation are heme peroxidases and multicopper oxidases (laccases), supported by auxiliary enzymes that enhance their activity and stability (Wang et al., 2018). Additionally, emerging enzyme families such as dye-decolorizing peroxidases (DyPs) and unspecific peroxygenases (UPOs) have been recognized for their significant contributions to lignin breakdown (Hofrichter et al., 2010; Lončar et al., 2016; Singh & Gupta, 2020).

TABLE 6.2
A Summary of bacterial taxa contributing to lignin biodegradation

Bacterial Taxa	Contribution to Lignin Decay	Ecological Niches	References
Proteobacteria (Pseudomonas, Sphingobium, Rhodopseudomonas, Comamonadaceae, and Caulobacteraceae)	High degradation of lignin substrates; extracellular laccases for initial breakdown; funneling pathways for aromatics such as vanillin and ferulic acid	Soils/forest litter (aerobic decay); termite guts (symbiotic aromatic metabolism); and wastewater (residual lignin)	Bugg et al. (2011); Tian et al. (2014); Kamimura et al. (2019)
Actinobacteria (Rhodococcus, Amycolatopsis, and Streptomyces)	Enzymatic depolymerization and aromatic catabolism; laccases and peroxidases	Deadwood/plant litter (associated with fungi); soils (enriched in advanced lignin decay)	Tian et al. (2014); Brown and Chang (2014); Kamimura et al. (2019)
Firmicutes (Clostridium)	Anaerobic degradation pathways, such as converging on benzoyl-CoA); contributes to partial lignin breakdown	Termite guts (symbiotic); anaerobic man-made environments in sludge or wastewater	Shi et al. (2013); Kamimura et al. (2019)
Acidobacteria (Holophaga; Bacteroidetes; and Spirochaetes)	Specialized aromatic degradation; less common but key in niche roles	High-saline environments; termite guts; forest ground layer	Kamimura et al. (2019)

6.3.1.1 Heme peroxidases: LiP, MnP, VP, and DyPs

Heme peroxidases utilize hydrogen peroxide (H_2O_2) as an oxidant and include LiP, manganese peroxidase (MnP), and VP. LiP (EC 1.11.1.14) was the first ligninolytic enzyme identified with the ability to oxidize non-phenolic lignin model compounds, which constitute the majority of lignin linkages (Gómez-Méndez et al., 2018; Hofrichter et al., 2010). Due to its high redox potential, LiP directly abstracts an electron from the aromatic rings of non-phenolic lignin units, generating aryl cation radicals. These radicals initiate lignin depolymerization through non-enzymatic secondary reactions, including β-O-4 ether bond cleavage, aromatic ring opening, and Cα-Cβ bond breakage (Ayuso-Fernández, 2018). Some WRF produce veratryl alcohol, a secondary metabolite that acts as a redox mediator, facilitating electron transfer from LiP to bulky lignin substrates. MnP (EC 1.11.1.13) primarily oxidizes phenolic lignin structures but is uniquely dependent on Mn^{2+} as a substrate. MnP catalyzes the H_2O_2-dependent oxidation of Mn^{2+} to Mn^{3+}, which forms a diffusible oxidant complex when chelated by organic acids such as malate or oxalate. This Mn^{3+}-chelate complex penetrates the lignocellulosic matrix to oxidize phenolic lignin units. Moreover, MnP indirectly extends its degradative capacity by initiating lipid peroxidation, generating peroxyl radicals capable of oxidizing non-phenolic lignin components (Wang et al., 2021). In contrast, VPs (VP; EC 1.11.1.16) integrate the catalytic properties of both LiP and MnP. They oxidize a broad spectrum of substrates, including high redox-potential non-phenolic compounds such as veratryl alcohol and dimethoxybenzene, as well as Mn^{2+} to Mn^{3+} (Ambatkar et al., 2022). Some VPs contain an exposed tryptophan residue that facilitates oxidation of bulky lignin substrates through long-range electron transfer, a feature shared with certain LiPs and regarded as an evolutionary adaptation for efficient lignin degradation (Grgas et al., 2023). The co-expression and complementary activities of these heme peroxidases in ligninolytic fungi enable effective degradation of diverse lignin polymer structures.

Dye-decolorizing peroxidases (DyPs) represent a distinct family of heme peroxidases found in fungi and bacteria that oxidize a wide range of lignin-related substrates, including both phenolic and non-phenolic compounds. DyPs utilize H_2O_2 to generate reactive radicals that cleave lignin bonds and have shown enhanced stability under variable environmental conditions, making them promising for industrial lignin valorization (Lončar et al., 2016). Unspecific peroxygenases (UPOs), mainly secreted by fungi, are heme-thiolate enzymes that catalyze oxygen transfer reactions using H_2O_2, functioning similarly to monooxygenases. UPOs can hydroxylate and oxidize non-phenolic lignin units and aromatic compounds, facilitating lignin degradation through selective oxyfunctionalization reactions under mild conditions (Hofrichter et al., 2010; Lončar et al., 2016; Singh & Gupta, 2020). Their unique catalytic versatility complements classical peroxidases and expands the enzymatic toolbox for lignin biorefineries.

6.3.1.2 Multicopper oxidases: laccases

Laccases (benzenediol: oxygen oxidoreductase; EC 1.10.3.2) are a prominent class of multicopper oxidases extensively involved in lignin biodegradation, predominantly produced by fungi but also by bacteria (Singh & Gupta, 2020). Their catalytic activity depends on four copper atoms arranged across three distinct sites within the enzyme, enabling the four-electron reduction of molecular oxygen to water coupled with the one-electron oxidation of a broad spectrum of phenolic substrates, including phenols, polyphenols, anilines, and aromatic amines (Debnath & Saha, 2020). Laccases primarily oxidize the phenolic hydroxyl groups in lignin, generating phenoxy radicals that undergo subsequent non-enzymatic reactions such as polymerization, repolymerization, alkyl-aryl cleavage, Cα-oxidation, and Cα-Cβ bond cleavage (Debnath & Saha, 2020; Yadav et al., 2019). The enzyme's oxidative capacity is significantly enhanced by low-molecular-weight redox mediators, such as 2,2'-azino-bis(3-ethylbenzothiazoline-6-sulfonic acid) (ABTS) and 1-hydroxybenzotriazole (HBT), which laccases oxidize to stable radicals or highly reactive species. These mediators diffuse into the complex lignin matrix, enabling oxidation of non-phenolic lignin units that are otherwise inaccessible to the enzyme alone. This laccase-mediator system is a powerful tool for lignin modification and depolymerization (Ma et al., 2016). Beyond depolymerization, laccases contribute to lignin polymerization processes relevant to humus formation and biotechnological applications such as fiberboard manufacturing (Tanveer et al., 2021). They also detoxify phenolic compounds generated during lignin degradation. Notably, bacterial laccases, such as the highly alkaline MetZyme® LIGNO™, exhibit remarkable stability and activity under harsh industrial conditions, including high-pH environments where lignin solubility is enhanced, making them promising candidates for lignin biorefinery applications (Hämäläinen et al., 2018). Furthermore, laccases functionalize lignin to produce novel compounds, including dispersion agents and bio-based additives, expanding lignin's industrial utility.

6.3.1.3 Auxiliary enzymes and synergistic interactions

Effective lignin biodegradation relies on the synergistic action of multiple enzymes that target diverse lignin structures or sequentially act on intermediate degradation products. Among these, auxiliary oxidoreductases such as aryl alcohol oxidase (AAO) and glyoxal oxidase (GLOX) play crucial roles by generating hydrogen peroxide (H_2O_2), the essential co-substrate for heme peroxidases including LiP, manganese peroxidase (MnP), and VP (Wang et al., 2021). These auxiliary enzymes do not degrade lignin directly but sustain the catalytic cycles of peroxidases, thereby amplifying the overall oxidative capacity of the ligninolytic system. In addition to H_2O_2 generation, laccases and MnPs exhibit cooperative interactions by producing new phenolic hydroxyl groups on lignin polymers. These newly formed phenolic sites serve as substrates for further oxidation by laccases, peroxidases, or other enzymes, creating a positive feedback loop that enhances lignin depolymerization efficiency (Debnath & Saha, 2020). This enzymatic interplay exemplifies the complex, multi-enzyme network nature employs to overcome lignin's recalcitrance. Enzyme classes from fungi and bacteria are compiled in Table 6.3.

TABLE 6.3
Summary of Lignin-degrading enzymes in fungi and bacteria

Enzyme Class	Source Organisms	Degradation Mechanism	Substrate Specificity	Reference
Lignin peroxidase (LiP)	White-rot fungi *Phanerochaete chrysosporium*, and *Pleurotus ostreatus*	Oxidative cleavage of non-phenolic lignin structures	Non-phenolic lignin units	Dong et al. (2013)
Manganese peroxidase (MnP)	White-rot fungi	Oxidation of manganese (II) to manganese (III), which oxidizes phenolic structures	Phenolic lignin units	Wang et al. (2018)
Laccase	White-rot fungi and some bacteria	Oxidative polymerization/ depolymerization via radical formation	Phenolic and some non-phenolic lignin units	Wu et al. (2023)

(Continued)

TABLE 6.3
(Continued)

Enzyme Class	Source Organisms	Degradation Mechanism	Substrate Specificity	Reference
Dye-decolorizing peroxidase (DyP)	Fungi (*Bjerkendra adusta*) and bacteria (*Rhodococcus jostii*)	Oxidative cleavage of aromatic rings; broad substrate specificity including dyes and lignin model compounds	Aromatic dyes, phenolic and non-phenolic lignin model compounds	Hofrichter et al. (2010); Alruwaili, Rashid and Bugg (2023)
versatile peroxidase (VP)	Fungi	Combines lignin peroxidase and manganese peroxidase activities	Broad substrate range	Weng et al. (2021)
Multicopper oxidase (*CopA*)	Bacteria (*Pseudomonas* and *Enterobacter*)	Oxidation of aromatic compounds	ABTS, 2,6-dimethoxyphenol, lignin-derived aromatics	Mandic et al. (2019)
Etherase (*LigD, E, F, G*)	Bacteria (*Sphingomonads*)	Cleavage of -O-4 ether bonds (a common linkage in lignin)	-O-4 lignin model compounds	Picart et al. (2015)
Mixed-function oxygenases/ dioxygenases	Fungi and bacteria	Demethylation, hydroxylation, ring-fission	Aromatic rings, lignin fragments	Kumar et al. (2018)
Hydrogenperoxide-requiring oxygenase	Fungi (*Phanerochaete chrysosporium*)	Oxidative cleavage requiring hydrogen peroxide	Lignin side chains	Jeyakumar et al. (2013)
Unspecific peroxygenase (UPO)	Fungi (e.g., *Agrocybe aegerita*)	Peroxygenase activity; oxygen transfer reactions using H_2O_2; hydroxylation and oxidation of lignin units	Phenolic and non-phenolic lignin units, aromatic compounds	de Gonzalo et al. (2016)

6.4 LIGNIN VALUABLE BIOPRODUCTS

A vital step in turning this complicated polymer into useful compounds that could be turned into high-value products is the breakdown of lignin by bacteria and fungi (Khan & Ahring, 2019). Lignin bio-depolymerization has the potential to make the lignocellulosic biorefinery model economically feasible by displacing the use of aromatic materials obtained from petrochemicals (Table 6.4). Central to this bioconversion are ligninolytic enzymes, particularly LiPs, manganese peroxidases (MnP),

TABLE 6.4
Examples of high-value bioproducts that can be derived from microbial-based lignin degradation

Product Category	Examples of Products	Microbial Enzyme(s) & Source Species	Applications/ Industrial Uses	References
Fine chemicals	Vanillin, syringaldehyde, guaiacol, catechol, phenol derivatives	Laccases from *Obba rivulosa, Schizophyllum commune, Pichia pastoris* (recombinant fungal laccases), bacterial laccase (e.g., *Bacillus*) LiP, MnP, VP from white-rot fungi (e.g., *Phanerochaete chrysosporium*)	Flavorings, fragrances, antioxidants, pharmaceutical intermediates, bio-based chemicals	Jia et al. (2022); Tovar-Herrera et al. (2018); Zhou et al. (2022)
High-performance materials	Carbon fibers, phenolic resins, polyurethane, polymeric foams	Laccases from bacterial species (e.g., MetZyme® from alkaliphilic bacteria), fungal laccases from *Obba rivulosa,* and others	Lightweight composites, automotive parts, construction materials, adhesives, coatings	Hämäläinen et al. (2018)
Biofuels and energy	Bioethanol, hydrogen-enriched biofuels, biogas	Fungal and bacterial laccases (e.g., *Pichia pastoris* expressed laccases), peroxidases from white-rot fungi	Renewable transportation fuels, biogas production, enhanced enzymatic hydrolysis of cellulose	Jia et al. (2022); Püllmann and Weissenborn (2021)
Bioplastics and biopolymers	Lignin-based bioplastics, coatings, and adhesives	Laccases from *Bacillus, Pseudomonas, Streptomyces* spp., and fungal laccases	Biodegradable plastics, environmentally friendly coatings, and sustainable adhesives	Grgas et al. (2023)
Dispersants and additives	Lignin-based dispersants, emulsifiers	Bacterial laccases and fungal laccases	Paper industry, agriculture, cosmetics, and paints	Trache et al. (2020)

(Continued)

TABLE 6.4
(Continued)

Product Category	Examples of Products	Microbial Enzyme(s) & Source Species	Applications/ Industrial Uses	References
Detoxification/ waste treatment	Degradation of phenolic pollutants, detoxification of lignin-derived compounds	Laccases and peroxidases from fungi (*Schizophyllum commune*, *Phanerochaete chrysosporium*) and bacteria (*Bacillus*)	Wastewater treatment, bioremediation, and reduction of toxic byproducts	Srikanth et al. (2022); Tovar-Herrera et al. (2018)
Chemical intermediates	Benzene, toluene, xylene (BTX), styrene, cyclohexane	Microbial depolymerization enzymes from white-rot fungi and bacteria	Precursors for plastics, synthetic rubbers, and solvents	Saira et al. (2022); Wu et al. (2023)
Technical lignins and lignosulfonates	Lignosulfonates, kraft lignin derivatives	Partial enzymatic modification with fungal laccases and peroxidases	Concrete additives, animal feed binders, and dust control	Abdelaziz and Hulteberg (2020)

VPs, and laccases, secreted predominantly by white-rot fungi such as *Phanerochaete chrysosporium*, *Ganoderma lucidum*, and *Irpex lacteus*, as well as bacterial species including *Rhodococcus jostii*, *Pseudomonas putida*, and *Bacillus ligniniphilus*. LiPs initiate oxidative cleavage of the polymer, generating smaller phenolic compounds such as vanillic acid, syringaldehyde, guaiacol, catechol, and hydroxybenzoic acid. These intermediates serve as precursors for the synthesis of fine chemicals, including toluene, xylene, benzene, muconic acid, and pyrogallol, high-value compounds used in flavorings, pharmaceuticals, polymers, and specialty chemicals (Asina et al., 2016; Bugg & Rahmanpour, 2015; Ma et al., 2016; Weiland et al., 2023). The broad substrate specificity and high redox potential of enzymes like LiP and VP enable the oxidation of both phenolic and recalcitrant non-phenolic lignin structures, enhancing depolymerization efficiency. Bacterial species such as *Rhodococcus jostii* and *Pseudomonas putida* have been genetically engineered to optimize lignin degradation pathways, improving yields of aromatic monomers and facilitating downstream biotransformations into biofuels and bioplastics (Grgas et al., 2023). For example, *Bacillus ligniniphilus* utilizes alkaline lignin as its sole carbon source, metabolizing it via multiple catabolic routes, including the β-ketoadipate, gentisate, and benzoic acid pathways, and producing a diverse array of aromatic compounds suitable for biorefinery valorization (Zhu et al., 2017). *Bacillus ligniniphilus* bacteria use lignin as their sole carbon source and undergo alkaline lignin biodegradation. This process produces three lignin degradation pathways, including the β-ketoadipate pathway, the gentisate pathway, and the benzoic acid pathway, as well as 15 aromatic compounds. Fungal biodegradation studies, such as those involving *Ganoderma lucidum* and *Irpex lacteus*, demonstrate the formation of phenolic, carboxylic acid, and aliphatic moieties during lignin breakdown. These modifications not only improve lignin's chemical reactivity but also enhance chromogenic properties relevant for applications in natural dyes and bio-based pigments (Rehman et al., 2022). Furthermore, the integration of bacterial and fungal ligninolytic systems into biorefinery processes is further supported by advances in enzyme engineering and synthetic biology, which have enabled the tailoring of enzyme cocktails for specific lignin feedstocks and product goals. For instance, bacterial laccases from *Bacillus* and *Pseudomonas* species exhibit high stability under industrial conditions, facilitating the functionalization of lignin into dispersants, adhesives, and bio-based additives (Hämäläinen et al., 2018). These enzymatic processes offer environmentally benign alternatives to harsh chemical treatments, reducing energy consumption and toxic byproducts. Figure 6.2 summarizes the diversity of microbial enzymatic systems with the potential to be utilized in lignin biorefineries to convert lignin into valuable fine chemicals, materials, biofuels, and other high-value products.

P-coumaryl, coniferyl, and sinapyl alcohols, the phenylpropanoid units found in lignin, form a heterogeneous aromatic polymer that is a flexible feedstock for microbial bioconversion, which produces value-added compounds. Through oxidative depolymerization and metabolic funneling routes, they facilitate the synthesis of valuable chemicals such as adipic acid, ferulic acid, syringic acid, muconic acid, and vanillin (Gu et al., 2024). Various monomers are channeled into central intermediates such as β-ketoadipate, protocatechuate, or catechol (Liu et al., 2024). Certain bacteria are essential to this process, such as *Streptomyces*, *Bacillus cereus*, *Rhodococcus opacus*, *Pseudomonas putida*, and *Sphingobium* species. To break the complicated β-O-4 and C-C bonds in lignin, they use extracellular enzymes such as laccases, manganese peroxidases, dye-decolorizing peroxidases (DyPs), and LiPs. Achieving conversion yields of up to 90% in engineered strains, these are followed by intracellular catabolic pathways (such as the β-ketoadipate or protocatechuate pathways) that transform aromatics into platform chemicals for biofuels, polyhydroxyalkanoates (PHAs), and nylon precursors (Gu et al., 2024; Hu et al., 2025). Systems biology optimizes pathways via CRISPR/Cas and AI-driven metabolic models to overcome lignin's recalcitrance and inhibitor toxicity. These bacteria, which are frequently employed in engineered consortia, increase efficiency through synergistic enzyme activities and metabolic complementarity (Hu et al., 2025; Reshmy et al., 2022). By creating chemicals like adipic acid for nylon and terephthalic acid for PET, this transformative bioconversion lowers the cost of biorefineries, valorizes lignin waste, and supports sustainable bioeconomy applications. It also lessens dependency on fossil-based feedstocks and makes it possible for scalable, environmentally friendly industrial processes (Liu et al., 2024).

6.5 CHALLENGES AND FUTURE PERSPECTIVES

Microbial degradation of lignin holds significant promise for the sustainable production of renewable chemicals and value-added bioproducts. However, several challenges must be addressed to maximize the efficiency and economic viability of these bioconversion processes. One major hurdle is the inherent complexity and heterogeneity of lignin, which complicates its separation from lignocellulosic biomass

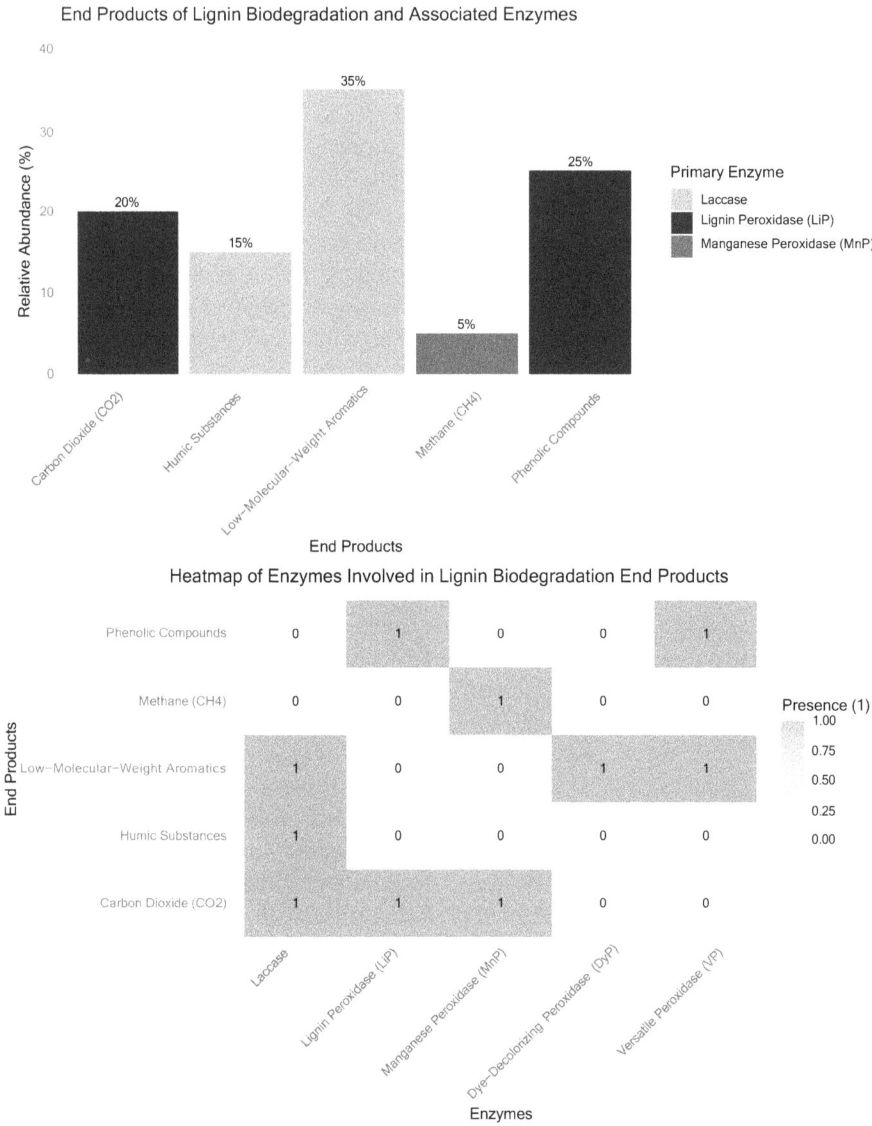

FIGURE 6.2 A summary of lignin's systemic effects, value-added byproducts, and involved enzymes

and subsequent enzymatic conversion. Current lignin extraction technologies often suffer from low selectivity and yield, limiting the availability of high-quality lignin feedstocks for biorefineries (Bugg & Rahmanpour, 2015). The catalytic potential of ligninolytic enzymes is also constrained by their relatively low productivity, stability under industrial conditions, and limited commercial availability. Moreover, the toxicity and inhibitory effects of lignin-derived aromatic compounds on microbial metabolism and enzyme activity pose additional bottlenecks. A deeper mechanistic understanding of enzyme synergism and the metabolic pathways involved in lignin biodegradation is essential to overcome these limitations and enhance conversion efficiency. Deciphering and harnessing these synergistic relationships is critical for the rational design of optimized enzyme cocktails tailored to specific lignin feedstocks and industrial bioprocesses. Such cocktails strategically combine enzymes with complementary substrate specificities and catalytic mechanisms to maximize lignin depolymerization and valorization. For example, pairing peroxidases with auxiliary oxidases and laccases can significantly improve the breakdown of both phenolic and non-phenolic lignin components (Wang et al., 2018). Recent advances in protein engineering, synthetic biology, and molecular genetics have accelerated the development of ligninolytic enzymes and microbial strains with enhanced catalytic efficiency, stability under industrial conditions, and refined substrate specificity. Engineered biocatalysts, integrated into customized enzyme mixtures, hold enormous promise for scalable and cost-effective lignin biorefineries aimed at producing biofuels, fine chemicals, and advanced materials, thereby supporting sustainable bioprocessing and circular bioeconomies (Fu et al., 2025; Weng et al., 2021). Additionally, engineering microbial consortia that mimic natural ligninolytic communities can provide synergistic interactions and metabolic complementarities beyond those of monocultures. Such consortia, isolated from diverse ecosystems, can be optimized for more efficient and stable lignin degradation through cooperative substrate utilization and cross-feeding (Liu & Qu, 2021). Scaling up microbial lignin valorization from laboratory to industrial scale also remains a critical challenge, requiring integrated bioprocess development that combines optimized pretreatment, enzyme production, lignin depolymerization, and downstream product recovery. Economic assessments and life cycle analyses are essential to evaluate sustainability and guide technology adoption (Liu & Qu, 2021). In conclusion, realizing the full potential of microbial lignin valorization demands multidisciplinary efforts in enzyme discovery and engineering, microbial strain development, process optimization, and ecosystem-level understanding. Continued research and innovation in these areas will enable lignin biorefineries to substantially contribute to the bio-based economy by sustainably producing fuels, chemicals, and materials from renewable lignocellulosic resources.

6.6 CONCLUSION

A key component of sustainable biomass valorization is the biological breakdown of lignin, which provides an environmentally beneficial substitute for energy-intensive chemical and thermal procedures. The intricacy of lignin as a highly heterogeneous and resistant aromatic biopolymer, as well as the special capacity of biological systems to overcome this resistance, have been emphasized in this chapter. By secreting oxidative enzymes, including laccases, LiPs, and manganese peroxidases, fungi especially white-rot and brown-rot species play a dominating role. By facilitating the downstream depolymerization, ring cleavage, and assimilation of aromatics produced from lignin into key metabolic pathways, bacteria enhance this process. Our knowledge of lignin-degrading pathways has been greatly enhanced by developments in enzymology, systems biology, and omics-based techniques, which have shown cooperative relationships between enzymes, mediators, and microbial consortia. These discoveries have led to the development of customized enzyme systems and engineering microorganisms with improved substrate selectivity and efficiency. Even with these developments, problems including low reaction rates, unstable enzymes, and insufficient lignin mineralization or valorization still exist. All things considered, biological lignin degradation has enormous potential for environmental remediation, biofuels, biochemicals, and biorefineries. To convert laboratory-scale achievements into commercially feasible and industrially scaled solutions, sustainable development and a circular bioeconomy will require ongoing integration of synthetic biology, metabolic engineering, and process optimization.

REFERENCES

Abdelaziz, O. Y., and C. P. Hulteberg. 2020. Lignin Depolymerization under Continuous-Flow Conditions: Highlights of Recent Developments. *ChemSusChem* 13 (17): 4382–84. https://doi.org/10.1002/cssc.202001225

Alruwaili, A., G. M. M. Rashid, and T. D. H. Bugg. 2023. Application of *Rhodococcus jostii* RHA1 Glycolate Oxidase as an Efficient Accessory Enzyme for Lignin Conversion by Bacterial Dyp Peroxidase Enzymes. *Green Chemistry* 25 (9): 3549–60. https://doi.org/10.1039/D3GC00475A

Ambatkar, N., D. D. Jadhav, S. Nandi, P. Kumbhar, and P. R. Kommoju. 2022. Optimized Process for the Production of Fungal Peroxidases and Efficient Saccharification of Pre-treated Rice Straw. *Bioresource Technology Reports* 17: 100913. https://doi.org/10.1016/j.biteb.2021.100913

Andlar, M., T. Rezić, N. Marđetko, D. Kracher, R. Ludwig, and B. Šantek. 2018. Lignocellulose Degradation: An Overview of Fungi and Fungal Enzymes Involved in Lignocellulose Degradation. *Engineering in Life Sciences* 18 (11): 768–78. https://doi.org/10.1002/elsc.201800039

Arantes, V., B. Goodell, and J. Jellison. 2018. Oxidative Damage Control during Decay of Wood by Brown Rot Fungus Using Antioxidants. *International Biodeterioration & Biodegradation* 135: 73–83. https://doi.org/10.1016/j.ibiod.2018.09.012

Arnold, F. H. 2018. Directed Evolution: Bringing New Chemistry to Life. *Angewandte Chemie International Edition* 57 (16): 4143–48. https://doi.org/10.1002/anie.201708408

Asina, F., I. Brzonova, K. Voeller, E. Kozliak, A. Kubátová, B. Yao, and Y. Ji. 2016. Biodegradation of Lignin by Fungi, Bacteria and Laccases. *Bioresource Technology* 220: 414–24. https://doi.org/10.1016/j.biortech.2016.08.016

Ayuso-Fernández, I., F. J. Ruiz-Dueñas, and A. T. Martínez. 2018. Evolutionary Convergence in Lignin-Degrading Enzymes. *Proceedings of the National Academy of Sciences* 115 (25): 6428–33. https://doi.org/10.1073/pnas.1802555115

Bautista-Zamudio, P. A., M. A. Flórez-Restrepo, X. López-Legarda, L. C. Monroy-Giraldo, and F. Segura-Sánchez. 2023. Biodegradation of Plastics by White-Rot Fungi: A Review. *Science of the Total Environment* 901: 165950. https://doi.org/10.1016/j.scitotenv.2023.165950

Beckham, G. T., C. W. Johnson, E. M. Karp, D. Salvachúa, and D. R. Vardon. 2016. Opportunities and Challenges in Biological Lignin Valorization. *Current Opinion in Biotechnology* 42: 40–53. https://doi.org/10.1016/j.copbio.2016.02.030

Brown, M. E., and M. C. Y. Chang. 2014. Exploring Bacterial Lignin Degradation. *Current Opinion in Chemical Biology* 19: 1–7. https://doi.org/10.1016/j.cbpa.2013.11.015

Bugg, T. D. H., M. Ahmad, E. M. Hardiman, and R. Rahmanpour. 2011. Pathways for Degradation of Lignin in Bacteria and Fungi. *Natural Product Reports* 28 (12): 1883–96. https://doi.org/10.1039/c1np00042j

Daniel, G. (2014). Fungal and bacterial biodegradation: White rots, brown rots, soft rots, and bacteria. In *Deterioration and Protection of Sustainable Biomaterials* (pp. 23–58). American Chemical Society. https://doi.org/10.1021/bk-2014-1158.ch002

de Boer, W., B. F. Larissa, R. C. Summerbell, and L. Boddy. 2005. Living in a Fungal World: Impact of Fungi on Soil Bacterial Niche Development. FEMS Microbiology Reviews 29(4): 795–811. https://doi.org/10.1016/j.femsre.2004.11.005

de Gonzalo, G., D. I. Colpa, M. H. M. Habib, and M. W. Fraaije. 2016. Bacterial Enzymes Involved in Lignin Degradation. *Journal of Biotechnology* 236: 110–19. https://doi.org/10.1016/j.jbiotec.2016.08.011

DeAngelis, K. M., D. Sharma, R. Varney et al. 2013. Evidence Supporting Dissimilatory and Assimilatory Lignin Degradation in *Enterobacter lignolyticus* SCF1. *Frontiers in Microbiology* 4: 280. https://doi.org/10.3389/fmicb.2013.00280

Debnath, R., and T. Saha. 2020. An Insight into the Production Strategies and Applications of the Ligninolytic Enzyme Laccase from Bacteria and Fungi. *Biocatalysis and Agricultural Biotechnology* 26: 101645. https://doi.org/10.1016/j.bcab.2020.101645

Dong, X. Q., J. S. Yang, N. Zhu, E. T. Wang, and H. L. Yuan. 2013. Sugarcane Bagasse Degradation and Characterization of Three White-Rot Fungi. *Bioresource Technology* 131: 443–51. https://doi.org/10.1016/j.biortech.2012.12.182

Falade, A. O., O. A. L. Eyisi, L. V. Mabinya, U. U. Nwodo, and A. I. Okoh. 2017. Peroxidase Production and Ligninolytic Potentials of Fresh Water Bacteria *Raoultella ornithinolytica* and *Ensifer adhaerens*. *Biotechnology Reports* 16: 12–17. https://doi.org/10.1016/j.btre.2017.10.001

Floudas, D., M. Binder, R. Riley et al. 2012. The Paleozoic Origin of Enzymatic Lignin Decomposition Reconstructed from 31 Fungal Genomes. *Science* 336 (6089): 1715–19. https://doi.org/10.1126/science.1221748

Fu, N., R. Y. Liu, Y. Zhou, B. Z. Li, Y. J. Yuan, and Z. H. Liu. 2025. Technological Advances in Ligninolytic Enzymes for the Biological Valorization of Lignin. *Green Chemistry* 27 (10): 4016–39. https://doi.org/10.1039/D4GC05724D

Gao, X., Z. Xu, T. Shi, C. Qi, L. D. Nghiem, G. Li, and W. Luo. 2024. Role of Lignocellulosic Biomass Composition to Regulate Microbial Mutualism for Organic Mineralization and Humification during Digestate Composting. ACS ES&T Engineering 4 (4): 771–82. https://doi.org/10.1021/acsestengg.3c00456

Gómez-Méndez, L. D., D. A. Moreno-Bayona, R. A. Poutou-Piñales, J. C. Salcedo-Reyes, A. M. Pedroza-Rodríguez, A. Vargas, and J. M. Bogoya. 2018. Biodeterioration of Plasma Pretreated LDPE Sheets by *Pleurotus ostreatus*. *PLoS ONE* 13 (9): 0203786. https://doi.org/10.1371/journal.pone.0203786

Goodell, B., Y. Zhu, S. Kim et al. 2017. Modification of the Nanostructure of Lignocellulose Cell Walls via a Non-enzymatic Lignocellulose Deconstruction System in Brown-Rot Wood-Decay Fungi. *Biotechnology for Biofuels* 10: 179. https://doi.org/10.1186/s13068-017-0865-2

Grand View Research. 2023. *Lignin Market Size, Share & Trends Analysis Report By Application, By Region, And Segment Forecasts,* 2023–30. Market Research Reports.

Grgas, D., M. Rukavina, D. Bešlo, T. Štefanac, V. Crnek, T. Šikić, M. Habuda-Stanić, and T. L. Dragičević. 2023. The Bacterial Degradation of Lignin—A Review. *Water* 15 (7): 1272. https://doi.org/10.3390/w15071272

Gu, J., Q. Qiu, Y. Yu, X. Sun, K. Tian, M. Chang, Y. Wang, F. Zhang, and H. Huo. 2024. Bacterial Transformation of Lignin: Key Enzymes and High-Value Products. *Biotechnology for Biofuels and Bioproducts* 17: 2. https://doi.org/10.1186/s13068-023-02447-4

Hämäläinen, V., T. Grönroos, A. Suonpää et al. 2018. Enzymatic Processes to Unlock the Lignin Value. *Frontiers in Bioengineering and Biotechnology* 6: 20. https://doi.org/10.3389/fbioe.2018.00020

Hofrichter, M., R. Ullrich, M. J. Pecyna, C. Liers, and T. Lundell. 2010. New and Classic Families of Secreted Fungal Heme Peroxidases. *Applied Microbiology and Biotechnology* 87 (3): 871–97. https://doi.org/10.1007/s00253-010-2633-0

Hu, J. N., B. C. Gao, Z. H. Liu, X. Li, B. Z. Li, and Y. J. Yuan. 2025. Prospecting Artificial Microbial Consortia toward Lignin Valorization. *Chemical Engineering Journal* 512: 162375. https://doi.org/10.1016/j.cej.2025.162375

Jardim, J. M., P. W. Hart, L. Lucia, and H. Jameel. 2020. Insights into the Potential of Hardwood Kraft Lignin to Be a Green Platform Material for Emergence of the Biorefinery. *Polymers* 12 (8): 1795. https://doi.org/10.3390/polym12081795

Jeffri, N. I., N. F. M. Rawi, M. H. B. M. Kassim, and C. K. Abdullah. 2024. Unlocking The Potential: Evolving Role Of Technical Lignin In Diverse Applications And Overcoming Challenges. *International Journal of Biological Macromolecules.* https://doi.org/10.1016/j.ijbiomac.2024.133506

Jeyakumar, D., J. Chirsteen, and M. Doble. 2013. Synergistic Effects of Pretreatment and Blending on Fungi Mediated Biodegradation of Polypropylenes. *Bioresource Technology* 148: 78–85. https://doi.org/10.1016/j.biortech.2013.08.074

Jha, V., N. A. Dafale, Z. Hathi, and H. Purohit. 2021. Genomic and Functional Potential of the Immobilized Microbial Consortium MCSt-1 for Wastewater Treatment. *Science of the Total Environment* 777: 146110. https://doi.org/10.1016/j.scitotenv.2021.146110

Jia, Y., Q. Huang, L. Zhu, and C. Pan. 2022. Characterization of a Recombinant Laccase B from *Trametes hirsuta* MX2 and Its Application for Decolorization of Dyes. *Molecules* 27 (5): 1581. https://doi.org/10.3390/molecules27051581

Jiju, P. S., A. K. Patel, N. S. Shruthy, S. Shalu, C. Di-Dong, and R. R. Singhania. 2025. Sustainability through Lignin Valorization: Recent Innovations and Applications Driving Industrial Transformation. *Bioresources and Bioprocessing* 12 (1): 88. https://doi.org/10.1186/s40643-025-00929-x

Kamimura, N., S. Sakamoto, N. Mitsuda, E. Masai, and S. Kajita. 2019. Advances in Microbial Lignin Degradation and Its Applications. *Current Opinion in Biotechnology* 56: 179–86. https://doi.org/10.1016/j.copbio.2018.11.011

Khan, M. U., and B. K. Ahring. 2019. Lignin Degradation under Anaerobic Digestion: Influence of Lignin Modifications—A Review. *Biomass and Bioenergy* 128: 105325. https://doi.org/10.1016/j.biombioe.2019.105325

Kumar, M., S. Verma, R. K. Gazara et al. 2018. Genomic and Proteomic Analysis of Lignin Degrading and Polyhydroxyalkanoate Accumulating β-Proteobacterium *Pandoraea* sp. ISTKB. *Biotechnology for Biofuels* 11: 154. https://doi.org/10.1186/s13068-018-1148-2

Liang, C., L. Lin, T. Xu, G. Kang, Z. H. Liu, and B. Z. Li. 2025. The *Pseudomonas* Ligninolytic Catalytic Network Reveals the Importance of Auxiliary Enzymes in Lignin Biocatalysts. *Proceedings of the National Academy of Sciences* 122 (40): e2417343122. https://doi.org/10.1073/pnas.2417343122

Lin, L. 2022. Bottom-up Synthetic Ecology Study of Microbial Consortia to Enhance Lignocellulose Bioconversion. *Biotechnology for Biofuels and Bioproducts* 15: 14. https://doi.org/10.1186/s13068-022-02113-1

Liu, G., and Y. Qu. 2021. Integrated Engineering of Enzymes and Microorganisms for Improving the Efficiency of Industrial Lignocellulose Deconstruction. *Engineering Microbiology* 1: 100005. https://doi.org/10.1016/j.engmic.2021.100005

Liu, Z. H., R. K. Le, Y. Shi et al. 2024. Microbial Valorization of Lignin: Pathways and Strategies. *Renewable and Sustainable Energy Reviews* 191: 114181. https://doi.org/10.1016/j.rser.2023.114181

Lončar, N., D. I. Colpa, and M. W. Fraaije. 2016. Exploring the Biocatalytic Potential of a DyP-Type Peroxidase by Profiling the Substrate Acceptance of *Thermobifida fusca* DyP Peroxidase. *Tetrahedron* 72 (46): 7276–81. https://doi.org/10.1016/j.tet.2016.09.055

Ma, R., M. Guo, K. T. Lin et al. 2016. Peracetic Acid Depolymerization of Biorefinery Lignin for Production of Selective Monomeric Phenolic Compounds. *Chemistry—A European Journal* 22 (31): 10884–91. https://doi.org/10.1002/chem.201600546

Mandic, M., L. Djokic, E. Nikolaivits et al. 2019. Identification and Characterization of New Laccase Biocatalysts from *Pseudomonas* Species Suitable for Degradation of Synthetic Textile Dyes. *Catalysts* 9 (7): 629. https://doi.org/10.3390/catal9070629

Nakamura, Y., M. G. Sungusia, T. Sawada, and M. Kuwahara. 1999. Lignin-Degrading Enzyme Production by *Bjerkandera adusta* Immobilized on Polyurethane Foam. *Journal of Bioscience and Bioengineering* 88 (1): 41–47. https://doi.org/10.1016/S1389-1723(99)80173-X

Park, G. W., G. Gong, J. C. Joo et al. 2022. Recent Progress and Challenges in Biological Degradation and Biotechnological Valorization of Lignin as an Emerging Source of Bioenergy: A State-of-the-Art Review. *Renewable and Sustainable Energy Reviews* 157: 112025. https://doi.org/10.1016/j.rser.2021.112025

Pei, Z., X. Liu, J. Chen, H. Wang, and H. Li. 2024. Research Progress on Lignin Depolymerization Strategies: A Review. Polymers 2024 (16): 2388. https://doi.org/10.3390/polym16172388

Penín, L., H. Lange, V. Santos, C. Crestini, and J. C. Parajó. 2020. Characterization of *Eucalyptus nitens* Lignins Obtained by Biorefinery Methods based on Ionic Liquids. *Molecules* 25 (2): 425. https://doi.org/10.3390/molecules25020425

Picart, P., P. D. De María, and A. Schallmey. 2015. From Gene to Biorefinery: Microbial β-Etherases as Promising Biocatalysts for Lignin Valorization. *Frontiers in Microbiology* 6: 916. https://doi.org/10.3389/fmicb.2015.00916

Pimviriyakul, P., T. Wongnate, R. Tinikul, and P. Chaiyen. 2020. Microbial Degradation of Halogenated Aromatics: Molecular Mechanisms and Enzymatic Reactions. *Microbial Biotechnology* 13 (1): 67–86. https://doi.org/10.1111/1751-7915.13488

Pollegioni, L., F. Tonin, and E. Rosini. 2015. Lignin-Degrading Enzymes. *FEBS Journal* 282 (7): 1190–213. https://doi.org/10.1111/febs.13224

Püllmann, P., and M. J. Weissenborn. 2021. Improving the Heterologous Production of Fungal Peroxygenases through an Episomal *Pichia pastoris* Promoter and Signal Peptide Shuffling System. *ACS Synthetic Biology* 10 (6): 1360–72. https://doi.org/10.1021/acssynbio.0c00641

Rabus, R., M. Boll, J. Heider et al. 2016. Anaerobic Microbial Degradation of Hydrocarbons: From Enzymatic Reactions to the Environment. *Journal of Molecular Microbiology and Biotechnology* 26 (1–3): 5–28. https://doi.org/10.1159/000443997

Rashid, G. M. M., V. Sodré, J. Luo, and T. D. H. Bugg. 2024. Overexpression of Endogenous Multi-Copper Oxidases mcoA and mcoC in *Rhodococcus jostii* RHA1 Enhances Lignin Bioconversion to 2,4-Pyridine-Dicarboxylic Acid. *Biotechnology and Bioengineering* 121 (4): 1365–69. https://doi.org/10.1002/bit.28620

Rehman, J. U., E. N. Joe, H. Y. Yoon et al. 2022. Lignin Metabolism by Selected Fungi and Microbial Consortia for Plant Stimulation: Implications for Biologically Active Humus Genesis. *Microbiology Spectrum* 10 (5): e02637–22. https://doi.org/10.1128/spectrum.02637-22

Reid, I. D. 1995. Biodegradation of Lignin. *Canadian Journal of Botany* 73 (S1): 1011–18. https://doi.org/10.1139/b95-351

Reshmy, R., P. Athiyaman Balakumaran, K. Divakar, E. Philip, A. Madhavan, A. Pugazhendhi, R. Sirohi, P. Binod, M. Kumar Awasthi, and R. Sindhu. 2022. Microbial Valorization of Lignin: Prospects and Challenges. *Bioresource Technology* 344(Part A): 126240. https://doi.org/10.1016/j.biortech.2021.126240

Riley, R., A. A. Salamov, D. W. Brown et al. 2014. Extensive Sampling of Basidiomycete Genomes Demonstrates Inadequacy of the White-Rot/Brown-Rot Paradigm for Wood Decay Fungi. *Proceedings of the National Academy of Sciences* 111 (27): 9923–28. https://doi.org/10.1073/pnas.1400592111

Saira, A., L. Maroof, M. Iqbal, S. Farman, and S. Faisal. 2022. Biodegradation of Low-Density Polyethylene (LDPE) Bags by Fungi Isolated from Waste Disposal Soil. *Applied and Environmental Soil Science* 2022: 8286344. https://doi.org/10.1155/2022/8286344

Sethupathy, S., G. M. Morales, L. Gao et al. 2022. Lignin Valorization: Status, Challenges and Opportunities. *Bioresource Technology* 347: 126696. https://doi.org/10.1016/j.biortech.2022.126696

Sharma, K. K., and R. C. Kuhad. 2010. Genetic Transformation of Lignin Degrading Fungi Facilitated by *Agrobacterium tumefaciens*. *BMC Biotechnology* 10: 67. https://doi.org/10.1186/1472-6750-10-67

Shi, Y., L. Chai, C. Tang, Z. Yang, H. Zhang, R. Chen, Y. Chen, and Y. Zheng. 2013. Characterization and Genomic Analysis of Kraft Lignin Biodegradation by the β-Proteobacterium Cupriavidus basilensis B-8. *Biotechnology for Biofuels* 6: 1. https://doi.org/10.1186/1754-6834-6-1

Sijinamanoj, V., T. Muthukumar, R. Muthuraja et al. 2021. Ligninolytic Valorization of Agricultural Residues by *Aspergillus nomius* and *Trichoderma harzianum* Isolated from Gut and Comb of *Odontotermes obesus* (Termitidae). *Chemosphere* 284: 131384. https://doi.org/10.1016/j.chemosphere.2021.131384

Singh, D., and N. Gupta. 2020. Microbial Laccase: A Robust Enzyme and Its Industrial Applications. *Biologia* 75 (8): 1183–93. https://doi.org/10.2478/s11756-019-00414-9

Srikanth, M., T. S. R. S. Sandeep, K. Sucharitha, and S. Godi. 2022. Biodegradation of Plastic Polymers by Fungi: A Brief Review. *Bioresources and Bioprocessing* 9: 42. https://doi.org/10.1186/s40643-022-00532-4

Tanveer, M. J., A. Khan, M. Nasir, and M. Asim. 2021. *Eco-Friendly Adhesives for Wood and Natural Fiber Composites*. Springer Nature.

Tian, J.-H., A.-M. Pourcher, T. Bouchez, E. Gelhaye and P. Peu. 2014. Occurrence of Lignin Degradation Genotypes and Phenotypes among Prokaryotes. *Applied Microbiology and Biotechnology* 98(23): 9527–44. https://doi.org/10.1007/s00253-014-6142-4 (PMID: 25343973)

Tovar-Herrera, O. E., A. M. Martha-Paz, Y. Pérez-Llano et al. 2018. *Schizophyllum commune*: An Unexploited Source for Lignocellulose Degrading Enzymes. *MicrobiologyOpen* 7 (5): 00637. https://doi.org/10.1002/mbo3.637

Trache, D., A. F. Tarchoun, M. Derradji et al. 2020. Nanocellulose: From Fundamentals to Advanced Applications. *Frontiers in Chemistry* 8: 392. https://doi.org/10.3389/fchem.2020.00392.

Wang, D., H. Li, X. Du, X. Wei, and Q. Zhang. 2021. Enhanced Lignin Biodegradation by Consortium of White Rot Fungi: Microbial Synergistic Effects and Product Mapping. *Biotechnology for Biofuels* 14: 162. https://doi.org/10.1186/s13068-021-02011-y

Wang, J., T. Suzuki, T. Mori et al. 2021. Transcriptomics Analysis Reveals the High Biodegradation Efficiency of White-Rot Fungus *Phanerochaete sordida* YK-624 on Native Lignin. *Journal of Bioscience and Bioengineering* 132 (3): 253–57. https://doi.org/10.1016/j.jbiosc.2021.05.009

Wang, X., B. Yao, and X. Su. 2018. Linking Enzymatic Oxidative Degradation of Lignin to Organics Detoxification. *International Journal of Molecular Sciences* 19 (11): 3373. https://doi.org/10.3390/ijms19113373

Weiland, F., N. Barton, M. Kohlstedt, J. Becker, and C. Wittmann. 2023. Systems Metabolic Engineering Upgrades *Corynebacterium glutamicum* to High-Efficiency *cis, cis*-Muconic Acid Production from Lignin-Based Aromatics. *Metabolic Engineering* 75: 153–69. https://doi.org/10.1016/j.ymben.2022.12.005

Weng, C., X. Peng, and Y. Han. 2021. Depolymerization and Conversion of Lignin to Value-Added Bioproducts by Microbial and Enzymatic Catalysis. *Biotechnology for Biofuels* 14: 84. https://doi.org/10.1186/s13068-021-01934-w

Wu, F., Z. Guo, K. Cui et al. 2023. Insights into Characteristics of White Rot Fungus during Environmental Plastics Adhesion and Degradation Mechanism of Plastics. *Journal of Hazardous Materials* 448: 130878. https://doi.org/10.1016/j.jhazmat.2023.130878

Wynands, B., F. Kofler, A. Sieberichs, N. da Silva, and N. Wierckx. 2023. Engineering a *Pseudomonas taiwanensis* 4-Coumarate Platform for Production of Para-Hydroxy Aromatics with High Yield and Specificity. *Metabolic Engineering* 78: 115–27. https://doi.org/10.1016/j.ymben.2023.05.004

Yadav, M., G. Bista, R. Maharjan et al. 2019. Secretory Laccase from *Pestalotiopsis* Species CDBT-F-G1 Fungal Strain Isolated from High Altitude: Optimization of Its Production and Characterization. *Applied Sciences* 9 (2): 340. https://doi.org/10.3390/app9020340

Zhang, B., G. Qiang, K. Barta et al. 2024. Bio-Based Polymers from Lignin. *The Innovation Materials* 2 (2): 100062. https://doi.org/10.59717/j.xinn-mater.2024.100062.

Zhang, J., G. N. Presley, K. E. Hammel, J.-S. Ryu, J. R. Menke, M. Figueroa, D. Hu, G. Orr, and J. S. Schilling. 2016. Localizing Gene Regulation Reveals a Staggered Wood Decay Mechanism for the Brown Rot Fungus Postia Placenta. *Proceedings of the National Academy of Sciences* 113(39): 10968–73. https://doi.org/10.1073/pnas.1608454113

Zhang, X., Y. He, B. Zhang, L. Qin, Q. Yang, and H. Huang. 2019. Factors Affecting Microbiological Quality of Household Drinking Water Supplied by Small-Scale Ultrafiltration Systems: A Field Study. *Science of the Total Environment* 689: 725–33. https://doi.org/10.1016/j.scitotenv.2019.06.327

Zhao, L., J. Zhang, D. Zhao et al. 2022. Biological Degradation of Lignin: A Critical Review on Progress and Perspectives. *Industrial Crops and Products* 188: 115715. https://doi.org/10.1016/j.indcrop.2022.115715

Zhou, N., W. P. D. W. Thilakarathna, Q. S. He, and H. P. V. Rupasinghe. 2022. A Review: Depolymerization of Lignin to Generate High-Value Bio-Products: Opportunities, Challenges, and Prospects. *Frontiers in Energy Research* 9: 758744. https://doi.org/10.3389/fenrg.2021.758744

Zhu, D., P. Zhang, C. Xie et al. 2017. Biodegradation of Alkaline Lignin by *Bacillus ligniniphilus* L1. *Biotechnology for Biofuels* 10: 44. https://doi.org/10.1186/s13068-017-0735-y

7 Lignin to renewable hydrocarbons and biofuels

Kopal Jain, Roshini Singh, Nikita Basant, Sachchida Nand Rai, Shruti Shukla, and Neeraj Tiwari

7.1 INTRODUCTION

Lignin is the second most abundant natural biopolymer after cellulose and constitutes approximately 15 and 30 wt% of lignocellulosic biomass. It is a promising source of natural arenes and plays a crucial role in providing rigidity and strength to plant cell walls (Hill, 2007; Xu & Ferdosian, 2017; Rathour et al., 2025). Due to its insolubility in water and alcohol, lignin presents challenges in conventional processing methods. Furthermore, the disposal of lignin waste, often through combustion or landfill dumping, constitutes a significant environmental issue (Kumari et al., 2024; Kumari & Singh, 2020). Consequently, advancements in lignin valorization through technological innovations are crucial for sustainable resource management (Sharma et al., 2020). Lignin, due to its high hydrogen and carbon content, holds significant potential as a promising raw material source for valuable petrochemical products. Its complex macromolecular structure enables its conversion into arenes, synthesis gas, and hydrogen through advanced thermochemical and catalytic processes. These transformations underscore lignin's viability as a sustainable resource for industrial applications, reducing reliance on traditional fossil-based resources (Arapova et al., 2020; Edwards et al., 2007). Lignin has demonstrated significant potential for the production of aromatic chemicals and biofuels through various conversion strategies, including hydrogenation, oxidation, pyrolysis, and catalytic pyrolysis (Bi et al., 2015). Further, the advancements in thermochemical transformation technologies have enabled the valorization of lignin into high-value chemicals. For instance, ethylbenzene serves as a crucial intermediate in organic synthesis, guaiacol plays a key role in phenolic resin production, and p-hydroxyacetophenone is employed in cholagogue drug formulation and perfume synthesis (Luo et al., 2018; Regmi et al., 2018; Stein et al., 2015). Despite the enormous availability of lignin, only a small fraction is utilized as a fuel or energy source. This underscores the necessity for advanced lignin valorization strategies to optimize its industrial applications and support sustainable resource management. However, the direct transformation of lignin into high-value chemicals presents significant challenges due to its inherent

DOI: 10.1201/9781003653189-7

hydrophobicity, poor solubility, and the complexity of its phenolic ring structures. Overcoming these obstacles requires innovative approaches in catalytic processing and thermochemical conversion to enable efficient lignin depolymerization and selective product formation. The continual growth in industrial, transportation, and energy sectors, along with stricter environmental regulations, has led to immense interest in developing biofuels (green fuels) from biomass. In this regard, lignin can play an important role in fulfilling the needs of all the above sectors and could act as a potent growth engine. Recently, researchers have shown keen interest in exploring the mechanisms and processes for effective conversion of lignin into renewable hydrocarbons and biofuels. The present chapter provides an understanding of lignin, its sources and structural chemistry, the conversion process, lignin-based biofuels, and challenges and limitations associated with the conversion process of lignin.

7.2 STRUCTURE, SOURCES, AND PROPERTIES OF LIGNIN

Lignin is a natural aromatic polymer of plant cell walls that contributes significantly to plant development and structural integrity. It primarily contributes by strengthening the structure of cell walls, especially in larger roots and stems. In addition to this, the presence of lignin provides additional tensile strength, ranging from 25 to 74 MPa, and exhibits a Young's modulus between 2.5 and 3.7 GPa, thereby improving the plant's overall rigidity and resistance to mechanical stress (Gibson, 2012). In addition to its structural functions, lignin also regulates the transport of water and solutes throughout the plant. Its biochemical properties also provide protection against biotic stressors by inhibiting enzymatic degradation of other cellular components (Boerjan et al., 2003). Furthermore, lignin serves as a protective barrier against microbial invasion and contributes to oxidative stress mitigation via its intrinsic antioxidant properties. In certain specialized plant cell types, lignin facilitates the waterproofing of cell walls, thereby preventing excessive water loss and enhancing plant adaptability to environmental fluctuations (Barros et al., 2015).

Lignin is predominantly composed of three typical 4-hydroxyphenylpropanoids known as monolignols, i.e., p-coumaryl alcohol, coniferyl alcohol, and sinapyl alcohols (Figure 7.1). All of these monolignols exhibit variations in aromatic ring methoxylation, yielding distinct p-hydroxyphenyl (H), guaiacyl (G), and syringyl (S) units within lignin macromolecules (Sun, 2020). Due to their intrinsic toxicity and instability, monolignols cannot accumulate in significant concentrations within living plant cells. To mitigate this, glycosylation of cinnamyl alcohols occurs, converting them into monolignol glucosides (Whetten & Sederoff, 1995). Following biosynthesis of monolignols, lignin monomers are translocated to the secondary plant cell wall, where they are oxidized for polymerization.

Lignin polymerization occurs through radical coupling reactions that establish various interunit linkages, determining the structural complexity and mechanical properties of lignin across different plant species. The degree of cross-linking and the relative composition of H, G, and S units contribute to the variations observed in lignin architecture among different plant species. In angiosperms, lignin predominantly consists of G and S units and low levels of H units. Conversely, lignin in gymnosperms is primarily composed of G units, with traces of H units. In grasses,

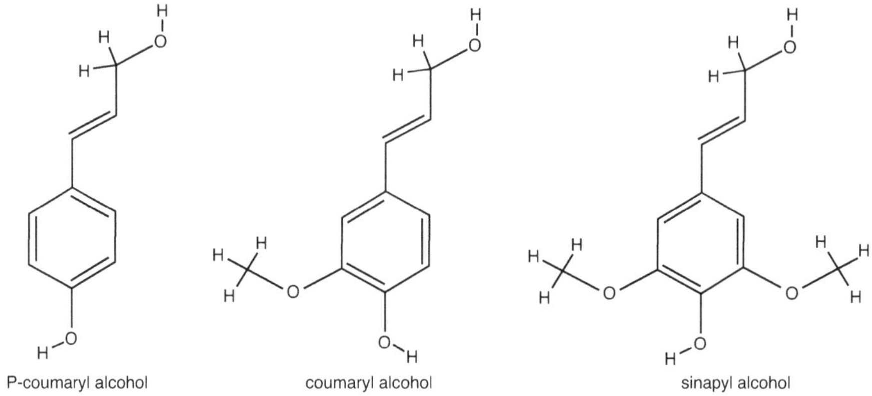

FIGURE 7.1 Types of monolignols

lignin exhibits a nearly equal distribution of G and S units, along with a higher proportion of H units compared to dicots, but varies by species and tissues. The higher concentration of H units contributes to a more heterogeneous and chemically diverse polymer structure, potentially enhancing the adaptability of monocots to diverse environmental conditions (Espiñeira et al., 2011; Loh et al., 2018).

7.3 PRETREATMENT TECHNIQUES ASSISTING THE CONVERSION OF LIGNOCELLULOSIC BIOMASS

The conversion of lignocellulosic biomass into commercially viable fuels and platform chemicals relies on a range of pretreatment processes, including physical, chemical, physico-chemical, and biological approaches (Figure 7.2). (Kumari & Singh, 2018; Hideno et al., 2009; Taherzadeh & Karimi, 2008). These methods can be applied as either single-step or multi-step treatments. One of the most fundamental preparatory steps in any biomass conversion process is the reduction of its physical size. The reduction in particle size is crucial for achieving high yields, as it enhances both biological and thermochemical conversion efficiency, thereby contributing to commercial viability. Additionally, reducing particle size significantly increases the surface area of biomass, which enhances accessibility to intra- and intermolecular bonds. This improved accessibility is particularly beneficial for biocatalysts operating in the bulk solution phase, allowing them to more effectively interact with the biomass and facilitate conversion (Garver & Liu, 2014). Overall, these pretreatment strategies are essential for maximizing efficiency and ensuring the success of lignocellulosic biomass conversion at an industrial scale (Bhatia et al., 2020).

7.3.1 Physical pretreatment methods

Physical pretreatment methods involve mechanical processes designed to reduce the particle size of biomass to a range of 0.2–30 mm, thereby increasing its surface area. In order to achieve this, techniques such as disk milling, ball milling, hammer milling, and colloid milling are commonly employed before thermochemical pretreatment to improve biomass accessibility (Kumari et al., 2025; Sun & Cheng, 2002).

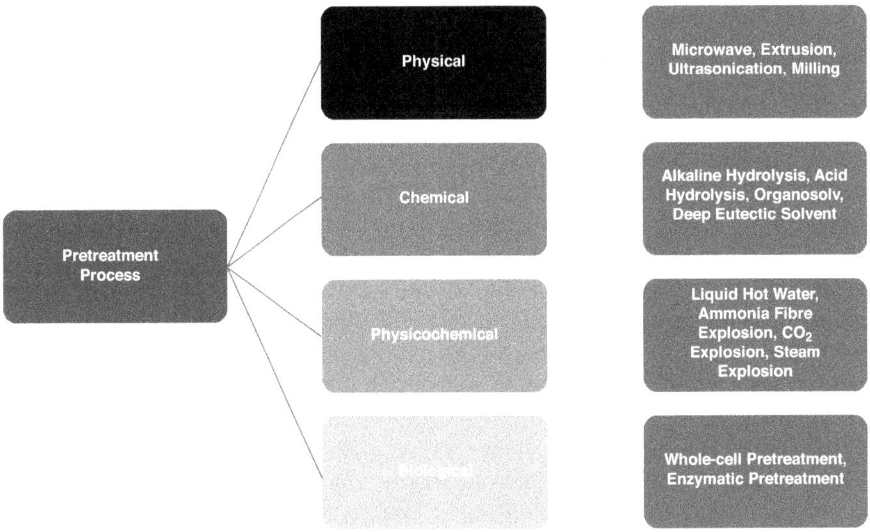

FIGURE 7.2 Pretreatment processes for the conversion of lignocellulosic biomass

7.3.2 Chemical pretreatment methods

Chemical pretreatment methods involve delignification of lignocellulosic biomass and target its major polymeric component materials, primarily cellulose and hemicellulose. The frequently used pretreatment methods are acidic, alkaline, and neutral hydrolytic techniques (Rathour et al., 2025). Acidic hydrolysis efficiently degrades hemicellulose and disrupts lignin, enhancing the release of fermentable sugars. Alkaline hydrolysis primarily targets lignin removal, improving cellulose accessibility for enzymatic breakdown. Neutral hydrolytic approaches provide a balanced chemical effect while minimizing structural alterations, offering a gentler method for biomass fractionation (Anwar et al., 2014).

7.3.3 Physico-chemical pretreatment methods

Physico-chemical pretreatment techniques integrate both physical and chemical approaches to enhance the efficiency of biomass conversion. These methods include ammonia-based pretreatment, which helps in lignin disruption and improves cellulose accessibility for enzymatic hydrolysis (Kumari & Singh, 2022; Balan et al., 2012). Auto-hydrolysis, on the other hand, utilizes high-temperature water or steam to break down hemicellulose and facilitate biomass fractionation without the addition of external chemical agents (Cara et al., 2008).

7.3.4 Biological pretreatment methods

Biological pretreatment offers eco-friendly approaches to biomass processing by utilizing microorganisms to degrade lignin and modify the chemical structure of lignocellulosic biomass with the help of ligninolytic enzymes such as peroxidases and laccases at mild conditions (Anwar et al., 2014).

7.4 DEPOLYMERIZATION STRATEGIES OF LIGNIN

Depolymerization of lignin is an important process, as it is involved in the controlled breakdown of its complex polymeric structure into low-molecular-weight aromatic compounds, which serve as precursors for hydrocarbon fuels and chemicals. There are various strategies among them; thermochemical depolymerization is one of the most established strategies (Chio et al., 2019). Pyrolysis, a non-catalytic method conducted in the absence of oxygen, thermally deconstructs lignin to produce a complex mixture of products, including phenols, catechols, and guaiacols. However, the resulting bio-oil often contains a high oxygen content and typically requires further upgrading. Alternatively, the use of supercritical fluids such as water, ethanol, or carbon dioxide under supercritical conditions can enhance lignin solubilization and promote cleavage of interunit linkages, leading to higher aromatic yields with reduced char formation.

In comparison, catalytic depolymerization offers more selective routes for producing desirable products. Reductive catalytic fractionation (RCF) stands out as an effective approach that simultaneously extracts and depolymerizes lignin using metal catalysts like nickel, ruthenium, or palladium under a hydrogen atmosphere. This method is highly selective for monomeric phenolics and has the added benefit of preserving cellulose and hemicellulose fractions for subsequent use (Bourbiaux et al., 2021). Oxidative depolymerization, which employs oxidants such as oxygen, hydrogen peroxide, or copper-based systems, can generate valuable aromatic aldehydes and acids, including vanillin. However, uncontrolled oxidation can result in over-degradation or repolymerization of products. Similarly, base- or acid-catalyzed depolymerization, utilizing agents like NaOH, KOH, or H_2SO_4, can effectively cleave ether bonds in lignin to yield a range of oligomeric and monomeric products, though these methods often risk condensation reactions or char formation (Zhou et al., 2022).

Enzymatic depolymerization using oxidative enzymes such as laccases and peroxidases represents an environmentally friendly alternative that operates under mild conditions. Despite its ecological benefits, this biological approach is limited by its slow reaction rates and narrow substrate specificity (Rathour et al., 2025). Together, these depolymerization strategies offer diverse pathways for lignin valorization, each with its own advantages and limitations depending on the desired end products and processing conditions (Liu et al., 2024).

7.5 CONVERSION PROCESS OF LIGNIN

The aromatic polymer, lignin, can be converted into renewable hydrocarbons and fuels through thermochemical, biological, and chemical processes. As part of a sustainable bioeconomy, lignin valorization into fuels and chemicals offers an environmentally friendly route for utilizing renewable biomass.

7.5.1 Thermochemical process

The transformation of lignin into high-value-added chemicals and materials through thermochemical methods has garnered significant research attention recently.

These methods are particularly attractive due to their industrial scalability, relative operational simplicity, and cost-effectiveness. The major thermochemical conversion pathways include pyrolysis, gasification, and hydrothermal liquefaction (HTL) (Prakash et al., 2015). Each of these processes exploits high temperatures and, in some cases, pressure to break down lignin into smaller, more valuable products (Mendu et al., 2011).

7.5.1.1 Pyrolysis

Pyrolysis involves the thermal decomposition of lignin in the absence of oxygen, typically at temperatures ranging from 400°C to 600°C. Pyrolysis is especially effective in converting lignin into bio-oil, along with solid biochar and non-condensable gases. Pretreatment is required to optimize lignin for pyrolysis because it removes impurities, reduces oxygen content, and improves thermal stability, thereby enhancing product yield and selectivity (Ghai et al., 2022). Fast pyrolysis is particularly effective in maximizing liquid yields, minimizing carbonization, and enhancing efficiency. Initially, weak ether bonds break down, initiating thermal degradation. As the temperature surpasses 400°C, decomposition intensifies and tar formation begins (Lu & Gu, 2022). Pyrolysis is generally classified into slow and fast modes. Slow pyrolysis primarily yields biochar, whereas rapid pyrolysis favors the production of bio-oil. Rapid heating rates (>100°C/s) are essential for achieving high bio-oil yields, which can reach up to ~70% in optimized systems. Lignin pyrolysis results in various thermally stable products, including CO, CO_2, H_2O, volatile liquids, phenolics, gaseous hydrocarbons, coke, and coal-like solids (Aboelela et al., 2023).

7.5.1.2 Gasification

Gasification converts lignin into a synthesis gas (syngas), which is a mixture of CO, H_2, CO_2, and small amounts of coke, ash, char, and short-chain gaseous hydrocarbons by reacting it with a gasifying agent (air, oxygen, or steam) at high temperatures (800–1200°C). Syngas can be further converted into liquid fuels via Fischer-Tropsch synthesis or methanol-to-gasoline (MTG) routes. The primary objective of gasification is to convert a low-value waste feedstock into a uniform and standardized gas mixture that can be further utilized for power generation (Bhatia et al., 2020). This serves as a feedstock in chemical processes to produce valuable compounds such as methanol and alkanes (Garcia et al., 2022). There are a large number of reactions and interactions that happen simultaneously during the lignin gasification process. However, these reactions take place at different temperatures, as shown in Figure 7.3.

Recent research has focused extensively on the thermochemical degradation of lignin to yield high-value fuels and aromatic chemicals. The intrinsic aromatic structure of lignin enables the generation of phenolic-rich products with significant industrial value. Efficient lignin utilization not only contributes directly to the reduction of greenhouse gas emissions through biofuel production but also enhances the overall sustainability and economic feasibility of integrated biorefinery systems (Farzad et al., 2016).

7.5.1.3 Hydrothermal Carbonization (HTC)

Hydrothermal carbonization refers to the chemical and physical transformation of lignocellulosic biomass under high pressure and temperature in aqueous conditions.

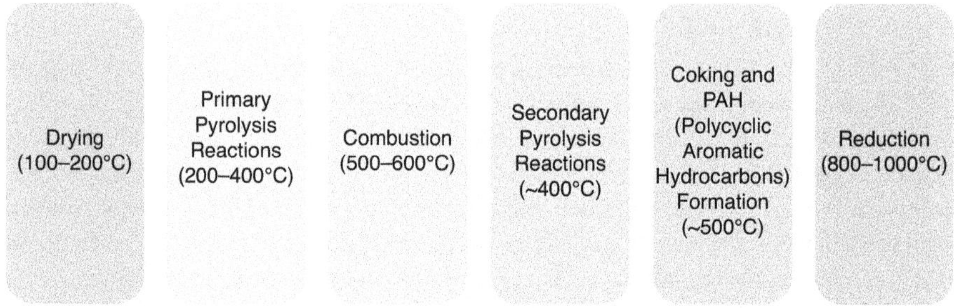

FIGURE 7.3 Lignin gasification process

During this process, lignin becomes more soluble, and the interaction with water on both physical and chemical levels also increases, ultimately forming carbon-rich structures (Khan et al., 2019). The process encompasses various reactions such as demethoxylation, alkylation, condensation, hydrolysis, and cleavage of C–O–C and C–C bonds. Notably, while aromatic rings generally remain intact, ether and C–C bonds are often broken during HTC. Unlike pyrolysis, HTC can process high-moisture lignin without the need for drying or significant energy input (Wang et al., 2020a). However, its industrial adoption remains limited due to factors such as high-pressure reactor requirements, elevated capital costs, and heat–mass transfer inefficiencies. Despite these limitations, HTC remains a promising approach for producing value-added chemicals and multifunctional materials from lignin derivatives.

7.5.1.4 Hydrothermal Liquefaction (HTL)

HTL is usually used in the case of wet biomass. This process converts the wet biomass into liquid fuels under high temperature and pressure conditions, typically below 400°C and up to 20 MPa, respectively. The major advantage of HTL is that it utilizes water as a solvent, eliminating the need for biomass drying and enabling efficient reactions at moderate temperatures. The resulting bio-oils are semi-liquid with dark coloration and a smoky odor (Fan et al., 2020). Numerous studies have demonstrated the influence of varying lignin types and reaction conditions on HTL efficiency. HTL operates under supercritical conditions (250–400°C, high pressure), where organic components in biomass are efficiently extracted. Solvents like ethanol, methanol, 2-propanol, and 2-butanol are commonly used to improve extraction efficiency and fuel quality (Mathanker et al., 2021). Although HTL effectively extracts phenolic compounds from lignin in aqueous media, it has limitations—such as the need for high temperatures, dilute feedstock concentrations, and additional hydroprocessing steps for fuel-grade bio-oil (Wörner et al., 2025).

7.5.2 Biological process

The biological conversion of lignin to biofuels primarily occurs through enzymatic and microbial degradation.

7.5.2.1 Enzymatic Degradation

The two primary categories of enzymes involved in lignin degradation are lignin-modifying enzymes (LME) and lignin-degrading auxiliary (LDA) enzymes. Although they cannot break down lignin on their own, LDA enzymes are required to complete the degradation process (Janusz et al., 2017). Heme-containing peroxidases, such as lignin peroxidase, manganese peroxidase, multifunctional peroxidase, and phenol oxidase (laccase), are the primary enzymes involved in lignin breakdown and have received significant attention (Chan et al., 2020). Because of its higher reduction potential, lignin peroxidase is the primary enzyme that breaks down lignin when compared to other peroxidases (Weng et al., 2021). Lignin peroxidase generated by liquid-state fermentation of *Aspergillus oryzae* CGMCC 5992 showed high activity against lignin in H_2O_2-pretreated maize stover. Additionally, gene modification and mineral supplementation were employed to increase lignin peroxidase yield (Zhang et al., 2015).

7.5.2.2 Microbial Degradation

Bacteria and fungi are the two major groups of microorganisms that are actively involved in the degradation or conversion process of lignin. White-rot and brown-rot fungi, which have the ability to mineralize lignin, have been the main focus of research on the microbial breakdown of lignin (Rathour et al., 2024). Oxidases and peroxidases are the main enzymes secreted by white-rot fungi for the degradation of lignin (Weng et al., 2021). Bacterial species like *Rhodococcus and Pseudomonas* are capable of efficiently breaking down lignin (Wang et al., 2022a). Lignin-degrading bacteria can be found in various common lignin-rich environments, such as compost, rotten wood, soil, termite guts, damaged bamboo slips, pulp and paper wastewater, etc. (Xu et al., 2018). A significant percentage of lignin-degrading genes have been found in the actinobacteria and proteobacteria phyla, according to bioinformatics studies. This suggests that there may be unidentified enzymes and processes. As a consequence, the breakdown of bacterial and fungal lignin may differ significantly (Grgas et al., 2023).

7.5.3 Chemical process

The chemical conversion of lignin involves the use of different chemical reagents like alcohols, bases, H_2O_2, etc. to break down the complex structure of lignin. Since lignin has a large number of hydroxyl groups in its structure, oxidative depolymerization or oxidation is a possible way to increase its utilization. This process is typically used for producing phenolic derivatives using oxidants, including oxygen, hydrogen peroxide, nitrobenzene, and metallic oxide, that aid in maintaining the aromatic rings of lignin. Various subsequent events, including the aromatic ring hydroxylation, phenolic oxidation, benzyl oxidation, ring-opening reactions, and demethylation, are typically linked to the process and generally proceed via electron transfer or hydrogen atom extraction from lignin. Aromatic aldehydes (such as vanillin and syringaldehyde) and associated acids (like vanillic acid and syringic acid) are the primary byproducts of oxidative depolymerization of lignin (Zhou et al., 2022).

7.6 PRODUCTION OF HYDROCARBONS FROM LIGNIN (AROMATIC AND ALIPHATIC)

Lignin is a complex biopolymer comprising phenylpropanoid units interlinked through ether linkages, predominantly ether bonds (e.g., β-O-4, α-O-4). This usually results in diverse streams of aromatic compounds, including phenolics, aromatic ethers, arenes, and other monomers. Lignin is an excellent substitute for petroleum in the production of arene products due to its rich aromaticity. The production of aromatic hydrocarbons is often achieved using the one-step catalytic pyrolysis of lignin conducted at high temperatures (Wang & Song, 2018).

7.6.1 Aromatic hydrocarbons from lignin

All around the world, aromatic hydrocarbons are used extensively as chemical intermediates. The majority of aromatic hydrocarbons produced today come from traditional petroleum refining methods, which are expensive and complicated. Green aromatic hydrocarbons have gained significant attention due to the utilization of renewable biomass resources comprising lignin. Phenylpropane units are prevalent in lignin, which is produced at a rate of about 50 million tons annually, in contrast to fossil fuels. These resources are plentiful, but they have not been used to their full potential and are linked to serious environmental contamination issues. Lignin is an excellent raw material for the production of renewable aromatic hydrocarbons since it is a special biomass resource (Wang et al., 2021). The various types of aromatic hydrocarbon produced from lignin are described in detail in the following section:

7.6.1.1 Phenols

There have been several documented techniques for producing phenols from lignin, including pyrolysis, alkaline oxidation, hydrocracking (hydrogenolysis), and solvolysis, which includes hydrolysis. A novel conversion method with two reaction stages can be used to produce phenols from lignin. The first stage involves utilizing a silica-alumina catalyst in a water/1-butanol solution to depolymerize lignin in an autoclave reactor. Under the right reaction conditions, the yield of the liquid product generated from lignin can reach 85–88 C-mol %. The liquid products from the first step are catalytically cracked in the second step utilizing an iron oxide catalyst and a fixed-bed flow reactor. The conversion of methoxy phenol and the total recovered fraction of phenols were 92–94% and 6.6–8.6%, respectively, using this approach (Yoshikawa et al., 2013). Phenols derived from lignin can be used as a building block in the production of different products like adhesives and resins, therefore offering numerous industrial applications.

7.6.1.2 Benzene, Toluene, Ethylbenzene, Xylenes (BTEX)

Lignin serves as an excellent source of BTEX. Both thermal and catalytic pyrolysis can convert lignin into BTEX. BTEX can be used in industries like rubber, plastic, dyes, textiles, and pesticide manufacturing.

7.6.1.3 Vanillin

Currently, vanillin is one of the few molecular phenolic chemicals that can be produced industrially from lignin. Vanillin thus serves as a crucial intermediate in the production of bio-based polymers, which require aromatic monomers to achieve acceptable thermo-mechanical characteristics (Fache et al., 2016). A variety of functional groups, including carbonyl, ether, and aromatic alcohol, are present in the aromatic aldehyde vanillin (4-hydroxy-3-methoxybenzaldehyde). It is a white substance that dissolves in water and is the primary fragrance component that gives real vanilla its creamy, sweet scent. Using chemical or microbiological techniques, the highly branched lignin skeleton is broken down into smaller aromatic molecules, which can subsequently be further transformed into vanillin. Moreover, lignin accounts for 15% of the 20,000 tons of vanillin produced annually (D'Arrigo et al., 2024).

7.6.2 Aliphatic hydrocarbons from lignin

The aliphatic hydrocarbons of lignin can be produced using methods like catalytic pyrolysis and hydrodeoxygenation. Although lignin can be pyrolyzed by a catalytic hydrogenation reaction to produce aliphatic hydrocarbons, a hydrogenation/hydrotreating phase is always necessary (Zabeti et al., 2016). The major aliphatic products include alkanes, alkenes, and other hydrocarbon components commonly found in gasoline. The industrial applications of these aliphatic hydrocarbons include the production of biofuels, chemicals, etc.

7.7 CHALLENGES IN LIGNIN VALORIZATION

Although lignin valorization has made some advancements, further studies and development will be needed to fully use the natural potential of lignin. The complex, heterogeneous, and recalcitrant nature of lignin is the greatest challenge of lignin valorization. Some of the challenges related to lignin valorization are summarized in Table 7.1.

7.8 LIGNIN-BASED BIOFUELS

Lignin-based biofuels are emerging as a promising class of renewable energy carriers due to lignin's abundance and high energy content. These biofuels can be produced

TABLE 7.1
Challenges and limitations associated with lignin valorization

Challenge	Explanation	Approach	References
Source of lignin and depolymerization ability	Inefficient depolymerization due to the recalcitrant nature of lignin; some bacterial species are not able to depolymerize lignin.	Use of targeted depolymerization techniques, selection of appropriate microbial strain for effective depolymerization	Li et al. (2016); Sun et al. (2018)
Selection of optimal microbial strain	Phenolic compounds and tannin derivatives from lignin cause toxicity for the growth of microorganisms	Screening and selection of tolerant and efficient microbial strains	Moraes et al. (2018)
Enzyme-microbe mutualism for lignin degradation	Synergistic effect of microbes and their enzymes for lignin degradation	Genetic manipulation to create a consortium of lignolytic enzymes and microbes	Hou et al. (2020); Zhao et al. (2015)
Optimization of metabolic pathways	Low yield of lignin-derived products	Engineering the metabolic pathway to increase the yield of desired products	Xie et al. (2016); Lee et al. (2019); Li and Zheng (2020)

through two primary pathways: thermochemical and catalytic routes. In the thermochemical route, lignin is exposed to processes such as pyrolysis, gasification, or hydrothermal liquefaction, breaking it down into energy-rich compounds such as bio-oil, syngas, and char. These products can be further upgraded for use as transportation fuels or industrial energy sources (Lu & Gu, 2022). In contrast, the catalytic route involves selective depolymerization of lignin using catalysts under controlled conditions to yield high-value hydrocarbons, alcohols, and aromatic compounds. This approach offers better control over product distribution and quality, making it suitable for producing drop-in fuels and petrochemical replacements. Both routes offer viable strategies for converting lignin into valuable biofuels, contributing to a more sustainable and circular bioeconomy. Some of the biofuels derived from biomass through different conversion processes and their respective applications are summarized in Table 7.2.

7.8.1 Lignin-based biofuels by catalytic route

Lignin-derived pyrolysis oil possesses numerous drawbacks, including thermal instability, low volatility, reduced calorific value, and a high tendency to form coke during processing. Due to these limitations, raw lignin pyrolysis oil is not suitable for direct application as biofuel. Consequently, catalytic upgrading methods are employed to enhance the quality of lignin-derived bio-oil. The most commonly applied catalytic strategies include hydrodeoxygenation (HDO), zeolite-assisted cracking,

TABLE 7.2
Biofuels obtained from biomass through different conversion processes

S.No.	Biofuel	Conversion Process	Application	References
1	Bio-oil	Pyrolysis	Heating fuel, crude for upgrading	Zhang et al. (2021)
2	Syngas	Gasification	Power generation	Santos and Alencar (2020)
3	Aromatic hydrocarbons	Catalytic depolymerization	Fuel blending, petrochemical feedstock	Kong et al. (2023)
4	Alcohols (e.g., ethanol, methanol)	Fermentation	Biofuel, solvent, chemical precursor	Schubert (2020)
5	Cycloalkanes	Hydrogenation	Jet fuel range hydrocarbons	Tang et al. (2019)
6	Phenolic compounds	Oxidative depolymerization	Fuel additives, intermediates	Lian et al. (2024)
7	Benzene, Toluene, Xylene (BTX)	Catalytic fast pyrolysis or zeolite cracking	Petrochemical fuels and solvents	Ma et al. (2022)

and hydrogenation. These techniques aim to improve the physico-chemical characteristics of the bio-oil and convert it into transportation-grade fuels. Table 7.2 outlines a comparative assessment of these catalytic pathways.

7.9 INDUSTRIAL APPLICATIONS OF LIGNIN

There are several industrial applications of lignin, as discussed in the following section:

7.9.1 Biopolymers and Biocomposites

Lignin is used in the production of eco-friendly materials such as bioplastics, which can be used as fillers. It also acts as a low-cost alternative for producing carbon fibers used in the aerospace and automobile industries. Additionally, lignin can also be used as an adhesive as a substitute for petroleum-derived phenols in phenol-formaldehyde resins, which are applied in the production of wood-based panels like plywood and blackboard (Wang et al., 2020b).

7.9.2 Energy and Fuel Applications

Due to its high carbon content, lignin is a valuable source of renewable energy. It can be thermochemically converted into biofuels such as bio-oil and syngas or used directly as a solid fuel (Kocaturk et al., 2023). Furthermore, lignin-derived carbon materials are being explored as components in energy storage devices like lithium-ion batteries and supercapacitors. Lignin-rich biodigester waste can be converted into activated carbon, offering a sustainable solution for both waste management and energy production (Mutuma et al., 2021).

7.9.3 Agricultural and environmental applications

In agriculture, lignin is used to develop controlled-release fertilizers and pesticides, ensuring a slow and steady supply of nutrients and active agents to crops. It also acts as a soil conditioner, which helps in enhancing the soil texture, water retention, and microbial activity, thus promoting sustainable farming practices (Boarino & Klok, 2023).

7.9.4 Pharmaceutical and biomedical applications

Due to its biocompatibility and functional properties, lignin can be used for drug delivery, providing controlled and targeted release of therapeutic compounds. Its antioxidant and antimicrobial properties can also be applied for biomedical applications like wound healing and the development of health supplements (Mukheja et al., 2023; Ullah et al., 2022).

7.9.5 Food packaging and coatings

Lignin has emerged as a promising material in the creation of functional surfaces, films, and coatings, particularly in enhancing the durability and functionality of packaging materials (Ruwoldt et al., 2023). In the context of food packaging, lignin is being incorporated as a sustainable additive to improve antioxidant and antimicrobial properties, thereby improving shelf life and ensuring product safety. These advancements position lignin as a valuable, eco-friendly component in the design of active packaging solutions (Priyadarshi et al., 2024).

7.9.6 Cosmetics and personal care

In the cosmetics industry, lignin has been used as a promising candidate due to its UV-absorbing capacity and antioxidant activity. It is used in products such as sunscreens, lotions, and hair care formulations to enhance product stability and offer protective benefits to the skin and hair (Antunes et al., 2023; Widsten, 2020).

7.9.7 Water treatment

Modified lignin materials are used as adsorbents to remove pollutants from wastewater, including heavy metals, dyes, and organic contaminants (Gouveia & de Sousa, 2025). Recently, numerous researchers have developed high-performance lignin-derived adsorbent materials through modification, carbonization, and composite formation. These materials demonstrate significant potential in wastewater treatment applications and contribute to environmental sustainability by valorizing lignin waste streams (Wang et al., 2022b; Rathour et al., 2024).

7.10 CHALLENGES AND LIMITATIONS OF LIGNIN BIOREFINERY

Lignin, the second most abundant natural polymer after cellulose, has garnered significant attention for its potential in various industrial applications. Its unique aromatic structure makes it a promising candidate for producing bio-based materials, chemicals, and fuels. However, despite its potential, several challenges and limitations hinder its widespread industrial utilization (Shorey et al., 2024). One of the primary issues is lignin's structural complexity and heterogeneity, which vary based on its botanical source and extraction method, making it difficult to achieve consistent quality and performance. The variability affects its reactivity and compatibility with other materials, making standardization difficult (Liu et al., 2024). The extraction processes, such as kraft or sulfite pulping, can introduce impurities and modify the lignin structure, reducing its suitability for high-value applications. These impurities can interfere with subsequent processing steps, limiting its reactivity and processability (Vishtal & Kraslawski, 2011). Furthermore, native lignin has low reactivity and limited compatibility with other materials, necessitating chemical modifications that can increase processing complexity and cost. The need for chemical modifications to enhance its reactivity can also introduce environmental and safety concerns (Bouhali et al., 2021). Economically, lignin remains undervalued, with the majority still being burned for energy rather than used in value-added products. Integration of lignin into existing industrial processes also presents technical and logistical barriers, often requiring process redesigns that may not be economically justifiable (Yoo & Ragauskas, 2021).

7.11 CONCLUSION

Lignin, the second most abundant aromatic biopolymer material found in nature, serves as an excellent raw material for the production of value-added products like hydrocarbons and biofuels. Lignin pretreatment and depolymerization play a crucial role in the refinery process. Nowadays, more research is focused on the use of microorganisms like bacteria and fungi for lignin pretreatment. The breakdown of lignin using enzymes like peroxidases, laccases, and other lignolytic enzymes is also an efficient way to depolymerize lignin. It can be converted into useful products like chemicals and fuels, thereby providing a new platform for the bioenergy sector. Fuels derived from lignin can provide a sustainable solution to environmental problems by reducing their dependency on fossil fuels. Biofuels like biodiesel, bioethanol, biogas, etc., serve as an excellent alternative to the conventional fuels. These biofuels can be produced from lignin through methods like pyrolysis and hydrothermal liquefaction. The economic viability of lignocellulosic biorefineries toward sustainability and circular economy can be greatly increased by the effective use of lignin as a resource. However, certain challenges related to the lignin biorefinery approach still need to be addressed, like the higher production cost of lignin-based fuels. Additionally, the depolymerization process is made more challenging by the recalcitrant and complex heterogeneous structure of lignin, which results in a lower product yield. Although there are constraints to be addressed, lignin-based fuels and the bioenergy industry at large appear to have a promising future.

REFERENCES

Aboelela, D., H. Saleh, A. M. Attia, et al. 2023. Recent Advances in Biomass Pyrolysis Processes for Bioenergy Production: Optimization of Operating Conditions. *Sustainability* 15 (14): 11238. https://doi.org/10.3390/su151411238

Antunes, F., I. Mota, J. F. Fangueiro, et al. 2023. From Sugarcane to Skin: Lignin as a Multifunctional Ingredient for Cosmetic Application. *International Journal of Biological Macromolecules* 234: 123592. https://doi.org/10.1016/j.ijbiomac.2023.123592

Anwar, Z., M. Gulfraz, and M. Irshad. 2014. Agro-Industrial Lignocellulosic Biomass a Key to Unlock the Future Bio-Energy: A Brief Review. *Journal of Radiation Research and Applied Sciences* 7 (2): 163–73. https://doi.org/10.1016/j.jrras.2014.02.003

Arapova, O. V., A. V. Chistyakov, M. V. Tsodikov, and I. I. Moiseev. 2020. Lignin as a Renewable Resource of Hydrocarbon Products and Energy Carriers (A Review). *Petroleum Chemistry* 60 (3): 227–43. https://doi.org/10.1134/S0965544120030044

Balan, V., L. da Costa Sousa, P. S. Shishir, et al. 2012. Overview to Ammonia Pretreatments for Lignocellulosic Biorefineries. *Dynamic Biochemistry Process Biotechnology and Molecular Biology* 6 (2): 1–11.

Barros, J., H. Serk, I. Granlund, and E. Pesquet. 2015. The Cell Biology of Lignification in Higher Plants. *Annals of Botany* 115 (7): 1053–74. https://doi.org/10.1093/aob/mcv046

Bhatia, R. K., G. Ramadoss, A. K. Jain, R. K. Dhiman, S. K. Bhatia, and A. K. Bhatt. (2020). Waste Biomass to Gaseous Fuels Process, Policies and Challenges in India: A Mini Review. *Bioenergy Research* 13: 1046108. https://doi.org/10.1007/s12155-020-10137-4

Bhatia, S. K., S. P. Jagtap, A. A. Bedekar, et al. 2020. Recent Advancement in Lignocellulosic Biomass Pretreatment and Deteoxificaion: Effect of Key Factors, Technological Improvements and Challenges. *Bioresource Technology* 300: 122724. https://doi.org/10.1016/j.biortech.2019.122724

Bi, P., J. Wang, Y. Zhang, et al. 2015. From Lignin to Cycloparaffins and Aromatics: Directional Synthesis of Jet and Diesel Fuel Range Biofuels Using Biomass. *Bioresource Technology* 183: 10–17. https://doi.org/10.1016/j.biortech.2015.02.023

Boarino, A., and H. A. Klok. 2023. Opportunities and Challenges for lignin Valorization in Food Packaging, Antimicrobial, and Agricultural Applications. *Biomacromolecules* 24 (3): 1065–77. https://doi.org/10.1021/acs.biomac.2c01385

Boerjan, W., J. Ralph, and M. Baucher. 2003. Lignin Biosynthesis. *Annual Review of Plant Biology* 54 (1): 519–46. https://doi.org/10.1146/annurev.arplant.54.031902.134938

Bourbiaux, D., J. Pu, F. Rataboul, et al. 2021. Reductive or Oxidative Catalytic Lignin Depolymerization: An Overview of Recent Advances. *Catalysis Today* 373: 24–37. https://doi.org/10.1016/j.cattod.2021.03.027

Cara, C., E. Ruiz, M. Ballesteros, P. Manzanares, M. J. Negro, and E. Castro. 2008. Production of Fuel Ethanol from Steam-Explosion Pretreated Olive Tree Pruning. *Fuel* 87 (6): 692–700. https://doi.org/10.1016/j.fuel.2007.05.008

Chan, J. C., M. Paice, and X. Zhang. 2020. Enzymatic Oxidation of Lignin: Challenges and Barriers toward Practical Applications. *Chem Cat Chem* 12 (2): 401–25. https://doi.org/10.1002/cctc.201901480

Chio, C., M. Sain, and W. Qin. 2019. Lignin Utilization: A Review of Lignin Depolymerization from Various Aspects. *Renewable and Sustainable Energy Reviews* 107: 232–49. https://doi.org/10.1016/j.rser.2019.03.008

D'Arrigo, P., L. Rossato, A. Strini, and S. Serra. 2024. From Waste to Value: Recent Insights into Producing Vanillin from Lignin. *Molecules* 29 (2): 2. https://doi.org/10.3390/molecules29020442

Edwards, P., V. Kuznetsov, and W. David. 2007. Hydrogen Energy. *Philosophical Transactions of the Royal Society A: Mathematical, Physical and Engineering Sciences* 365 (1853): 1043–56. https://doi.org/10.1098/rsta.2006.1965

El Bouhali, A., P. Gnanasekar, and Y. Habibi. 2021. Chemical Modifications of Lignin. In *Lignin-Based Materials for Biomedical Applications,* ed. Santos, H. and P. Figueiredo, 159–94. Elsevier. https://doi.org/10.1016/B978-0-12-820303-3.00012-6

Espiñeira, U. E., G. Ros, F. Pomar, et al. 2011. Distribution of Lignin Monomers and the Evolution of Lignification among Lower Plants. *Plant Biology* 13 (1): 59–68. https://doi.org/10.1111/j.1438-8677.2010.00345.x

Fache, M., B. Boutevin, and S. Caillol. 2016. Vanillin Production from Lignin and Its Use as a Renewable Chemical. *ACS Sustainable Chemistry & Engineering* 4 (1): 35–46. https://doi.org/10.1021/acssuschemeng.5b01344

Fan, L., H. Zhang, J. Li, et al. 2020. Algal Biorefinery to Value-Added Products by Using Combined Processes based on Thermochemical Conversion: A Review. *Algal Research* 47: 101819. https://doi.org/10.1016/j.algal.2020.101819

Farzad, S., M. Mandegari, and J. Görgens. 2016. A Critical Review on Biomass Gasification, Co-Gasification, and Their Environmental Assessments. *Biofuel Research Journal* 3 (4): 483–95. https://doi.org/10.18331/BRJ2016.3.4.3

Garcia, C. A., S. Cheng, and J. Cross. 2022. Lignin Gasification: Current and Future Viability. *Energies* 15 (23): 23. https://doi.org/10.3390/en15239062

Garver, M. P., and S. Liu. 2014. Development of Thermochemical and Biochemical Technologies for Biorefineries. In Bioenergy Research: Advances and Applications, ed. Gupta, V. K., Tuohy, M. G., Kubicek, C. P., Saddler, J., and Feng X. (457–88). Elsevier. https://doi.org/10.1016/B978-0-444-59561-4.00027-9

Ghai, H., D. Sakhuja, S. Yadav, et al. 2022. An Overview on Co-Pyrolysis of Biodegradable and Non-Biodegradable Wastes. *Energies* 15: 4168. https://doi.org/10.3390/en15114168

Gibson, L. J. 2012. The Hierarchical Structure and Mechanics of Plant Materials. *Journal of the Royal Society Interface* 9 (76): 2749–66. https://doi.org/10.1098/rsif.2012.0341

Gouveia, J. R., and F. de Sousa. 2025. Lignin Polymers for Water Treatment. In *Biopolymers for Water Purification*, ed. Thomas, S., H. J. Maria, and G. J. Vadakkekara (181–210). Wiley. https://doi.org/10.1002/9783527835904.ch7

Grgas, D., M. Rukavina, D. Bešlo, et al. 2023. The Bacterial Degradation of Lignin—A Review. *Water* 15 (7): 7. https://doi.org/10.3390/w15071272

Hideno, A., H. Inoue, K. Tsukahara, et al. 2009. Wet Disk Milling Pretreatment without Sulfuric Acid for Enzymatic Hydrolysis of Rice Straw. *Bioresource Technology* 100 (10): 2706–11. https://doi.org/10.1016/j.biortech.2008.12.057

Hill, C. 2007. *Wood Modification: Chemical, Thermal and Other Processes.* John Wiley & Sons.

Hou, L., D. Ji, W. Dong, et al. 2020. The Synergistic Action of Electro-Fenton and White-Rot Fungi in the Degradation of Lignin. *Frontiers in Bioengineering and Biotechnology* 8: 99. https://doi.org/10.3389/fbioe.2020.00099

Janusz, G., A. Pawlik, J. Sulej, et al. 2017. Lignin Degradation: Microorganisms, Enzymes Involved, Genomes Analysis and Evolution. *FEMS Microbiology Reviews* 41 (6): 941–62. https://doi.org/10.1093/femsre/fux049

Khan, A. T., A. Saud, S. Jamari, et al. 2019. Hydrothermal Carbonization of Lignocellulosic Biomass for Carbon Rich Material Preparation: A Review. *Biomass and Bioenergy* 130: 105384. https://doi.org/10.1016/j.biombioe.2019.105384

Kocaturk, E., T. Salan, O. Ozcelik, M. Alma, and Z. Candan. 2023. Recent Advances in Lignin-Based Biofuel Production. *Energies* 16 (8): 8. https://doi.org/10.3390/en16083382

Kong, L., L. Dai, and Y. Wang. 2023. Enhancing Aromatic Hydrocarbon Formation via Catalytic Depolymerization of Lignin Waste over Ru/WOx/N-C Catalyst. *Fuel* 332: 126263. https://doi.org/10.1016/j.fuel.2022.126263

Kumari, D., and R. Singh. 2018. Pretreatment of Lignocellulosic Wastes for Biofuel Production: A Critical Review. *Renewable and Sustainable Energy Reviews* 90: 877–91. https://doi.org/10.1016/j.rser.2018.03.111

Kumari, D., and R. Singh. 2020. Ultrasonic Assisted Petha Waste Water Pretreatment of Rice Straw for Optimum Production of Methane and Ethanol Using Mixed Microbial Culture. *Renew Energy* 145: 682–90. https://doi.org/10.1016/j.renene.2019.06.082

Kumari, D., and R. Singh. 2022. Rice Straw Structure Changes Following Green Pretreatment with Petha Wastewater for Economically Viable Bioethanol Production. *Scientific Reports* 12: 10443. https://doi.org/10.1038/s41598-022-14627-7

Kumari, D., G. S. Maurya, E. Tarif, and Y. Jain. 2025. Sustainable Biohythane Production from Lignocellulosic Biomass and Industrial by-Products: A Review of Pretreatment Technologies and Economic Feasibility. *Journal of Environmental Chemical Engineering* 13 (6): 119936. https://doi.org/10.1016/j.jece.2025.119936

Kumari, D., G. S. Maurya, and R. Singh. 2024. Coupled Pretreatment of Rice Straw with Dairy Wastewater and Ultrasonication for Biofuel Production Using Mixed Microbial Source. *Bioresource Technology Reports* 25: 101738. https://doi.org/10.1016/j.biteb.2023.101738

Lee, S., M. Kang, J. Bae, J. Sohn, and B. Sung. 2019. Bacterial Valorization of Lignin: Strains, Enzymes, Conversion Pathways, Biosensors, and Perspectives. *Frontiers in Bioengineering and Biotechnology* 7: 209. https://doi.org/10.3389/fbioe.2019.00209

Li, Mi, Y. Pu, and A. Ragauskas. 2016. Current Understanding of the Correlation of Lignin Structure with Biomass Recalcitrance. *Frontiers in Chemistry* 4. https://doi.org/10.3389/fchem.2016.00045

Li, X., and Y. Zheng. 2020. Biotransformation of Lignin: Mechanisms, Applications and Future Work. *Biotechnology Progress* 36 (1): e2922. https://doi.org/10.1002/btpr.2922

Lian, P., S. Liu, Z. Ma, X. Wang, and Y. Han. 2024. Lignin Oxidative Depolymerization to Phenolic Compounds in a Continuous-Flow Manufacturing Platform. *Industrial Crops and Products* 212: 118376. https://doi.org/10.1016/j.indcrop.2024.118376

Liu, E., F. Segato Mercado, and M. Wilkins. 2024. A Green Pathway for Lignin Valorization: Enzymatic Lignin Depolymerization in Biocompatible Ionic Liquids and Deep Eutectic Solvents. *Enzyme and Microbial Technology* 174: 110392. https://doi.org/10.1016/j.enzmictec.2023.110392

Liu, R., S. Willför, and C. Xu. 2024. New Insights into the Effects of Structural and Constituent Heterogeneities in Lignin on the Formation of Nanoparticles. *Industrial Crops and Products* 220: 119149. https://doi.org/10.1016/j.indcrop.2024.119149

Loh, J. X., D. Kai, and Zibiao. 2018. Functional Materials from Lignin: Methods and Advances. 3. Sustainable Chemistry Series. World Scientific (Europe). https://doi.org/10.1142/q0153

Lu, X., and X. Gu. 2022. A Review on Lignin Pyrolysis: Pyrolytic Behavior, Mechanism, and Relevant Upgrading for Improving Process Efficiency. *Biotechnology for Biofuels and Bioproducts* 15 (1): 106. https://doi.org/10.1186/s13068-022-02203-0

Luo, M., H. Lin, B. Li, et al. 2018. A Novel Modification of Lignin on Corncob-Based Biochar to Enhance Removal of Cadmium from Water. *Bioresource Technology* 259: 312–18. https://doi.org/10.1016/j.biortech.2018.03.075

Ma, C., S. Kumagai, Y. Saito, et al. 2022. Production of BTX via Catalytic Fast Pyrolysis of Printed Circuit Boards and Waste Tires using Hierarchical ZSM-5 Zeolites and Biochar. *ACS Sustainable Chemistry & Engineering* 10 (45): 14775–82. https://doi.org/10.1021/acssuschemeng.2c04031

Mathanker, A., S. Das, D. Pudasainee, et al. 2021. A Review of Hydrothermal Liquefaction of Biomass for Biofuels Production with a Special Focus on the Effect of Process Parameters, Co-Solvents, and Extraction Solvents. *Energies* 14 (16): 16. https://doi.org/10.3390/en14164916

Mendu, V., A. Ware, M. Crocker, et al. 2011. Identification and Thermochemical Analysis of High-Lignin Feedstocks for Biofuel and Biochemical Production. *Biotechnology for Biofuels* 4 (1): 43. https://doi.org/10.1186/1754-6834-4-43

Moraes, E., T. Alvarez, G. Persinoti, et al. 2018. Lignolytic-Consortium Omics Analyses Reveal Novel Genomes and Pathways Involved in Lignin Modification and Valorization. *Biotechnology for Biofuels* 11: 75. https://doi.org/10.1186/s13068-018-1073-4

Mukheja, Y., J. Kaur, K. Pathania, et al. 2023. Recent Advances in Pharmaceutical and Biotechnological Applications of Lignin-Based Materials. *International Journal of Biological Macromolecules* 241: 124601. https://doi.org/10.1016/j.ijbiomac.2023.124601

Mutuma, B., N. Sylla, A. Bubu, et al. 2021. Valorization of Biodigestor Plant Waste in Electrodes for Supercapacitors and Microbial Fuel Cells. Version 1. https://doi.org/10.48550/ARXIV.2111.01434

Prakash, A., R. Singh, B. Balagurumurthy, et al. 2015. Thermochemical Valorization of Lignin. In *Recent Advances in Thermo-Chemical Conversion of Biomass*, ed. Pandey, A., T. Bhaskar, M. Stöcker, and R. K. Sukumaran (455–78). Elsevier. https://doi.org/10.1016/B978-0-444-63289-0.00016-8

Priyadarshi, R., T. Ghosh, S. Purohit, V. Prasannavenkadesan, and J. Rhim. 2024. Lignin as a Sustainable and Functional Material for Active Food Packaging Applications: A Review. *Journal of Cleaner Production* 469: 143151. https://doi.org/10.1016/j.jclepro.2024.143151

Rathour, R. K., M. Behl, D. Sakhuja, et al. 2025. Recent Updates in Biomass Pretreatment Techniques for a Sustainable Biofuel Industry: A Comprehensive Review. *Waste and Biomass Valorization* 1–26. https://doi.org/10.1007/s12649-025-03016-6

Rathour, R. K., N. Rana, V. Sharma, N. Sharma, A. K. Bhatt, and R. K. Bhatia. 2024. Combatting Synthetic Dye Toxicity through Exploring the Potential of Lignin Peroxidase from Pseudomonas Fluorescence LiP RL5. *Environmental Science and Pollution Research* 1–11. https://doi.org/10.1007/s11356-024-34400-9

Regmi, Y., J. Mann, J. McBride, et al. 2018. Catalytic Transfer Hydrogenolysis of Organosolv Lignin Using B-Containing FeNi Alloyed Catalysts. *Catalysis Today*, Selected Contributions to the Symposium on Catalysts & Catalytic Technologies for Conversion of Biomass & Its Derivatives (252nd ACS Meeting and Exposition, Philadelphia, PA) 302: 190–95. https://doi.org/10.1016/j.cattod.2017.05.051

Ruwoldt, J., F. Blindheim, and G. Carrasco. 2023. Functional Surfaces, Films, and Coatings with Lignin – a Critical Review. *RSC Advances* 13 (18): 12529–53. https://doi.org/10.1039/D2RA08179B

Santos, D. R., and A. Alencar. 2020. Biomass-Derived Syngas Production via Gasification Process and Its Catalytic Conversion into Fuels by Fischer Tropsch Synthesis: A Review. *International Journal of Hydrogen Energy* 45 (36): 18114–32. https://doi.org/10.1016/j.ijhydene.2019.07.133

Schubert, T. 2020. Production Routes of Advanced Renewable C1 to C4 Alcohols as Biofuel Components – a Review. *Biofuels, Bioproducts and Biorefining* 14 (4): 845–78. https://doi.org/10.1002/bbb.2109

Sharma, S., A. Sharma,S. Mulla, et al. 2020. Lignin as Potent Industrial Biopolymer: An Introduction. In *Springer Series on Polymer and Composite Materials, Lignin*, ed. Swati, S. and A. Kumar (1–15). Springer International Publishing. https://doi.org/10.1007/978-3-030-40663-9_1

Shorey, R., A. Salaghi, P. Fatehi, and T. Mekonnen. 2024. Valorization of Lignin for Advanced Material Applications: A Review. *RSC Sustainability* 2 (4): 804–31. https://doi.org/10.1039/D3SU00401E

Stein, V. T., T. Hartog, J. Buendia, et al. 2015. Ruthenium-Catalyzed C-C Bond Cleavage in Lignin Model Substrates. *Angewandte Chemie* 127 (20): 5957–61. https://doi.org/10.1002/ange.201410620

Sun, R. C. 2020. Lignin Source and Structural Characterization. *Chem Sus Chem* 13 (17): 4385–93. https://doi.org/10.1002/cssc.202001324

Sun, Y., and J. Cheng. 2002. Hydrolysis of Lignocellulosic Materials for Ethanol Production: A Review. *Bioresource Technology Reviews* 83 (1): 1–11. https://doi.org/10.1016/S0 960–8524(01)00212-7

Sun, Z., B. Fridrich, A. Santi, S. Elangovan, and K. Barta. 2018. Bright Side of Lignin Depolymerization: Toward New Platform Chemicals. *Chemical Reviews* 118 (2): 614–78. https://doi.org/10.1021/acs.chemrev.7b00588

Taherzadeh, M. J., and K. Karimi. 2008. Pretreatment of Lignocellulosic Wastes to Improve Ethanol and Biogas Production: A Review. *International Journal of Molecular Sciences* 9 (9): 1621–51. https://doi.org/10.3390/ijms9091621

Tang, H., N. Li, G. Li, et al. 2019. Synthesis of Gasoline and Jet Fuel Range Cycloalkanes and Aromatics from Poly (Ethylene Terephthalate) Waste. *Green Chemistry* 21 (10): 2709–19. https://doi.org/10.1039/C9GC00571D

Ullah, I., Z. Chen, Y. Xie, et al. 2022. Recent Advances in Biological Activities of Lignin and Emerging Biomedical Applications: A Short Review. *International Journal of Biological Macromolecules* 208: 819–32. https://doi.org/10.1016/j.ijbiomac.2022.03.182

Vishtal, A., and A. Kraslawski. 2011. Challenges in Industrial Applications of Technical Lignins. *BioResources* 6(3): 3547–68. https://doi.org/10.15376/biores.6.3.3547-3568

Wang, A., and H. Song. 2018. Maximizing the Production of Aromatic Hydrocarbons from Lignin Conversion by Coupling Methane Activation. *Bioresource Technology* 268: 505–13. https://doi.org/10.1016/j.biortech.2018.08.026

Wang, H., X. Peng, H. Li, A. Giannis, and Chao. 2022a. Recent Biotechnology Advances in Bio-Conversion of Lignin to Lipids by Bacterial Cultures. *Frontiers in Chemistry* 10. https://doi.org/10.3389/fchem.2022.894593

Wang, S., Z. Li, W. Yi, et al. 2021. Renewable Aromatic Hydrocarbons Production from Catalytic Pyrolysis of Lignin with Al-SBA-15 and HZSM-5: Synergistic Effect and Coke Behaviour. *Renewable Energy* 163: 1673–81. https://doi.org/10.1016/j.renene.2020.10.108

Wang, T., M. Jiang, X. Yu, N. Niu, and L. Chen. 2022b. Application of Lignin Adsorbent in Wastewater Treatment: A Review. *Separation and Purification Technology* 302: 122116. https://doi.org/10.1016/j.seppur.2022.122116

Wang, H. Y., X. Hao, P. Peng, et al. 2020a. Hydrothermal Synthesis and Applications of Advanced Carbonaceous Materials from Biomass: A Review. *Advanced Composites and Hybrid Materials* 3 (3): 267–84. https://doi.org/10.1007/s42114-020-00158-0

Wang, Y., X. Meng, Y. Pu, and A. Ragauskas. 2020b. Recent Advances in the Application of Functionalized Lignin in Value-Added Polymeric Materials. *Polymers* 12 (10): 10. https://doi.org/10.3390/polym12102277

Weng, C., X. Peng, and Y. Han. 2021. Depolymerization and Conversion of Lignin to Value-Added Bioproducts by Microbial and Enzymatic Catalysis. *Biotechnology for Biofuels* 14 (1): 84. https://doi.org/10.1186/s13068-021-01934-w

Whetten, R., and R. Sederoff. 1995. Lignin Biosynthesis. *The Plant Cell* 1: 1001–13. https://doi.org/10.1105/tpc.7.7.1001

Widsten, P. 2020. Lignin-Based Sunscreens—State-of-the-Art, Prospects and Challenges. *Cosmetics* 7 (4): 85. https://doi.org/10.3390/cosmetics7040085

Wörner, M., U. Hornung, S. Karagöz, T. Zevaco, and N. Dahmen. 2025. Focus on Hydrochars Produced from Hydrothermal Liquefaction of Beech Wood, Soda Lignin and Black Liquor. *European Journal of Wood and Wood Products* 83 (2): 61. https://doi.org/10.1007/s00107-025-02214-2

Xie, S., A. Ragauskas, and J. Yuan. 2016. Lignin Conversion: Opportunities and Challenges for the Integrated Biorefinery. *Industrial Biotechnology* 12: 161–67. https://doi.org/10.1089/ind.2016.0007

Xu, C., and F. Ferdosian. 2017. Structure and Properties of Lignin. In *Conversion of Lignin into Bio-Based Chemicals and Materials*, ed. Xu, C. and F. Ferdosian (1–12). Springer. https://doi.org/10.1007/978-3-662-54959-9_1
Xu, R., K. Zhang, Liu, et al. 2018. Lignin Depolymerization and Utilization by Bacteria. *Bioresource Technology* 269: 557–66. https://doi.org/10.1016/j.biortech.2018.08.118
Yoo, G. C., and A. Ragauskas. 2021. Opportunities and Challenges of Lignin Utilization. In *ACS Symposium Series*, ed. Yoo, C. G. and A. Ragauskas, 1377. American Chemical Society. https://doi.org/10.1021/bk-2021-1377.ch001
Yoshikawa, T., T. Yagi, S. Shinohara, et al. 2013. Production of Phenols from Lignin via Depolymerization and Catalytic Cracking. *Fuel Processing Technology*, Special Issue of APCRE11 108: 69–75. https://doi.org/10.1016/j.fuproc.2012.05.003
Zabeti, M., J. Baltrusaitis, and K. Seshan. 2016. Chemical Routes to Hydrocarbons from Pyrolysis of Lignocellulose Using Cs Promoted Amorphous Silica Alumina Catalyst. *Catalysis Today*, Transformations of Biomass and its derivatives to Fuels and Chemicals, 269: 156–65. https://doi.org/10.1016/j.cattod.2015.11.024
Zhang, M., Y. Hu, H. Wang, et al. 2021. A Review of Bio-Oil Upgrading by Catalytic Hydrotreatment: Advances, Challenges, and Prospects. *Molecular Catalysis* 504: 111438. https://doi.org/10.1016/j.mcat.2021.111438
Zhang, Z., L. Xia, F. Wang, et al. 2015. Lignin Degradation in Corn Stalk by Combined Method of H_2O_2 Hydrolysis and Aspergillus Oryzae CGMCC5992 Liquid-State Fermentation. *Biotechnology for Biofuels* 8 (1): 183. https://doi.org/10.1186/s13068-015-0362-4
Zhao, C., S. Xie, P. Yunqiao, et al. 2015. Synergistic Enzymatic and Microbial Conversion of Lignin for Lipid. *Green Chem.* 18. https://doi.org/10.1039/C5GC01955A
Zhou, N., W. Thilakarathna, Q. Sophia, and V. Rupasinghe. 2022. A Review: Depolymerization of Lignin to Generate High-Value Bio-Products: Opportunities, Challenges, and Prospects. *Frontiers in Energy Research* 9. https://doi.org/10.3389/fenrg.2021.758744

8 Lignin-based high-value biochemicals for sustainable industries

Amandeep Singh, Tanima Bhattacharya, and Chhavi Sharma

8.1 INTRODUCTION

The plant bark serves as a protective external layer for the stems as well as the roots and is a complicated lignocellulosic biomass mainly made up of cellulose, lignin, and hemicellulose (Gibson, 2012), as shown in Figure 8.1. However, the composition percentage of all these components varies from case to case. The cellulose contains the fibrous framework and is accountable for delivering strength as well as structural rigidity to the inner delicate tissues. It may further be transformed into different high-value biochemicals and composites for the sustainable industries (Singh et al., 2021, 2022a). The hemicellulose is an amorphous hetero-polysaccharide in nature and works as a matrix substance that holds the cellulose microfibrils together and imparts the enhanced flexibility. Lignin is a third primary constituent of the lignocellulosic biomass and is a heterogeneous aromatic polymer. It imparts the hydrophobic features, offers good mechanical support, and provides resistance to the microbial degradation. The synergistic interaction of these three constituents of the lignocellulosic biomass makes bark an important renewable resource with possible uses in biofuel production, biocomposites, value-added biochemicals, and composites for the sustainable industries. The obtained value-added biochemicals by valorization of lignocellulosic biomass recommend the idea of generation of green and sustainable materials for the sustainable industries.

Since the lignin is considered a side product in pulp and paper productions, the aromatic properties of the lignin allow its transformation into numerous high-value biochemicals, including biofuels, bioplastics, fine chemicals, and pharmaceuticals, as well as composites for sustainable industrial applications. Yearly, around 50 million tons of lignin are formed as an agricultural side product; thus far, only 2% is converted into high-value biochemicals (Stewart, 2008; Borchert et al., 2022). However, because of the chemical composition of the phenylpropionate subunits, the lignin is a potential sustainable feedstock as a precursor for high-value biochemicals and composites for the sustainable industries. Though the depolymerization of lignin into an intense stream of scientifically valued biochemicals remains always a problem, as the aromatic chemicals become toxic to microbial cell reactors working for their

DOI: 10.1201/9781003653189-8

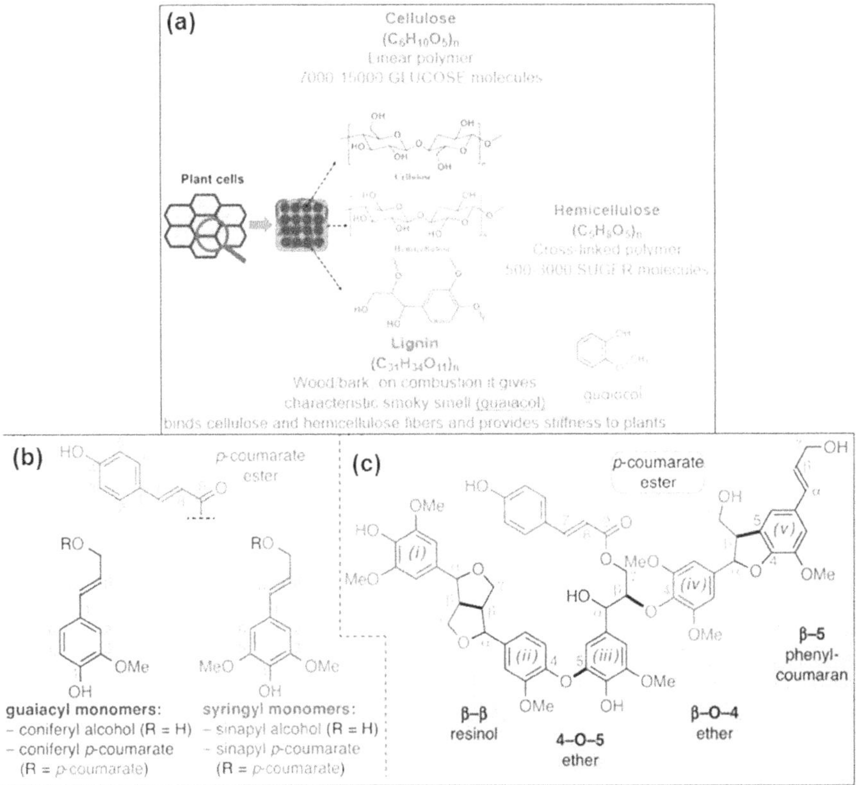

FIGURE 8.1 (a) Constituents of a plant cell. (b) Segmental parts of lignin. (c) Guaiacyl (blue) and syringyl (red) units along pendant *p*-coumarates (green) in lignin

Source: Reproduced with due permission from (G2017)

chemical transformation into value-added biomaterials. Subsequently, to advance the valorization of lignin, current scientific advancements are mainly focused on the following areas. (i) Biological, chemical, and thermal degradation of lignin to precursors. (ii) Designing of the bioconversion plants. (iii) Bioscience behind the microbes for direct bioconversion of lignin into value-added biochemicals. (iv) Systems biology research to analyze the metabolic process to increase the valorization efficiency. (v) Progress of translational synthetic biology tools intended for the valorization process. Recent advancements in green catalytic systems, microbial engineering, and solvent-based fractionation technologies are facilitating controlled lignin depolymerization and improving product selectivity. Furthermore, the recycling and circular economy concern regarding the uses and reuses of the materials emphasizes the importance of sustainable production, their application, and recycling to decrease the environmental influence (Singh et al., 2022b). The use of lignin is reliable within the aspects of the circular economy, as it transforms the discarded industrial side product into high-value biochemicals (Singh & Vijayan, 2023). By using the designed catalytic, biotechnological, and materials production methods, the valorization of

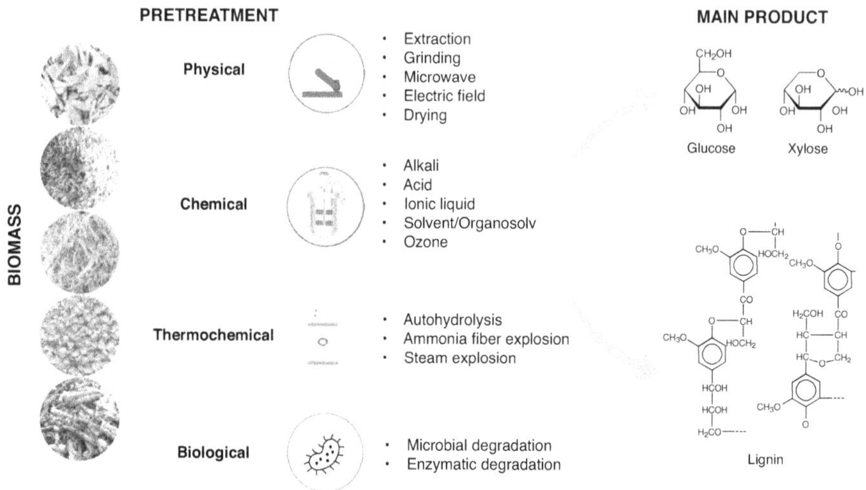

FIGURE 8.2 Methods for pretreating the lignocellulosic biomass

Source: Reproduced from an open access research paper (Ingle et al., 2025)

lignin improves the efficiency of the lignin, reduces the requirement of fossil-derived resources, decreases the waste generation, and raises the sustainable paths to prepare bioenergy, chemicals, and environmentally friendly biomaterials. Furthermore, before being utilized to produce value-added biomaterials, the biomasses are pretreated using several methods, as shown in Figure 8.2.

The successful use of lignin necessitates the adoption of effective and focused valorization strategies. It is essential to comprehend the advantages and drawbacks of each method to assess their appropriateness for different lignin sources and industrial applications. Table 8.1 provides a summary of the various lignin valorization techniques, emphasizing their individual benefits and the challenges they present.

8.2 HIGH-VALUE CHEMICALS FROM LIGNIN

Being a substantial component of the lignocellulosic feedstock, it serves as an appreciable pool for the phenolic and aromatic biochemicals. Subsequently, it is important to develop an approach for the selective depolymerization of the lignin into fine biochemicals for the sustainable industries. Factually, several prospects exist for the preparation of value-added biochemicals and drugs derived from the lignocellulosic biomass. Lignin has a significantly high aromatic component; therefore, it is characterized as a potential natural source for producing several aromatic biochemicals, including phenol, guaiacol, and vanillin.

8.2.1 Aromatic monomers

8.2.1.1 Phenol

Phenol is an important commodity biochemical within the manufacturing sector, with a universal yield of about 11 million tons in 2024, expected to increase at a

TABLE 8.1
Advantages and disadvantages of lignin valorization methods

Valorization Techniques	Feedstock	Advantages	Disadvantages	Biomass Source and Lignin Recovery		References
				Biomass (Lignin %)	Extraction yield (%)	
Alkaline hydrolysis	Agricultural residuals: Rice straw, wheat straw, sunflower stalk	Operates under lower temperatures	Requires an expensive catalyst and the formation of inhibitors	De-oiled Jatropha waste (40–50)	93.4	Oruganti et al. (2024); Novakovic et al. (2021)
Acid hydrolysis (dilute)	Wheat straw	Acid recycling is not required	High temperature and pressure are needed	Corn stover (17–21)	75.7	Tu and Hallett (2019); Sheng and Xu (2022); Gill et al. (2021)
Acid hydrolysis (concentrated)	Wheat straw and bagasse	Operates under mild temperatures	Acid recovery is expensive and requires corrosion-free equipment	Sugar cane (23.7)	91	Momayez et al. (2022); Ávila and Goldbeck (2022)
Oxidative cleavage	Wheat straw, bagasse, and poplar sawdust	Effective lignin degradation	Expensive reagent requirement	Toonna sinensis branches (13)	77	Wei et al. (2021); Lin et al. (2024)
Ionic liquid-assisted depolymerization	Wheat straw, bagasse and peanut	Sufficient dissolution	Expensive ionic liquids are required in higher amounts	Miscanthus x giganteus (17–25)	82	Wang and Qian (2020); Taylor et al. (2024); Radhakrishnan et al. (2021)
Pyrolysis	Various lignin sources, such as dry impregnated lignin and agricultural residues	Simple and less-expensive	Higher temperature is required for the initiation of decomposition, which results in reduced yield of bio-oil	Tobacco waste (18.6)	85	Wang et al. (2023); Tan et al. (2024); Lu and Gu (2022)

(Continud)

TABLE 8.1
(Continued))

Valorization Techniques	Feedstock	Advantages	Disadvantages	Biomass Source and Lignin Recovery		References
				Biomass (Lignin %)	Extraction yield (%)	
Microbial depolymerization by fungi	Rice straw, wheat straw, and softwood	Moderate conditions and minimal energy are sufficient, cost-effective	A wide sterile area, low hydrolysis rate, and more time are required	Sugarcane baggase (20–30)	87.6	Hermosilla et al. (2018); Giri and Sharma (2020); Dong et al. (2013)
Microbial depolymerization by bacteria	Rice straw, wheat straw, and softwood	Moderate conditions are sufficient, cost-effective		Paper mill waste	89	Nurika et al. (2022); Kumar et al. (2022); Hemati et al. (2023); Gu et al. (2024)
Enzymatic depolymerization	Rice straw, wheat straw, and softwood	Moderate conditions and minimal energy are sufficient		Paper mill waste	90	Zhang et al. (2021); Weiss et al. (2020); Kumar et al. (2022)

Source: Reproduced with due permission from Jiju et al. (2025).

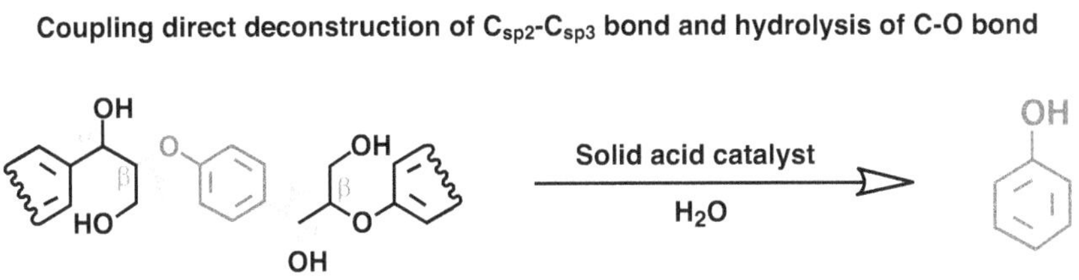

FIGURE 8.3 Chemical reaction showing phenol production by cleavage of Csp2-Csp3 and Cβ–O bonds of lignin

Source: Reproduced from open access research paper (Yan et al., 2020). Copyright © 2020 The Authors

compound yearly growth rate (CAGR) of 3.45% during the forecast period ranging to 2035. The global phenol marketplace is anticipated to observe a constant evolution because of the trade viability and the cumulative demand for its widespread applications as an organic chemical intermediate and precursor element to produce various different industrial goods, predominantly to prepare phenolic resin, nylon, bisphenol A, epoxy, cyclohexanol, pharmaceuticals, polycarbonate, etc. In recent research, other researchers (Yan et al., 2020) developed a method to obtain phenol from the lignin using direct cleavage of the Csp2-Csp3 bond and the C-O bond, as shown in Figure 8.3. The research showed that a zeolite catalyst can efficiently enable the direct breaking of the Csp2-Csp3 bond to remove the propyl structure and hydrolysis of the aliphatic β carbon–oxygen bond, resulting in the formation of an OH group on the aromatic ring. The phenol yield was achieved at 10.9% along with 91.8% selectivity. Furthermore, a large-scale research experiment showed that about 4.1 g of phenol can be obtained from the 50.0 g of lignin precursor.

Similarly, other research (Feghali et al., 2015) has used the reductive depolymerization method without using any metal. By using the hydrosilanes as reducing agents and B(C6F5)3 as a Lewis acid catalyst, the wood lignin was efficiently converted into various phenol derivatives at normal reaction conditions. Furthermore, these phenolic biochemicals were successfully synthesized from lignin with 7 to 24% yield and from wood with 0.5 to 2.4%. This well-established approach was recognized as a reliable one and is being used across 15 different wood species, including gymnosperm and angiosperm types. In other research, researchers (Sriprom et al., 2020) have advanced the working protocols to synthesize the phenol using the partial oxidation method. In this method, 4 to 20 g/L aqueous NaOH solution along with 140 to 180°C reaction temperatures and 15 to 45 minutes reaction time was used. Under the optimized parameters, the concentration of obtained phenol was achieved as 30 mg/L. The results recommend that partial oxidation of an alkali solution of lignin can efficiently produce the phenol under a normal reaction setup. Since breaking the 5-5′ bonds in lignin required a proper strategy because of the significant high bonding energy. Therefore, biphenyl can be considered as a substantial compound for lignin for this purpose, focusing on the requirements to develop a method for the cleavage. To tackle this challenge, researchers (Wang et al., 2024) have presented an advanced technique for biphenyl conversion that comprises selective hydrogenation to cyclohexylbenzene using a Ni/SiO_2 catalyst and oxidation to phenol and cyclohexanone through a radical method. The results showed that the Ni/SiO_2 with small particle size showed greater catalytic activity, even though the selectivity was not changed. Thus, high selectivity for the cyclohexylbenzene is considered for limited adsorption of cyclohexylbenzene on the catalyst. This very work presents a new method to break the C-C bonds in lignin.

8.2.1.2 Guaiacol

Guaiacol (1-hydroxy-2-methoxybenzene) is an important biochemical and is being used widely to synthesize a range of different high-value biochemicals for the sustainable industries. However, presently, the synthesis of guaiacol largely depends upon fossil resources. Furthermore, the use of lignin as a reasonable and renewable feedstock for various selective syntheses of guaiacol offers substantial possibilities. However, it also attributes the challenges to the heterogeneity and inert reactivity. To overcome this issue, other researchers (Shen et al., 2020) have analyzed that lanthanum triflate [$La(OTf)_3$] can efficiently catalyze the transformation of lignin into guaiacol as the only liquid product, as shown in Figure 8.4. The catalyst $La(OTf)_3$ enables the hydrolysis of lignin ether linkages, which results in the creation of alkyl-syringol and alkyl-guaiacol. These intermediates then undergo decarbonization and demethoxylation procedures to obtain guaiacol as a liquid product, while the leftover residue remains as a solid, which is easily separated out.

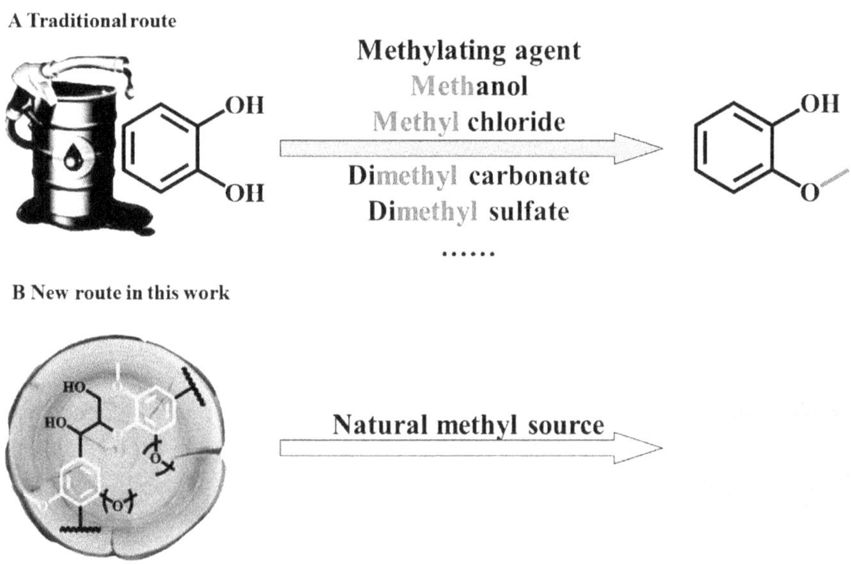

FIGURE 8.4 Synthesis methods for guaiacol: (a) traditional method and (b) novel method using a natural methyl source

Source: Reproduced from Shen et al. (2020). Open access

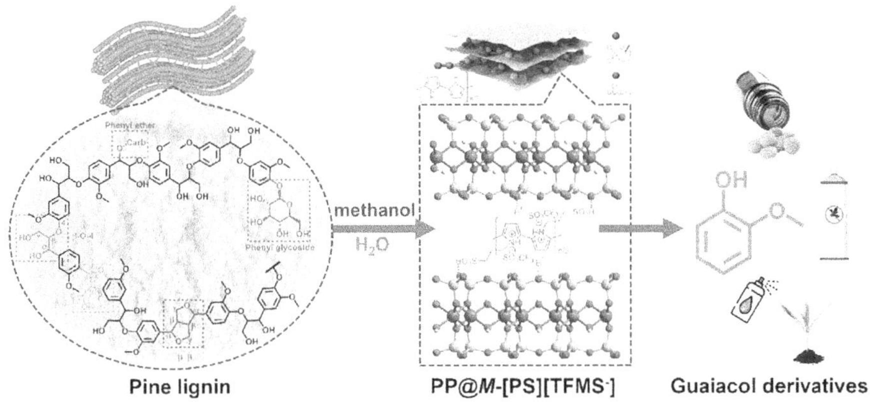

FIGURE 8.5 Acid-catalyzed depolymerization of pine-derived lignin to synthesize guaiacol biomolecule

Source: Reproduced with due permission from Li et al. (2023). © 2023 Elsevier B.V. All rights reserved

Similarly, other researchers (Li et al., 2023) have presented a method for the direct breaking of Car-Cα and Cβ-O bonds of pine-derived guaiacol using an acid catalyst, as shown in Figure 8.5. The results showed that an approximately 18.2% yield of guaiacol was achieved using a 90% methanol aqueous solution. Results also showed that the breaking of the Car-Cα bond is related to the deacetylation, whereas the formation of the Cα=O bond changes charge distribution, thus enabling the breaking of the Car-Cα bond. The inclusion of water increases the decomposition of methanol, which leads to the creation of active hydrogen species. This study offers a feasible technique to produce guaiacol using the direct cleavage method of Car-Cα and Cβ-O bonds within the lignin biomolecules.

8.2.1.3 Vanillin

Hydroxybenzaldehydes (HBAs) represent a category of widely distributed compounds that are present in reaction mixtures derived from various forms of lignin degradation. These compounds illustrate the substitution patterns of the aromatic nuclei found in lignin and primarily consist of vanillin, syringaldehyde, and p-hydroxybenzaldehyde. The generation of HBAs is especially significant during oxidative degradation reactions of lignin, which aligns with their structural characteristics. For example, in the alkaline nitrobenzene oxidation technique utilized for lignin analysis, the yield of HBAs can attain levels between 30% and 50%, contingent upon the specific lignin sample employed. The oxidative breakdown of lignin to generate HBAs, including vanillin, holds considerable industrial significance. Vanillin (4-hydroxy-3-methoxybenzaldehyde) is an extensively used aroma worldwide. Its synthesis mainly depends on the fossil feedstock. Lignin, which is a discarded byproduct of the pulp industry, is being widely used to synthesize the natural vanillin. Vanillin is an aromatic aldehyde with several functional groups, including carbonyl, ether, and aromatic alcohol. It appears as a white solid and is soluble in water. It is the main aromatic compound present in the natural vanilla, imparting its sweet and creamy smell. Since it is known as one of the most predominant and valued flavoring agents globally, it is the primary component of vanilla bean extract, which is extracted from seed pods. Due to the important demand in the confectionery industry, perfume industry, and medicine, the majority of vanillin being used today is prepared in the lab from fossil resources. Less than 1% of the vanilla flavoring market is sourced from the natural compounds. Thus, the increasing global demand for vanillin requires the search for alternative and sustainable resources and manufacturing approaches using bio-based materials such as lignin, which offers an environmentally friendly alternative over using traditional fossil-based precursors (D'Arrigo et al., 2024).

Since the sulfite pulping technique presently is the only source of lignin to synthesize vanillin, it imparts only around 10% of total lignin extracted universally (Fache et al., 2016). On the other hand, the lignin produced using the kraft method contributes about 90% of global production and is mainly burned to produce the energy required for the pulping process (Rodrigues et al., 2012). Yearly, about 50 million tons of lignin are obtained worldwide, and most of it are incinerated to produce steam (Demuner et al., 2019). Alongside, there remains a continuous interest in the depolymerization of different types of lignin, along with those that are obtained by the kraft method. However, other lignin sources vary based on the local availability, i.e., bamboo (McCallum et al., 2021), marigold residues (Poveda-Giraldo & Cardona Alzate, 2021a), eucalyptus (Poveda-Giraldo & Cardona Alzate, 2021b), etc. Similarly, other researchers (Lu et al., 2020) prepared an economical catalyst (Fe-Zn/A12O3) in order to transform alkaline lignin to the phenolic monomers in the presence of formic acid. This catalyst was found to be efficiently synthesized using the co-impregnation method and showed selectivity toward generating the phenolic monomers. Using optimum reaction parameters, a formic acid (used as the in situ hydrogen source) to lignin ratio of 4:1 and a reaction temperature of 180°C for 6 hours, the maximum production of the bio-oil achieved was up to 28%. The composition analysis of bio-oil showed that the main phenolic monomers are 2-methoxy-phenol (2.86%), vanillin (2.83%), and apocynin (2.06%). Eventually, the recyclability tests confirmed that the catalytic activity of the Fe–Zn/A12O3 catalyst remained operative up to 5 cycles. In other research, Fache et al. (2016) have recognized the uses of vanillin as an important constituent for various chemical industries. The vanillin is considered as one of the limited molecular phenolic chemicals being synthesized on an industrial level from the discarded material of the paper and pulp industry to produce high-value biochemicals and composites for the sustainable industries. Subsequently, it has substantial capabilities to be considered as an important intermediate compound for the synthesis of bio-derived polymers, which need aromatic monomers to get optimum thermo-mechanical features. Furthermore, the alkaline oxidation method to transform lignin into vanillin is predominantly important, as it includes both advantages and restrictions, different techniques for the product separation following depolymerization, and finally, an assessment of the future for wider applications of vanillin as a building block for further synthesis. Furthermore, in similar research, other researchers (Dhar & Vinu, 2018) have explored the kinetics of

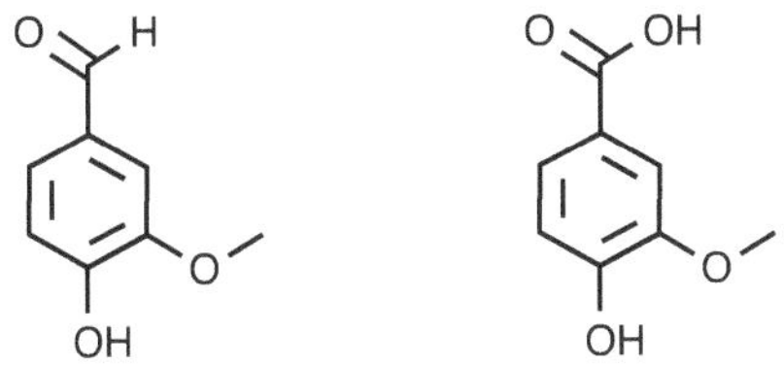

FIGURE 8.6 Chemical structures of vanillin and vanillic acid

lignin depolymerization and generation of phenols using a microwave-enabled catalytic procedure carried out at 80°C temperature in DMSO/water solvent. The 4 different catalysts used for this research are named Fe2O3, LaFeO3, ZrO2, and zeolite-Y hydrogen (ZYH). The vanillin was obtained at about 11 mg/g using both the Fe_2O_3 and ZYH catalysts. The selectivity for monomeric phenols was better with these two catalysts, whereas ZYH and ZrO_2 catalysts showed excellent selectivity for the reaction. Furthermore, the recycling of such catalysts as well as the influence of the catalyst amount on the kinetics of lignin depolymerization was found to be impressive. Likewise, other researchers (Rinesch et al., 2017) depolymerize kraft lignin using a Cu-V hydrotalcite-like catalyst in conjunction with V(acac)3/Cu(NO3)2.3H2O mixtures. This catalyst was used to obtain monomeric aromatic aldehydes and aromatic acids, including vanillin and vanillic acid (structures are shown in Figure 8.6). The obtained products were found to be aligned with model-based tests having erythro-1-(3,4-dimethoxyphenyl)-2-(2-methoxyphenoxy)-1,3-propanediol. The kinetic study experiment showed that a synergistic effect of V(acac)3 and Cu(NO3)2.3H2O is responsible for increasing the catalytic performance along with increasing the selectivity. The synthesis of veratraldehyde was found to be enabled by the C–C bond breaking via the retro-aldol reaction mechanism or a single electron transfer phenomenon. Similarly, other researchers (Zhu et al., 2020) offered an active technique for delignifying the pine wood using oxidative biorefining under an aqueous alkali solution, producing both vanillin and cellulose as side products. Using optimum parameters (stirring at 400 rpm, reaction temperature of 160°C, reaction duration of 1 hour, 7.5% aq. NaOH solution, 1 MPa O2 gas), about 90% of lignin is transformed and isolated to monophenolics (29%) with a vanillin yield of about 21% along with a small amount of oligophenolics and carboxylic acids. The similar reaction conditions with substitute feedstocks, i.e., eucalyptus (hardwood), grasses, and bagasse waste, showed the feasibility of the defined oxidative catalytic fractionation process and highlighted the robust effect of biomass varieties.

8.2.1.4 Syringaldehyde

Syringaldehyde (4-Hydroxy-3,5-dimethoxybenzaldehyde) is a refined product obtained from the lignin (Figure 8.7). It owns various characteristic applications in the medical field because of its several unique features (Pearl & Dickey, 1951). In different experiments, the syringaldehyde is used as a starting material to synthesize

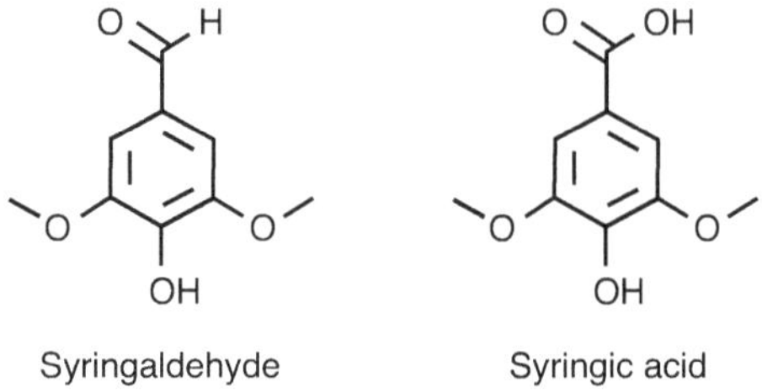

FIGURE 8.7 Chemical structure of syringaldehyde and syringic acid

the dendrimers from alkynes, enabled by the solid Cu catalysts, which improves the oxidation resistance up to a substantial level (Lee et al., 2013). Furthermore, the research concerning the intravenous administration of syringaldehyde in the diabetic rats discovered its ability to increase the glucose consumption in the bodies of affected rats and contribute a significant role to decreasing the blood sugar levels (Huang et al., 2012). Subsequently, syringaldehyde embraces the potential as an alternative treatment for the diabetic patients in the future (Deng et al., 2000). Additionally, the research revealed that the syringaldehyde has the capability to indirectly break the DNA strands via certain procedures and mechanisms. This property of syringaldehyde exhibits its potential as a nature-based resource having anticancer capabilities.

8.2.1.5 Catechol derivatives

Lignin is considered a viable source of various aromatic compounds and chemicals, serving as a substitute for petrochemicals. Catechol and its derivatives are significant platform chemicals utilized across numerous industrial sectors, including agrochemicals and pharmaceuticals. The global market for catechols is anticipated to expand in response to the increasing demand from end-use industries. Given that the current production of catechols is dependent on fossil fuels, it is imperative to explore alternative renewable sources to address sustainability challenges. In this context, lignin is recognized as a promising material due to its abundance and renewable nature. Furthermore, its structural similarity to catechols positions lignin as a potential feedstock for the synthesis of bio-catechols. Nevertheless, the conversion of lignin to catechols is still in the early stages due to the recalcitrant nature of lignin and the complexity of the depolymerized mixture. In this direction, Kim et al. (2021) reported a targeted production of catechols through the tandem conversion of lignin, specifically involving demethylation followed by catalytic hydrogenolysis (DFCH). The DFCH process yielded some liquid rich in catechols and their derivatives, which constituted 80% of the identified products. Additionally, the reaction mechanisms were investigated using quantum calculations to improve our understanding of the conversion process. The findings demonstrate that the selective modification of lignin's structure represents a promising method for generating specific platform chemicals, thereby offering valuable insights into an effective strategy for lignin valorization.

8.2.2 Organic acids and aldehydes

8.2.2.1 p-coumaric acid (pCA)

The coumarins signify a class of polyphenolic biochemicals that are part of the benzopyrone domain. The structure of coumarins contains a pyrone nucleus attached to the aromatic benzene ring. A diversity of coumarins is being documented as very significant secondary metabolites in various living organisms, i.e., bacteria, plants, and fungi. Provided their occurrence in nature, the coumarins are used in numerous production industries, i.e., pesticides, pharmaceuticals, cosmetics, and food. Recently, a report showed the 'clip-off' method to produce *p*CA as shown in Figure 8.8.

Furthermore, *p*CA is considered a nature-derived bioactive phenolic acid that has important potential for various fields, including pharmaceuticals, nutraceuticals, and cosmetics, as well as serving as a fundamental component in organic synthesis. For example, recent studies have shown its application as a precursor to 4-hydroxybenzoic acid (Tinikul et al., 2018). One compound that has the potential to enhance the value of lignin product streams, predominantly those derived from the grasses, is *p*CA. Serving as an important intermediate in different biosynthetic processes, the *p*CA is found naturally in all plant species, having a wide range of usable plants, i.e., fermented beverages, legumes, various vegetables, berries, and fruits (Yadav et al., 2017; Boz, 2015).

8.2.2.2 Ferulic acid

Ferulic acid, scientifically known as (2*E*)-3-(4-hydroxy-3-methoxyphenyl)prop-2-enoic acid or 4-hydroxy-3-methoxycinnamic acid, is produced in annual quantities of several hundred tons. It is a primary source that includes side products from the processing of rice, sugar beets, and sugarcane. In the plant cell wall, it contains ether and ester linkages with lignin and polysaccharides, respectively, thus supporting the cell wall and preventing its expansion. The term 'ferulic acid' originated from the plant genus *Ferula*, from which it was originally extracted. The rising interest in ferulic acid stems from its uses in cosmetics and its potential therapeutic benefits in fields such as neurology, gastroenterology, and oncology. The biotransformation of ferulic acid into vanillin has garnered substantial interest, as it signifies a promising route to producing 'biovanillin.' In fact, ferulic acid serves as the natural precursor to vanillin

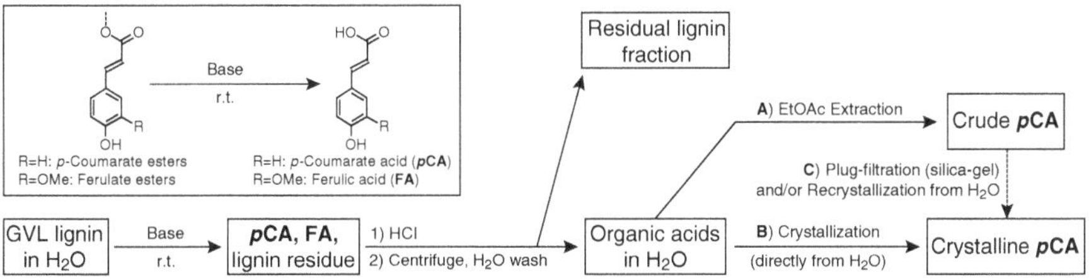

FIGURE 8.8 Schematic diagram showing the 'clip-off' method to synthesize *p*CA

Source: Reproduced from open access research paper (Timokhin et al., 2020). Copyright © 2020 American Chemical Society

found in *vanilla* plants. Furthermore, ferulic acid can be enzymatically transformed into vanillic acid and synthetically valuable vinyl bromides. In bacterial species, 5 different pathways for the catabolism of ferulate can be identified, depending on the specific reactions that simplify the activation of ferulic acid. In *Pseudomonas* and its closely related species, the transformation of ferulic acid into vanillin occurs through a two-step mechanism, which includes the generation of a high-energy thioester that serves as a crucial intermediate (Gallage & Møller, 2015; Calisti et al., 2008), as shown in Figure 8.9.

Also, the ferulic acid can be converted into the intermediate vanillic acid instead of 4-hydroxybenzoate through 4 different metabolic pathways: the non-oxidative decarboxylation pathway, the CoA-dependent β-oxidation pathway, the CoA-dependent non-β-oxidation pathway, and the side chain reduction pathway (Xu et al., 2019).

8.2.2.3 Muconic acid

Muconic acid serves as a bio-derived platform chemical that can be transformed into essential petrochemical building blocks such as adipic acid or terephthalic acid or utilized in innovative materials that offer performance advantages. Muconic acid functions as a metabolic intermediate within the β-ketoadipate pathway and can be synthesized from carbohydrates or other conventional carbon sources through the shikimate pathway. Additionally, muconic acid can be generated from lignin-derived aromatic compounds with significant atom efficiency via aromatic-catabolic pathways (Salvachúa et al., 2018). In a study, Chen et al. (Chen et al., 2021) prepared *E. coli*-based whole-cell biocatalysis to transform the lignin-derived isoeugenol into *cis,cis*-muconic acid (ccMA) (Figure 8.10). The recombinant *E. coli* expressing the ADO (aromatic dioxygenase from *T. thermophila*) and HFD1 (aldehyde dehydrogenase from *S. cerevisiae*) could efficiently transform the isoeugenol to vanillic acid (VAN). However, the researchers also observed that demethylation of VAN by VanAB (vanillic acid O-demethylase from *P. putida*) was strictly subdued by a high concentration of the isoeugenol that encouraged them to prepare a two-step method to produce the ccMA using isoeugenol. After the modification of the feeding approach and optimization of the pH parameter, 43.8 mM ccMA was obtained from 50 mM isoeugenol-derived VAN by recombinant *E. coli* expressing VanAB, GDC (gallate decarboxylase from *T. atroroseus*), and CatA (catechol dioxygenase from *A. radioresistens*). The study set an example toward the assessment of *E. coli*-based whole-cell biocatalysis for the lignin valorization.

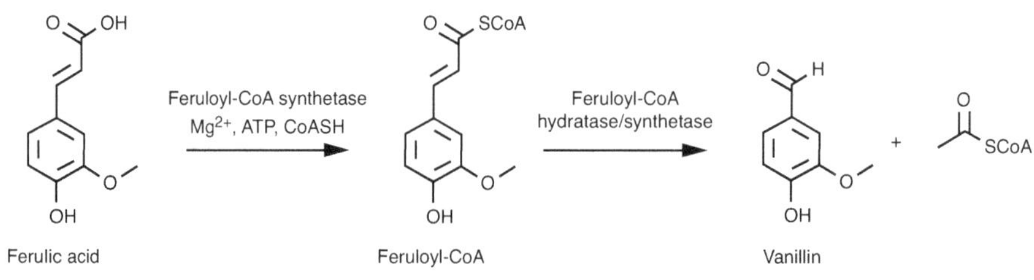

FIGURE 8.9 Chemical reaction showing non-β-oxidative route to transform ferulic acid into vanillin

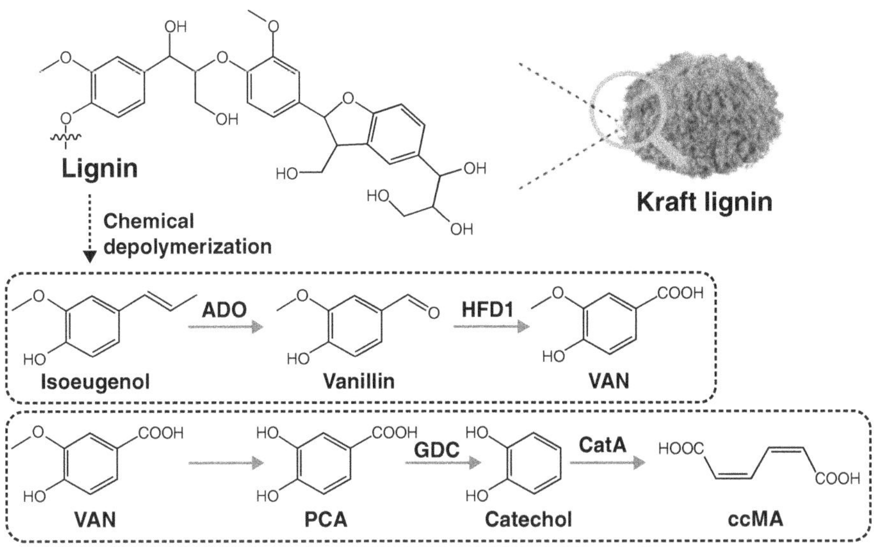

FIGURE 8.10 Synthesis of cis,cis-muconic acid (ccMA) from lignin-derived isoeugenol

Source: Reproduced with due permission from Chen et al. (2021)

8.2.2.4 Eugenol and isoeugenol

Eugenol (2-methoxy-4-prop-2-enylphenol) and isoeugenol (2-methoxy-4-propenylphenol) are guaiacol derivatives featuring propenyl side chains. A current summary highlighted the uses of eugenol as a precursor for different polymer building blocks. Furthermore, eugenol is applicable in the development of coatings and composite biomaterials. Eugenol as well as isoeugenol are suitable for the biocatalyzed synthesis of vanillin. The transformation of eugenol to vanillin by *Bacillus safensis* produces vanillin with a moderate conversion rate at pH 7 and temperature 37°C (Singh et al., 2019), and another fungus named *Trichosporon asahii* (yeast) converts isoeugenol into vanillin and vanillic acid at pH 5.8 and temperature 28°C (Ashengroph & Amini, 2017), as shown in Figure 8.11.

8.2.3 Polyols and polymers

8.2.3.1 Lignin-based polyols

The direct application of lignin as a bio-polyol to produce bio-based materials continues to be hindered by its naturally low reactivity, which is primarily linked to steric effects that restrict access to its active -OH groups. In this context, there are two approaches to improve the reactivity of lignin hydroxyl groups: (1) depolymerization into smaller fragments to yield more reactive hydroxyphenyl derivatives and (2) functionalization, such as through anionic ring-opening polymerization (AROP) of epoxides, utilizing lignin as an initiator in the presence of a base. Unfortunately, both methods typically necessitate severe reaction conditions; for instance, base-catalyzed lignin oxypropylation is conducted at temperatures

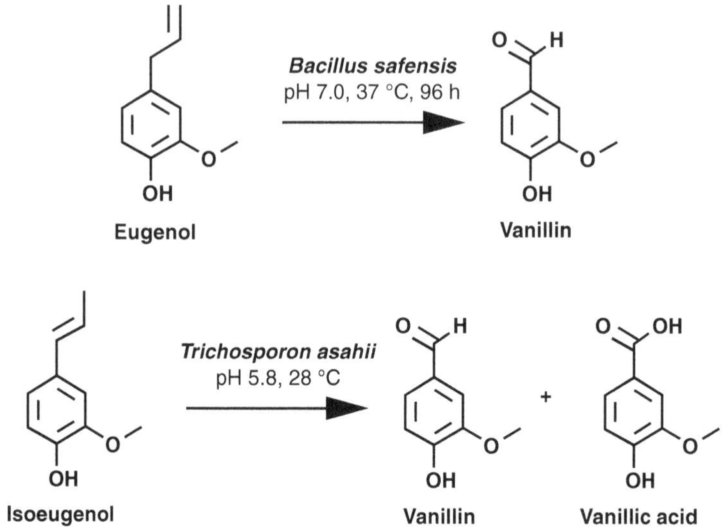

FIGURE 8.11 Synthesis routes for vanillin

ranging from 150 to 330°C and pressures between 6 and 40 bar. These extreme operating conditions lead to the generation of significant amounts of side products and unpleasant odor polyols due to the creation of phenolic moieties, which impede the final application of lignin-based polyols (LBPs). In order to circumvent the limitations associated with the existing base-catalyzed lignin oxypropylation technique for the production of LBPs, which include the generation of significant quantities of by-products, severe reaction conditions (elevated temperatures exceeding 150°C and high pressure) leading to lignin degradation, and the formation of phenolic compounds that emit unpleasant odors, thereby restricting the application of LBPs, the author has devised an alternative synthetic pathway. To illustrate the adaptability of the newly proposed synthetic method, various LBPs were synthesized utilizing butylene oxide, propylene oxide, glycidol, and epichlorohydrin as oxiranes (Perez-Arce et al., 2020). The resulting LBPs exhibited lignin contents ranging from 11.3 to 30.1 wt%, weight average molecular weights between 2730 and 30258 g/mol, and hydroxyl values spanning from 53 to 253 mg KOH/g. Additionally, the reactivity of these LBPs with diisocyanates was demonstrated, and a plausible reaction mechanism has been suggested. This newly developed approach holds significant industrial potential, particularly within the pulp industry, where lignin is a primary byproduct, as well as in the bio-polyols and biopolymers sectors. Furthermore, another approach for synthesizing lignin-based liquid polyols has been established (Duval et al., 2022), as shown in Figure 8.12. Organosolv lignin was reacted with ethylene carbonate in polyethylene glycol as the solvent, resulting in a complete conversion of the phenolic OH groups into primary aliphatic OH groups. These aromatic polyols are produced in a single step, without the need for purification. By modifying the molar mass of polyethylene glycol, a broad spectrum of hydroxyl values can be achieved. The polyols containing up to 30% lignin

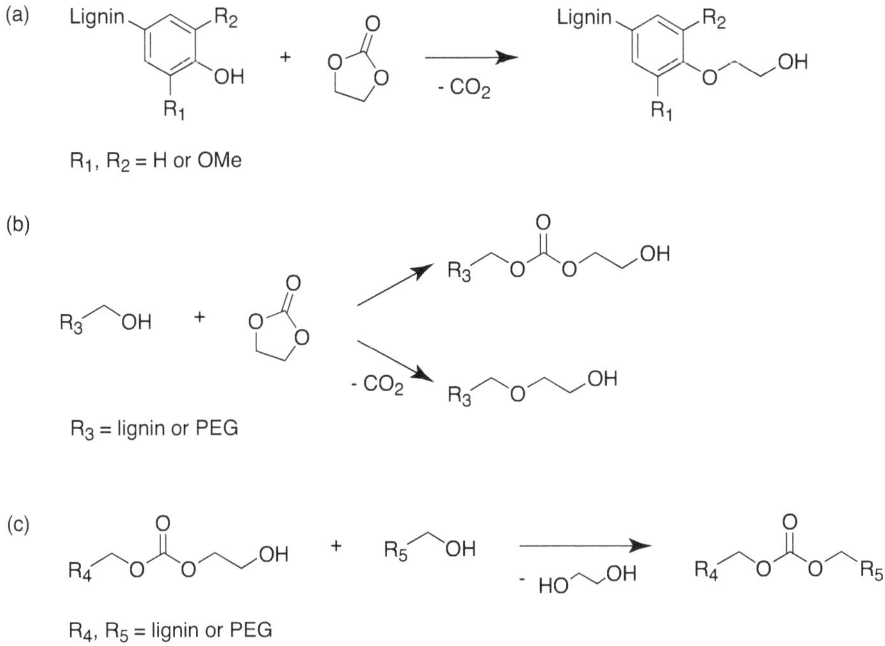

FIGURE 8.12 Chemical reactions of ethylene carbonate (EC) with lignin and polyethylene glycol (PEG). (a) chemical reaction with phenolic OH groups of lignin, (b) chemical reaction with aliphatic OH groups of lignin or PEG, and (c) chain coupling by transcarbonation

Source: Open access (Duval et al., 2022)

exhibit a viscosity that is suitable for the direct formulation of polyurethane foams. This method offers considerable advantages over oxypropylation, which is the most prevalent technique for generating LBPs, as it operates at ambient pressure, avoids toxic chemicals, eliminates the need for purification or post-treatment, and enables the production of polyols with adjustable properties. Subsequently, four distinct aromatic polyols were synthesized to create rigid polyurethane foams, substituting up to 100% of a conventional polyether polyol. The polyols developed demonstrated exceptionally high reactivity, facilitating a reduction in the catalyst content in the polyurethane formulation by 75%. Rigid polyurethane foams produced with a 25% substitution of the standard polyol exhibited properties comparable to those of commercial polyurethane foams, featuring over 90% closed cells and a thermal conductivity of approximately 25 mWm^{-1}/K, making them highly suitable for thermal insulation applications.

8.2.3.2 Carbon fibers

Lignin, recognized as the most prevalent polyphenol in nature, contains a significant number of aromatic backbones with a carbon content of up to 60%, rendering it a promising candidate for carbon fiber precursors. Carbon fibers derived from lignin possess distinctive characteristics: they are regarded as value-added products,

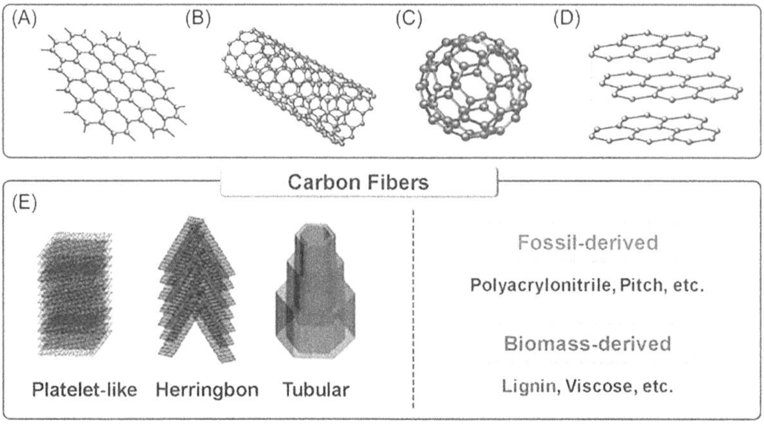

FIGURE 8.13 Representing images of (a) graphene, (b) carbon nanotubes, (c) carbon spheres, (d) graphite, and (e) carbon fibers with various morphologies and classifications of carbon fibers according to raw materials (fossil-derived and biomass-derived)

Source: Open access (Yan et al., 2025)

which help to lower preparation expenses and mitigate the costs linked to biomass management and disposal. Furthermore, lignin is rich in functional groups that can be utilized to enhance the properties of lignin-derived carbon fibers through various modifications and combinations with other materials (Figure 8.13). The microstructure and porosity of these carbon fibers are influenced by the intrinsic structure of their precursors, enabling the selection of different precursors to achieve the desired microstructure and morphology for specific applications (Yan et al., 2025).

8.2.4 Bioplastics and composites

8.2.4.1 Lignin-polylactic acid (PLA) composites

Currently, the global community is confronted with a significant issue of plastic pollution, primarily due to the non-degradable characteristics of conventional plastics. Polymer nanocomposites (PNCs) represent an innovative category of reinforced hybrid materials, created through the distribution of nanoscale clay particles within a polymer matrix (Singh et al., 2024, 2025; Kumar et al., 2015; Pattanayak et al., 2021; Kumar & Singh, 2013). In contrast, PLA, a biodegradable and bio-based polymer, is emerging as a viable alternative to traditional non-biodegradable plastics. Nevertheless, its elevated cost, restricted properties, and singular functionality limit its broader application. Lignin, a natural and sustainable biomass sourced from plant cell walls, has surfaced as a promising filler for PLA. The incorporation of lignin into PLA composites presents the opportunity to achieve a trifecta of affordability, high performance, and multifunctional attributes while adhering to the principles of biodegradation and sustainability. However, the inadequate compatibility between

PLA and lignin significantly undermines their overall performance, posing a substantial challenge for the advancement of PLA/lignin composites. Recent years have seen significant progress in addressing this issue (Nandhini et al., 2023; Ye et al., 2024). In a separate study, cellulose nanofibrils containing lignin (LCNF) derived from recycled newspaper were combined with a biodegradable PLA matrix to create multifunctional composites that exhibit both strength and ductility (Wang et al., 2021). The presence of lignin in LCNF enhanced the interfacial interaction and compatibility with the PLA matrix through hydrogen bonding and van der Waals forces, leading to a significant improvement in the mechanical properties of the PLA/LCNF composite films. Following the incorporation of LCNF into PLA, the resulting PLA/LCNF composite films demonstrated outstanding UV-shielding capabilities, effective water vapor barrier properties, and commendable thermal stability. These findings suggest a viable approach for developing fully biodegradable multifunctional composites utilizing environmentally friendly LCNF and biodegradable PLA. In a similar way, composites of PLA incorporating 0.5% lignin or nanolignin were fabricated using two distinct methodologies: (a) traditional melt-mixing and (b) in situ ROP through reactive processing (Makri et al., 2023). The synthesis of the composites was accomplished swiftly via reactive processing, taking less than 20 minutes. When the quantity of catalyst was increased twofold, the duration of the reaction was shortened to under 15 minutes. The advantages of reducing the size of lignin and employing in situ ROP through reactive processing were evident, as the composites containing nanolignin produced via reactive processing exhibited enhanced crystallization, mechanical strength, and antioxidant capabilities. These enhancements were ascribed to the involvement of nanolignin in the ROP of lactide acting as a macroinitiator, leading to the formation of PLA-grafted nanolignin particles that facilitated improved dispersion.

8.2.4.2 Polyhydroxyalkanoates *(PHA)*

Comprising as much as 30% of the lignocellulosic structure, effective lignin bioprocessing has the potential to enhance the feasibility of biorefineries and achieve comprehensive valorization of lignocellulosic waste. Recent developments have demonstrated the capacity of aerobic microorganisms, including *Pseudomonas putida* KT2440, to break down depolymerized lignin and generate intracellular energy storage compounds, particularly medium-chain-length polyhydroxyalkanoates (mcl-PHA), which may be utilized in the production of bioplastics. Analyzing the reported results of lignin bioconversion under varying nitrogen supplementation conditions, it was expected that lignin degradation would correlate with nitrogen presence, leading to an increase in bacterial biomass. In contrast, PHA production was associated with nitrogen-limited conditions, which coincided with a halt in cell growth. By utilizing the advantages of both scenarios, the potential for one-pot lignin bioconversion that merges high bacterial biomass with nitrogen-limited conditions is currently under investigation. Researchers have demonstrated that *Pseudomonas putida* was able to degrade 35% of the compounds present in alkali-pretreated lignin liquor when nitrogen was abundant, although this was accompanied by a low production of PHA. In contrast, a reduction in nitrogen supplementation led to an enhancement in PHA content (PHA/dry cell weight) to 43%, albeit at the cost of a decrease in lignin degradation, which fell to 22%. Furthermore, an increase in the initial cell biomass (ranging from 0.1 to 1.5 g/L) consistently enhanced lignin degradation from 22 to 33% under nitrogen-limited conditions (Sun & Loh, 2025).

8.2.5 Advanced materials

8.2.5.1 Activated carbon-based materials

Lignin-based activated carbon (LAC) represents a highly promising functional material, characterized by its significant surface area, extensive porous structure, and diverse functional groups. The modification process is crucial for enhancing the functionality of LAC by changing its porous and chemical characteristics. This review provides an overview of the latest modification technologies applicable to LAC for advanced applications. Various promising modification strategies are examined to illustrate their impact on the preparation of LAC. A comprehensive discussion is presented regarding the multiscale alterations in the porosity and surface chemistry of LAC. Subsequently, advanced applications are introduced to demonstrate the potential uses of LAC in supercapacitor electrodes, catalyst supports, hydrogen storage, and carbon dioxide capture. Finally, the mechanistic relationships between the structure and function of LAC are thoroughly explained. These findings underscore the significant role of modification technologies in transforming the properties and defining the functionalities of LAC, positioning it as a promising porous carbon material for industrial applications (Chen et al., 2022). In another work (Yang et al., 2020), activated carbon (AC) with an exceptionally high surface area exceeding 2,000 m^2/g was synthesized from low-sulfur acid hydrotropic lignin (AHL) derived from poplar wood using H_3PO_4 at a moderate temperature of 450°C (AHL-AC6). ACs exhibiting comparable surface areas were also produced under identical activation conditions from commercial hardwood alkali lignin and lignosulfonate. An initial assessment of AC performance was conducted utilizing nitrogen adsorption-desorption and dye adsorption techniques. AHL-AC6 demonstrated the most favorable specific surface area and dye adsorption capabilities. Moreover, the adsorption results for Congo red and methylene blue (MB) indicated that AHL-AC6 possessed a higher adsorption capacity than those documented in existing literature. The dye adsorption data aligned well with the Langmuir model. The fitting parameters imply that the adsorption is significantly strong and nearly irreversible, particularly for MB. This study presents, for the first time, a method for generating AC from lignin with a BET surface area exceeding 2,000 m^2/g using cost-effective and environmentally friendly H_3PO_4 at moderate temperatures.

8.2.5.2 Lignin-derived nanoparticles (NPs)

Regarding other materials and biopolymers, including cellulose and chitosan, the motivation behind the development of lignin nanomaterials lies in the acquisition of novel and appealing properties that emerge exclusively when substances are structured at the nanoscale. This phenomenon may be attributed to the so-called 'quantum effect,' which introduces new tunable characteristics at the nanoscale, or it may simply result from the increased surface-to-volume ratio of nanomaterials.

In comparison to conventional composite materials, nanocomposites present clear advantages, particularly those derived from biorenewable and sustainable resources. Given the compatibility, biodegradability, and eco-friendliness of bionanocomposites, their potential uses in the food packaging sector and pharmaceuticals are currently being explored. Moreover, given that PLA exhibits limited antioxidant and UV-barrier capabilities, blending PLA with lignin emerges as a promising approach to mitigate these shortcomings. Boarino et al. (2022) investigated that the incorporation of lignin NPs enhances the characteristics of PLA-based films. A significant challenge in fabricating PLA-lignin NP blend films is preventing the aggregation of NPs, which could adversely affect optical properties and antioxidant efficacy, among other factors. To circumvent NP aggregation within the PLA matrix, PLA-grafted lignin NPs were synthesized through organocatalyzed lactide ring-opening polymerization. Unlike lignin and unmodified lignin NPs, these PLA-grafted lignin NPs demonstrated uniform dispersion in PLA for lignin concentrations up to 10%. The introduction of merely 1% of these NPs effectively obstructed the transmission of 280 nm UV light. Concurrently, these blend films maintained satisfactory visible light transmittance. The optical characteristics of the PLA-lignin blend films were further enhanced by the well-dispersed nature of the PLA-grafted NPs, as indicated by a significantly higher visible light transmittance in blends of PLA and PLA-grafted NPs compared to those made from PLA with lignin or unmodified lignin NPs. Ultimately, the combination of PLA with PLA-grafted lignin NPs substantially improves the antioxidant activity of these films. A schematic representation of PLA-grafted lignin NPs is shown in Figure 8.14.

The perceptions derived from wood science underscore the significance of the complex molecular arrangement of lignin and crystalline cellulose in enhancing the strength of wood. In this direction, Pan et al. (2024) immersed pre-stretched PVA hydrogels in a solution containing nanosized lignosulfonate sodium (LS), a water-soluble anionic polyelectrolyte, to innovatively recreate a similar molecular structure within the hydrogels. The nanosized LS effectively stabilized and aggregated the pre-stretched PVA while promoting the development of dense crystalline

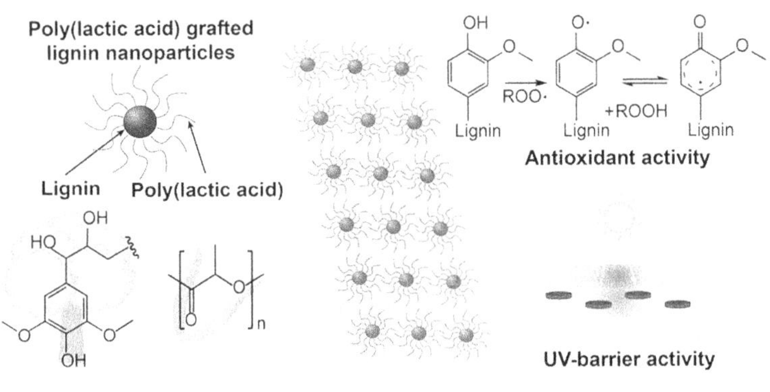

FIGURE 8.14 Schematic representation of PLA-grafted lignin NPs

Source: Open access (Boarino et al., 2022)

domains within the polymer matrix. This hydrogel significantly surpassed earlier hydrogels made solely from lignin and PVA. Furthermore, the composite hydrogels displayed remarkable anti-freezing capabilities (below −80°C). Importantly, the LS-assisted reconstruction technology presents avenues for the secondary fixation of PVA hydrogel shapes and the robust welding of hydrogel components. This research introduces a method for the high-value application of LS, a sustainable byproduct of pulp production. The profound biomimetic strategy of LS will be utilized in various multifunctional hydrogel applications.

8.2.5.3 Other porous materials

In the past ten years, porous biomaterials derived from lignin have demonstrated significant potential for various applications, including drug delivery, tissue engineering, wound dressing, pharmaceutical excipients, biosensors, and medical devices. These biomaterials incorporate lignin sourced from lignocellulosic biomass. As an aromatic compound, lignin is expected to enhance the mechanical and thermal properties of these materials, as well as their antioxidant and antibacterial characteristics, biodegradability, and biocompatibility. The characteristics of pore size, shape, and distribution are critical in defining the porous structure, porosity, surface area, permeability, water solubility, and adsorption capacity of the materials. Such attributes are particularly advantageous for medical applications, notably in controlled drug delivery systems, wound dressing, and tissue engineering. This review aims to present a comprehensive overview of the current landscape and prospects of lignin-based porous materials in medical and pharmaceutical contexts, emphasizing the types of materials, essential properties, methods of modification and fabrication, and promising applications in the medical field (Nan et al., 2022). Lignin serves as a precursor to produce porous carbon materials via thermochemical methods, including the development of lignin-based carbon for targeted drug delivery (Saha et al., 2013). Furthermore, lignin-derived natural adhesives can be utilized in materials that require adhesives in medical contexts, such as particleboard, which can function as a tissue-substitute phantom material (Zuber et al., 2022). Moreover, lignin can act as a pharmaceutical excipient, enhancing the bioavailability of medications, a crucial aspect in the formulation of oral dosages, and facilitating the release rate of drugs (for instance, aspirin) from tablets. Aquasolv lignin is deemed safe for consumption and exhibits antioxidant and antidiabetic properties, characteristics that can be beneficial in solid dosage forms within the pharmaceutical industry.

8.2.6 Nitrogenous-functionalized lignin materials

The nitrogenous functionalization of the lignin is carried out using different methods. Main methods include ammoxidation, the Mannich reaction, nitration, reductive amination, and reaction with quaternary ammonia, as shown in Figure 8.15.

8.2.6.1 Ammoxidation of lignin

The ammoxidation of lignin states a process in which lignin reacts with ammonia at designated pressure and temperature conditions using different oxidants, such as oxygen and hydrogen peroxide, to yield nitrogen-modified lignin. Ling et al. (1988)

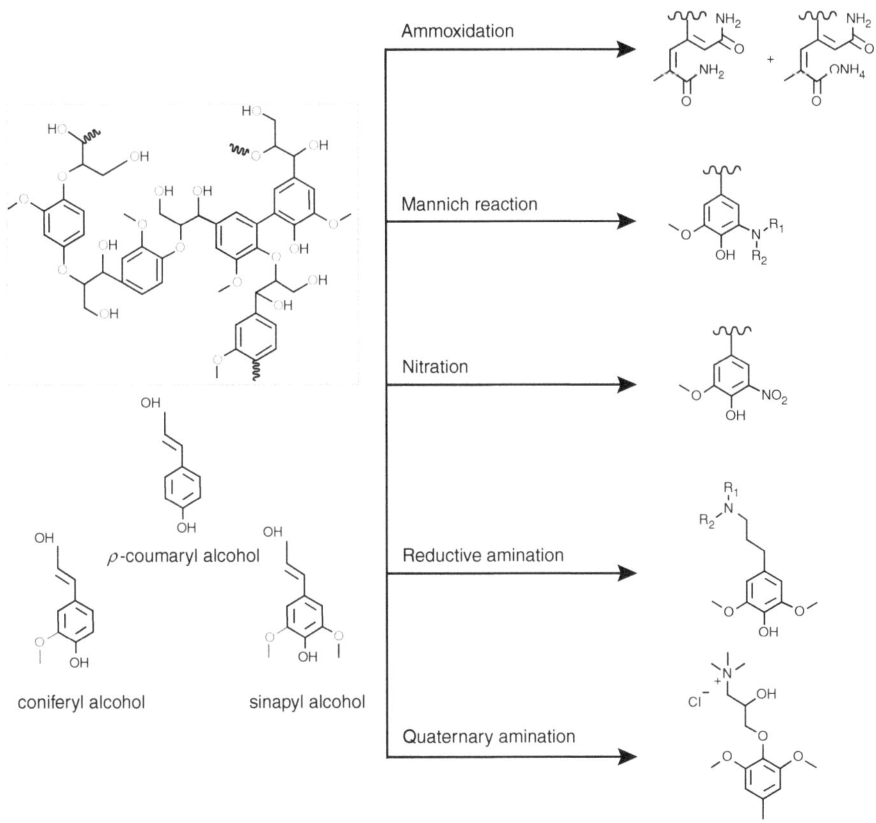

FIGURE 8.15 Nitrogenous-functionalized lignin materials

have conducted a comparative analysis of the chemical structures of three different lignosulfonates prior to and following the process of ammoxidation. It was observed that the molecular weights of the resultant products diminished, the structures of the aromatic rings underwent partial degradation, and the nitrogen content increased; however, the C/N ratios experienced a decline. Concurrently, nitrogen predominantly manifested in the forms of amine, amide, and ring-bonded nitrogen.

8.2.6.2 Mannich reaction

The Mannich reaction involving lignin is a process where lignin containing active hydrogen is concurrently condensed with formaldehyde and an amine, resulting in the substitution of the active hydrogen with either a substituted or unsubstituted aminomethyl group. Recently, the Mannich reaction has emerged as a significant and frequently employed technique for the incorporation of amino groups into lignin, extensively applied in the modification of lignin and the development of lignin-based functional materials (Jiao et al., 2019). The amination process of industrial softwood kraft lignin was performed utilizing the Mannich reaction as described by Du et al. (2014). Through phenolation pretreatment, the nitrogen content in the aminated lignin was elevated from 2.5 to 4.8%.

8.2.6.3 Nitration

The process of nitrating lignin typically involves the use of a non-aqueous solvent that contains nitric acid, which facilitates the introduction of NO or NO_2 functional groups onto the aromatic ring of lignin. The most employed nitrating agent is a combination of nitric acid with concentrated acetic acid, acetic anhydride, or sulfuric acid. Khabarov et al. (Khabarov et al., 2016) have conducted a study on the nitration of sulfate lignin using nitric acid in a dioxane-water solution under homogeneous conditions. This process yielded a significant degree of nitration at a temperature of 50°C with a minimal amount of nitric acid. They suggested that the resulting nitro-lignin could possess one nitro group for every 1.5–2 phenyl-propane units.

8.2.6.4 Reductive amination

The reductive amination involves the reaction of ketones and aldehydes with amines to create carbon–nitrogen double bonds, which are subsequently reduced using a reducing agent. In this direction, researchers have presented a novel single-step catalytic approach for the direct synthesis of phenolic amines from poplar lignin via reductive fractionation in an aqueous ammonia–alcohol mixture utilizing an Ru catalyst in the presence of H_2 (Ma et al., 2023). Depending on the type of alcohol employed, it is possible to produce primary, secondary, and tertiary phenolic amines. The highest yield of phenolic amines achieved was 26.6% relative to the lignin content. Throughout the processes of delignification, amination, and alkylation, native lignin was selectively extracted from woody biomass, resulting in the retention of cellulose, hemicellulose, and other components that remained largely unaffected. Incorporating the reductive amination reaction into the lignin reduction and fractionation process to generate valuable organonitrogen chemicals during the lignin treatment phase could represent a significant progression in the lignin-first strategy (Zhang et al., 2023).

8.2.6.5 Quaternary amination

The modification of lignin followed by pyrolysis presents a promising approach for its high-value conversion, with lignin amination emerging as a viable method to enhance lignin valorization during pyrolysis. To elucidate the effects of amination on the lignin pyrolysis mechanism, alkali lignin was examined using ethylenediamine and formaldehyde, and the pyrolysis behavior along with the formation of lignin derivatives was thoroughly investigated. The findings indicated that amination resulted in the creation of a C-N structure, alongside the substitution of Na+ to form C-N+ within the lignin, achieving a peak nitrogen content of 5.81%. Consequently, this led to a reduction in activation energy and an increased weight loss of 50.4% during the pyrolysis process (Liu et al., 2024).

8.3 CHALLENGES AND FUTURE SCOPES

Lignin, an intricate aromatic biopolymer, encounters obstacles such as structural heterogeneity, recalcitrance, and ineffective depolymerization, which restrict its transformation into valuable products. Innovative catalytic, biotechnological, and environmentally friendly processing methods are being developed to address these challenges. The future potential resides in creating customized lignin valorization pathways aimed at sustainable biofuels, chemicals, and high-performance biomaterials. Lignin, recognized as one of the most prevalent natural polymers, possesses significant potential as a sustainable precursor for the generation of value-added products, including biofuels, fine chemicals, resins, advanced biomaterials, and other high-value biochemicals and composites for the sustainable industries. Nevertheless, its application is fraught with challenges due to various inherent limitations. The foremost challenge stems from the highly heterogeneous and complex structure that defines lignin, which varies considerably based on its botanical source and extraction method, leading to inconsistent physicochemical properties. This complexity complicates the process of achieving selective depolymerization and acquiring well-defined monomeric units, which are essential for the synthesis of high-value products. Furthermore, the recalcitrant nature of lignin, characterized by robust carbon–carbon and carbon–oxygen linkages, along with the absence of effective catalytic or enzymatic systems for large-scale valorization, further obstructs its industrial use. Existing processing techniques, such as kraft pulping or organosolv, frequently break down lignin into irregular fragments, thereby diminishing its applicability for fine chemical synthesis. In spite of these obstacles, the potential for lignin valorization in the future remains optimistic. Recent progress in green catalytic systems, microbial engineering, and solvent-based fractionation technologies is paving the way for controlled lignin depolymerization and enhanced product selectivity. The incorporation of lignin valorization into current biorefineries has the potential to enhance overall process economics and resource efficiency. Furthermore, the creation of customized lignin-based nanomaterials, carbon fibers, adhesives, and bioactive compounds offers a sustainable alternative to petroleum-derived products. With ongoing research in catalytic science, bioprocessing, and material design, lignin can evolve from being an underutilized byproduct of the pulp and paper industry into a fundamental component of the circular bioeconomy, facilitating the global transition toward renewable and environmentally friendly material production.

8.4 CONCLUSION

Lignin serves as the primary natural source of aromatic compounds essential to the chemical industry. While the idea of leveraging lignin for aromatic production is not a recent development, it is becoming increasingly important in light of current petroleum shortages and rising prices. The application of lignin in the chemical sector encounters several challenges, such as the production and depolymerization of lignin, the improvement of depolymerization mixtures, the extraction of particular compounds, and their conversion into a wide range of high-value biochemicals and composites for the sustainable industries. Lignin, being an abundant and renewable aromatic polymer, is increasingly acknowledged as a promising precursor for generating value-added products. Its distinctive structure, rich in phenolic and aromatic components, facilitates the production of biofuels, fine chemicals, resins, carbon fibers, and advanced biomaterials. Nevertheless, obstacles such as structural heterogeneity, robust interunit linkages, and ineffective depolymerization hinder its large-scale application. Recent progress in catalytic conversion, microbial engineering, and fractionation methods presents opportunities to address these challenges. By incorporating lignin valorization into biorefineries, it can act as a sustainable substitute for fossil resources, thereby promoting the advancement of a circular and green bioeconomy.

REFERENCES

Ashengroph, M., and J. Amini. 2017. Bioconversion of Isoeugenol to Vanillin and Vanillic Acid Using the Resting Cells of *Trichosporon asahii. 3 Biotech* 7 (6): 358. https://doi.org/10.1007/s13205-017-0998-9

Ávila, P. F., and R. Goldbeck. 2022. Fractionating Process of Lignocellulosic Biomass for the Enzymatic Production of Short Chain Cello-Oligosaccharides. *Industrial Crops and Products* 178 (April): 114671. https://doi.org/10.1016/j.indcrop.2022.114671

Boarino, A., S. Aigoul, L. Yves, and H.-A. Koster. 2022. Uniformly Dispersed Poly(Lactic Acid)-Grafted Lignin Nanoparticles Enhance Antioxidant Activity and UV-Barrier Properties of Poly(Lactic Acid) Packaging Films. *ACS Applied Polymer Materials* 4 (7): 4808–17. https://doi.org/10.1021/acsapm.2c00420

Borchert, A. J., W. R. Henson, and G. T. Beckham. 2022. Challenges and Opportunities in Biological Funneling of Heterogeneous and Toxic Substrates beyond Lignin. *Current Opinion in Biotechnology* 73: 1–13. https://doi.org/10.1016/j.copbio.2021.06.007

Boz, H. 2015. *p*-Coumaric Acid in Cereals: Presence, Antioxidant and Antimicrobial Effects. *International Journal of Food Science & Technology* 50 (11): 2323–28. https://doi.org/10.1111/ijfs.12898

Calisti, C., A. G. Ficca, P. Barghini, and M. Ruzzi. 2008. Regulation of Ferulic Catabolic Genes in *Pseudomonas fluorescens* BF13: Involvement of a MarR Family Regulator. *Applied Microbiology and Biotechnology* 80 (3): 475–83. https://doi.org/10.1007/s00253-008-1557-4

Chen, Y., B. Fu, G. Xiao, et al. 2021. Bioconversion of Lignin-Derived Feedstocks to Muconic Acid by Whole-Cell Biocatalysis. *ACS Food Science & Technology* 1 (3): 382–87. https://doi.org/10.1021/acsfoodscitech.1c00023

Chen, K., H. Zi-Jing, L. Zhi-Hua, A. J. Ragauskas, L. Bing-Zhi, and Y. Ying-Jin. 2022. Emerging Modification Technologies of Lignin-Based Activated Carbon toward Advanced Applications. *ChemSusChem* 15 (21). https://doi.org/10.1002/cssc.202201284

D'Arrigo, P., L. A. M. Rossato, A. Strini, and S. Serra. 2024. From Waste to Value: Recent Insights into Producing Vanillin from Lignin. *Molecules* 29 (2): 442. https://doi.org/10.3390/molecules29020442

Demuner, I. F., J. L. Colodette, A. J. Demuner, and C. M. Jardim. 2019. Biorefinery Review: Wide-Reaching Products through Kraft Lignin. *BioResources* 14 (3): 7543–81. https://doi.org/10.15376/biores.14.3.Demuner

Deng, J. Z., D. J. Newman, and S. M. Hecht. 2000. Use of Compare Analysis to Discover Functional Analogues of Bleomycin. *Journal of Natural Products* 63 (9): 1269–72. https://doi.org/10.1021/np000084p

Dhar, P., and R. Vinu. 2018. Microwave-Assisted Catalytic Solvolysis of Lignin to Phenols: Kinetics and Product Characterization. *ACS Omega* 3 (11): 15076–85. https://doi.org/10.1021/acsomega.8b01509

Dong, X. Q., J. S. Yang, N. Zhu, E. T. Wang, and H. L. Yuan. 2013. Sugarcane Bagasse Degradation and Characterization of Three White-Rot Fungi. *Bioresource Technology* 131 (March): 443–51. https://doi.org/10.1016/j.biortech.2012.12.182

Du, X., J. Li, and M. E. Lindström. 2014. Modification of Industrial Softwood Kraft Lignin Using Mannich Reaction with and without Phenolation Pretreatment. *Industrial Crops and Products* 52: 729–35. https://doi.org/10.1016/j.indcrop.2013.11.035

Duval, A., D. Vidal, A. Sarbu, W. René, and L. Avérous. 2022. Scalable Single-Step Synthesis of Lignin-Based Liquid Polyols with Ethylene Carbonate for Polyurethane Foams. *Materials Today Chemistry* 24 (June): 100793. https://doi.org/10.1016/j.mtchem.2022.100793

Fache, M., B. Boutevin, and S. Caillol. 2016. Vanillin Production from Lignin and Its Use as a Renewable Chemical. *ACS Sustainable Chemistry & Engineering* 4 (1): 35–46. https://doi.org/10.1021/acssuschemeng.5b01344

Feghali, E., G. Carrot, P. Thuéry, C. Genre, and T. Cantat. 2015. Convergent Reductive Depolymerization of Wood Lignin to Isolated Phenol Derivatives by Metal-Free Catalytic Hydrosilylation. *Energy & Environmental Science* 8 (9): 2734–43. https://doi.org/10.1039/C5EE01304F

Gallage, N. J., and B. L. Møller. 2015. Vanillin – Bioconversion and Bioengineering of the Most Popular Plant Flavor and Its *De Novo* Biosynthesis in the Vanilla Orchid. *Molecular Plant* 8 (1): 40–57. https://doi.org/10.1016/j.molp.2014.11.008

Gibson, L. J. 2012. The Hierarchical Structure and Mechanics of Plant Materials. *Journal of the Royal Society Interface* 9 (76): 2749–66. https://doi.org/10.1098/rsif.2012.0341

Gill, M. K., G. S. Kocher, and A. S. Panesar. 2021. Optimization of Acid-Mediated Delignification of Corn Stover, an Agriculture Residue Carbohydrate Polymer for Improved Ethanol Production. *Carbohydrate Polymer Technologies and Applications* 2 (December): 100029. https://doi.org/10.1016/j.carpta.2020.100029

Giri, R., and R. K. Sharma. 2020. Fungal Pretreatment of Lignocellulosic Biomass for the Production of Plant Hormone by Pichia Fermentans under Submerged Conditions. *Bioresources and Bioprocessing* 7 (1): 30. https://doi.org/10.1186/s40643-020-00319-5

Gu, J., Q. Qiu, Y. Yu, et al. 2024. Bacterial Transformation of Lignin: Key Enzymes and High-Value Products. *Biotechnology for Biofuels and Bioproducts* 17 (1): 2. https://doi.org/10.1186/s13068-023-02447-4

Hemati, A., N. Aliasgharzad, R. Khakvar, N Delangiz, B. A. Lajayer, and E. D. van Hullebusch. 2023. Bioaugmentation of Thermophilic Lignocellulose Degrading Bacteria Accelerate the Composting Process of Lignocellulosic Materials. *Biomass Conversion and Biorefinery* 13 (17): 15887–901. https://doi.org/10.1007/s13399-021-02238-7

Hermosilla, E., O. Rubilar, H. Schalchli, A. Sant'Ana da Silva, V. Ferreira-Leitao, and M. C. Diez. 2018. Sequential White-Rot and Brown-Rot Fungal Pretreatment of Wheat Straw as a Promising Alternative for Complementary Mild Treatments. *Waste Management* 79 (September): 240–50. https://doi.org/10.1016/j.wasman.2018.07.044

Huang, C. H., M. F. Chen, H. H. Chung, and J. T. Cheng. 2012. Antihyperglycemic Effect of Syringaldehyde in Streptozotocin-Induced Diabetic Rats. *Journal of Natural Products* 75 (8): 1465–68. https://doi.org/10.1021/np3003723

Ingle, A. P., S. Saxena, M. P. Moharil, J. D. Rivaldi, L. Ramos, and A. K. Chandel. 2025. Production of Biomaterials and Biochemicals from Lignocellulosic Biomass through Sustainable Approaches: Current Scenario and Future Perspectives. *Biotechnology for Sustainable Materials* 2 (1): 3. https://doi.org/10.1186/s44316-025-00025-2

Jiao, G.-J., P. Peng, S.-L. Sun, Z.-C. Geng, and D. She. 2019. Amination of Biorefinery Technical Lignin by Mannich Reaction for Preparing Highly Efficient Nitrogen Fertilizer. *International Journal of Biological Macromolecules* 127: 544–54. https://doi.org/10.1016/j.ijbiomac.2019.01.076

Jiju, P. S., A. K. Patel, N. S. Shruthy, S. Shalu, C.-D. Dong, and R. R. Singhania. 2025. Sustainability through Lignin Valorization: Recent Innovations and Applications Driving Industrial Transformation. *Bioresources and Bioprocessing* 12 (1): 88. https://doi.org/10.1186/s40643-025-00929-x

Khabarov, Y. G., N. Y. Kuzyakov, V. A. Veshnyakov, G. V. Komarova, and A. Yu. Garkotin. 2016. Nitration of Sulfate Lignin under Homogeneous Conditions Studied by Electron Spectroscopy. *Russian Chemical Bulletin* 65 (12): 2925–31. https://doi.org/10.1007/s11172-016-1679-2

Kim, K. H., K. Jeong, J. Zhuang, et al. 2021. Tandem Conversion of Lignin to Catechols via Demethylation and Catalytic Hydrogenolysis. *Industrial Crops and Products* 159: 113095. https://doi.org/10.1016/j.indcrop.2020.113095

Kumar, P., A. D. Singh, V. Kumar, and P. P. Kundu. 2015. Incorporation of Nano-Al_2O_3 within the Blend of Sulfonated-PVdF-Co-HFP and Nafion for High Temperature Application in DMFCs. *RSC Advances* 5 (78): 63465–72. https://doi.org/10.1039/C5RA07992F

Kumar, R., A. Singh, A. Maurya, et al. 2022. Effective Bioremediation of Pulp and Paper Mill Wastewater Using Bacillus Cereus as a Possible Kraft Lignin-Degrading Bacterium. *Bioresource Technology* 352 (May): 127076. https://doi.org/10.1016/j.biortech.2022.127076

Kumar, V., and A. Singh. 2013. Polypropylene Clay Nanocomposites. *Reviews in Chemical Engineering* 29 (6). https://doi.org/10.1515/revce-2013-0014

Lee, C. Y., R. Held, A. Sharma, et al. 2013. Copper-Granule-Catalyzed Microwave-Assisted Click Synthesis of Polyphenol Dendrimers. *The Journal of Organic Chemistry* 78 (22): 11221–28. https://doi.org/10.1021/jo401603d

Li, Z., X. Bai, X.-Y. Wei, et al. 2023. A Solid Acid-Catalyzed Depolymerization of Pine Lignin to Obtain Guaiacol Using a Hydrogen-Free Strategy. *Fuel Processing Technology* 249: 107843. https://doi.org/10.1016/j.fuproc.2023.107843

Lin, J., P. Wen, W. Ying, J. Yu, and J. Zhang. 2024. Comparison of Lactic and Propionic Acid Hydrolysis for Production of Xylo-Oligosaccharides and Ethanol from Polysaccharides in Toona Sinensis Branch. *International Journal of Biological Macromolecules* 270 (June): 132339. https://doi.org/10.1016/j.ijbiomac.2024.132339

Ling, C., J. Y. Quan, and Z. Z. Li. 1988. A Study on the Structure of Ammoxidative Lignin. *Journal of Nanjing Forestry University* 22: 68–72. https://eurekamag.com/research/003/028/003028160.php

Liu, Z., J. Zeng, Z. Dong, Y. Chen, H. Yang, and H. Chen. 2024. Insight into the Mechanism of Lignin Amination Pretreatment on Lignin Structure and Its Pyrolysis Property for Lignin Valorization. *Chemical Engineering Journal* 499: 156386. https://doi.org/10.1016/j.cej.2024.156386

Lu, X., and X. Gu. 2022. A Review on Lignin Pyrolysis: Pyrolytic Behavior, Mechanism, and Relevant Upgrading for Improving Process Efficiency. *Biotechnology for Biofuels and Bioproducts* 15 (1): 106. https://doi.org/10.1186/s13068-022-02203-0

Lu, X., X. Zhu, H. Guo, et al. 2020. Efficient Depolymerization of Alkaline Lignin to Phenolic Compounds at Low Temperatures with Formic Acid over Inexpensive Fe–Zn/Al_2O_3 Catalyst. *Energy & Fuels* 34 (6): 7121–30. https://doi.org/10.1021/acs.energyfuels.0c00742

Ma, J., D. Le, and N. Yan. 2023. Single-Step Conversion of Wood Lignin into Phenolic Amines. *Chem* 9 (10): 2869–88. https://doi.org/10.1016/j.chempr.2023.05.030

Makri, S. P., X. Eleftheria, M. A. Valera, et al. 2023. Poly(Lactic Acid) Composites with Lignin and Nanolignin Synthesized by In Situ Reactive Processing. *Polymers* 15 (10): 2386. https://doi.org/10.3390/polym15102386

McCallum, C. S., W. Wang, W. J. Doran, et al. 2021. Life Cycle Thinking Case Study for Catalytic Wet Air Oxidation of Lignin in Bamboo Biomass for Vanillin Production. *Green Chemistry* 23 (4): 1847–60. https://doi.org/10.1039/D0GC03685D

Momayez, F., M. Hedenström, S. Stagge, L. J. Jönsson, and C. Martín. 2022. Valorization of Hydrolysis Lignin from a Spruce-Based Biorefinery by Applying γ-Valerolactone Treatment. *Bioresource Technology* 359 (September): 127466. https://doi.org/10.1016/j.biortech.2022.127466

Nan, N., W. Hu, and J. Wang. 2022. Lignin-Based Porous Biomaterials for Medical and Pharmaceutical Applications. *Biomedicines* 10 (4): 747. https://doi.org/10.3390/biomedicines10040747

Nandhini, R., B. Sivaprakash, N. Rajamohan, and D. V. N. Vo. 2023. Lignin and Polylactic Acid for the Production of Bioplastics and Valuable Chemicals. *Environmental Chemistry Letters* 21 (1): 403–27. https://doi.org/10.1007/s10311-022-01505-x

Novakovic, J., N. Kontogianni, E. M. Barampouti, et al. 2021. Towards Upscaling the Valorization of Wheat Straw Residues: Alkaline Pretreatment Using Sodium Hydroxide, Enzymatic Hydrolysis and Biogas Production. *Environmental Science and Pollution Research* 28 (19): 24486–98. https://doi.org/10.1007/s11356-020-08905-y

Nurika, Ir., E. N. Shabrina, N. Azizah, S. Sartini, T. D. H. Bugg, and G. C. Barker. 2022. Application of Ligninolytic Bacteria to the Enhancement of Lignocellulose Breakdown and Methane Production from Oil Palm Empty Fruit Bunches (OPEFB). *Bioresource Technology Reports* 17 (February): 100951. https://doi.org/10.1016/j.biteb.2022.100951

Oruganti, R. K., M. P. Gungupalli, and D. Bhattacharyya. 2024. Alkaline Hydrolysis for Yield of Glucose and Kraft Lignin from De-Oiled Jatropha Curcas Waste: Multiresponse Optimization Using Response Surface Methodology. *Biomass Conversion and Biorefinery* 14 (10): 11093–111. https://doi.org/10.1007/s13399-022-03204-7

Pan, X., X. Li, Z. Wang, Y. Ni, and Q. Wang. 2024. Nanolignin-Facilitated Robust Hydrogels. *ACS Nano* 18 (35): 24095–104. https://doi.org/10.1021/acsnano.4c04078

Pattanayak, P., F. Papiya, V. Kumar, A. Singh, and P. P. Kundu. 2021. Performance Evaluation of Poly (Aniline-co-Pyrrole) Wrapped Titanium Dioxide Nanocomposite as an Air-Cathode Catalyst Material for Microbial Fuel Cell. *Materials Science and Engineering:* C 118: 111492. https://doi.org/10.1016/j.msec.2020.111492

Pearl, I. A., and E. E. Dickey. 1951. Studies on Lignin and Related Products. IV. Separation of Syringaldehyde and Vanillin by Chromatography. *Journal of the American Chemical Society* 73 (2): 863–64. https://doi.org/10.1021/ja01146a521

Perez-Arce, J., A. Centeno-Pedrazo, J. Labidi, J. R. Ochoa-Gómez, and E. J. Garcia-Suarez. 2020. A Novel and Efficient Approach to Obtain Lignin-Based Polyols with Potential Industrial Applications. *Polymer Chemistry* 11 (46): 7362–69. https://doi.org/10.1039/D0PY01142H

Poveda-Giraldo, J. A., and C. A. Cardona Alzate. 2021a. A Biorefinery for the Valorization of Marigold (*Calendula officinalis*) Residues to Produce Biogas and Phenolic Compounds. *Food and Bioproducts Processing* 125: 91–104. https://doi.org/10.1016/j.fbp.2020.10.015

Poveda-Giraldo, J. A., and C. A. Cardona Alzate. 2021b. Biorefinery Potential of *Eucalyptus grandis* to Produce Phenolic Compounds and Biogas. *Canadian Journal of Forest Research* 51 (1): 89–100. https://doi.org/10.1139/cjfr-2020-0201

Radhakrishnan, R., P. Patra, M. Das, and A. Ghosh. 2021. Recent Advancements in the Ionic Liquid Mediated Lignin Valorization for the Production of Renewable Materials and Value-Added Chemicals. *Renewable and Sustainable Energy Reviews* 149 (October): 111368. https://doi.org/10.1016/j.rser.2021.111368

Rinesch, T., J. Mottweiler, M. Puche, P. Concepción, A. Corma, and C. Bolm. 2017. Mechanistic Investigation of the Catalyzed Cleavage for the Lignin β-O-4 Linkage: Implications for Vanillin and Vanillic Acid Formation. *ACS Sustainable Chemistry & Engineering* 5 (11): 9818–25. https://doi.org/10.1021/acssuschemeng.7b01725

Rodrigues Pinto, P. C., E. A. Borges da Silva, and A. E. Rodrigues. 2012. Lignin as Source of Fine Chemicals: Vanillin and Syringaldehyde. In *Biomass Conversion*, ed. Baskar, C., S. Baskar, and R. Dhillon, 381–420. Springer Berlin Heidelberg. https://doi.org/10.1007/978-3-642-28418-2_12

Saha, D., E. A. Payzant, A. S. Kumbhar, and A. K. Naskar. 2013. Sustainable Mesoporous Carbons as Storage and Controlled-Delivery Media for Functional Molecules. *ACS Applied Materials & Interfaces* 5 (12): 5868–74. https://doi.org/10.1021/am401661f

Salvachúa, D., C. W. Johnson, C. A. Singer, et al. 2018. Bioprocess Development for Muconic Acid Production from Aromatic Compounds and Lignin. *Green Chemistry* 20 (21): 5007–19. https://doi.org/10.1039/C8GC02519C

Shen, X., Q. Meng, Q. Mei, et al. 2020. Selective Catalytic Transformation of Lignin with Guaiacol as the Only Liquid Product. *Chemical Science* 11 (5): 1347–52. https://doi.org/10.1039/C9SC05892C

Sheng, Y., and Y Xu. 2022. Nuclear Magnetic Resonance Analysis of Ascorbic Acid Assisted Lignocellulose Decomposition in Dilute Acid Pretreatment and Its Stimulation on Enzymatic Hydrolysis. *Bioresource Technology* 343 (January): 126147. https://doi.org/10.1016/j.biortech.2021.126147

Singh, A., J. G. Vijayan, and K. G. Moodley. 2021. Surface Functionalizations of Nanocellulose for Wastewater Treatment. In *Handbook of Nanocelluloses*, ed. Barhoum, A., 1–48. Springer International Publishing. https://doi.org/10.1007/978-3-030-62976-2_49-1

Singh, A., and J. G. Vijayan. 2023. Biochar: A Green Material for Wastewater Treatment. In *Encyclopedia of Green Materials*, Baskar, C., S. Ramakrishna, and A. D. L. Rosa, 1–14. Springer Nature Singapore. https://doi.org/10.1007/978-981-16-4921-9_34-1

Singh, A., S. L. Banerjee, K. Kumari, and P. P. Kundu. 2022a. Recent Innovations in Chemical Recycling of Polyethylene Terephthalate Waste: A Circular Economy Approach toward Sustainability. In *Handbook of Solid Waste Management*, ed. Baskar, C., S. Ramakrishna, S. Baskar, R. Sharma, A. Chinnappan, and R. Sehrawat, 1149–76. Springer Nature Singapore. https://doi.org/10.1007/978-981-16-4230-2_53

Singh, A., K. Kumari, and P. P. Kundu. 2022b. Nanocellulose Biocomposites for Bone Tissue Engineering. In *Handbook of Nanocelluloses*, ed. Barhoum, A., 597–647. Springer International Publishing. https://doi.org/10.1007/978-3-030-89621-8_39

Singh, A., K. Mukhopadhyay, and S. G. Sachan. 2019. Biotransformation of Eugenol to Vanillin by a Novel Strain *Bacillus safensis* SMS1003. *Biocatalysis and Biotransformation* 37 (4): 291–303. https://doi.org/10.1080/10242422.2018.1544245

Singh, A., P. P. Kundu, and K. Kumari. 2025. Biomedical Engineering: Nanowires for Organ and Bone Tissue Engineering. In *One- and Two-Dimensional Nanomaterials*, ed. Joshi, S., P. Dua, H. Sarma, and P. Velmurugan, 199–221. Elsevier. https://doi.org/10.1016/B978-0-443-23703-4.00008-X

Singh, A., V. Dhiman, K. Kumari, and P. P. Kundu. 2024. Quantum Dots/Bioconjugates. In *Quantum Dots Based Nanocomposited,* ed. Thomas, S., P. Das, and S. Ganguly, 193–206. Springer. https://doi.org/10.1007/978-3-031-54779-9_10

Sriprom, P., P. Leephisuth, P. Assawasaengrat, A. Neramittagapong, and S. eramittagapong. 2020. Optimization of Lignin Conversion to Phenol via Partial Oxidation of Synthesized Wastewater Containing Lignin. *Energy Reports* 6: 144–49. https://doi.org/10.1016/j.egyr.2020.11.272

Stewart, D. 2008. Lignin as a Base Material for Materials Applications: Chemistry, Application and Economics. *Industrial Crops and Products* 27 (2): 202–07. https://doi.org/10.1016/j.indcrop.2007.07.008

Sun, J., and K. C. Loh. 2025. One-Pot Lignin Bioconversion to Polyhydroxyalkanoates based on Hierarchical Utilization of Heterogeneous Compounds. *Bioresource Technology* 419 (March): 132056. https://doi.org/10.1016/j.biortech.2025.132056

Tan, Z., Y. Li, F. Chen, et al. 2024. Challenges and Perspectives of the Conversion of Lignin Waste to High-Value Chemicals by Pyrolysis. *Processes* 12 (3): 589. https://doi.org/10.3390/pr12030589

Taylor, B. R., N. Kumar, D. K. Mishra, B. A. Simmons, H. Choudhary, and K. L. Sale. 2024. Computational Advances in Ionic Liquid Applications for Green Chemistry: A Critical Review of Lignin Processing and Machine Learning Approaches. *Molecules* 29 (21): 5073. https://doi.org/10.3390/molecules29215073

Timokhin, V. I., M. Regner, A. H. Motagamwala, et al. 2020. Production of *p*-Coumaric Acid from Corn GVL-Lignin. *ACS Sustainable Chemistry & Engineering* 8 (47): 17427–38. https://doi.org/10.1021/acssuschemeng.0c05651

Tinikul, R., P. Chenprakhon, S. Maenpuen, and P. Chaiyen. 2018. Biotransformation of Plant-Derived Phenolic Acids. *Biotechnology Journal* (1700632). https://doi.org/10.1002/biot.201700632

Tu, W. C., and J. P. Hallett. 2019. Recent Advances in the Pretreatment of Lignocellulosic Biomass. *Current Opinion in Green and Sustainable Chemistry* 20 (December): 11–17. https://doi.org/10.1016/j.cogsc.2019.07.004

Wang, L., X. Cheng, M. Dong, et al. 2024. Robust Carbon–Carbon Cleavage in Lignin to Produce Phenol and Cyclohexanone. *Chemical Research in Chinese Universities* 40 (1): 29–35. https://doi.org/10.1007/s40242-024-3279-2

Wang, S., Z. Wan, Y. Han, et al. 2023. A Review on Lignin Waste Valorization by Catalytic Pyrolysis: Catalyst, Reaction System, and Industrial Symbiosis Mode. *Journal of Environmental Chemical Engineering* 11 (1): 109113. https://doi.org/10.1016/j.jece.2022.109113

Wang, X., and E. W. Qian. 2020. Extraction and Modification of Lignin from Red Pine Using Ionic Liquid. *Journal of the Japan Petroleum Institute* 63 (2): 102–5. https://doi.org/10.1627/jpi.63.102

Wang, Y., S. Liu, Q. Wang, et al. 2021. Strong, Ductile and Biodegradable Polylactic Acid/Lignin-Containing Cellulose Nanofibril Composites with Improved Thermal and Barrier Properties. *Industrial Crops and Products* 171 (November): 113898. https://doi.org/10.1016/j.indcrop.2021.113898

Wei, X. Z., J. Liu, and L. Ma. 2021. Cleavage via Selective Catalytic Oxidation of Lignin or Lignin Model Compounds into Functional Chemicals. *ChemEngineering* 5 (4): 74. https://doi.org/10.3390/chemengineering5040074

Weiss, R., G. M. Guebitz, A. Pellis, and G. S. Nyanhongo. 2020. Harnessing the Power of Enzymes for Tailoring and Valorizing Lignin. *Trends in Biotechnology* 38 (11): 1215–31. https://doi.org/10.1016/j.tibtech.2020.03.010

Xu, Z., P. Lei, R. Zhai, Z. Wen, and M. Jin. 2019. Recent Advances in Lignin Valorization with Bacterial Cultures: Microorganisms, Metabolic Pathways, and Bio-Products. *Biotechnology for Biofuels* 12 (1): 32. https://doi.org/10.1186/s13068-019-1376-0

Yadav, S. K., S. Singh, H. B. Singh, and B. K. Sarma. 2017. Compatible Rhizosphere-Competent Microbial Consortium Adds Value to the Nutritional Quality in Edible Parts of Chickpea. *Journal of Agricultural and Food Chemistry* 65 (30): 6122–30. https://doi.org/10.1021/acs.jafc.7b01326

Yan, J., Q. Meng, X. Shen, et al. 2020. Selective Valorization of Lignin to Phenol by Direct Transformation of Csp2–Csp3 and C–O Bonds. *Science Advances* 6 (45). https://doi.org/10.1126/sciadv.abd1951

Yan, L., H. Liu, Y. Yang, L. Dai, and C. Si. 2025. Lignin-Derived Carbon Fibers: A Green Path from Biomass to Advanced Materials. *Carbon Energy* 7 (3). https://doi.org/10.1002/cey2.662

Yang, Z., R. Gleisner, D. H. Mann, J. Xu, J. Jiang, and J. Y. Zhu. 2020. Lignin-Based Activated Carbon Using H_3PO_4 Activation. *Polymers* 12 (12): 2829. https://doi.org/10.3390/polym12122829

Ye, H., T. You, H. Nawaz, and F. Xu. 2024. A Comprehensive Review on Polylactic Acid/Lignin Composites—Structure, Synthesis, Performance, Compatibilization, and Applications. *International Journal of Biological Macromolecules* 280 (November): 135886. https://doi.org/10.1016/j.ijbiomac.2024.135886

Zhang, H., L. Han, and H. Dong. 2021. An Insight to Pretreatment, Enzyme Adsorption and Enzymatic Hydrolysis of Lignocellulosic Biomass: Experimental and Modeling Studies. *Renewable and Sustainable Energy Reviews* 140 (April): 110758. https://doi.org/10.1016/j.rser.2021.110758

Zhang, P., S. Y. Dai, and J. S. Yuan. 2023. Improve Value for Lignin-First Strategy: Producing Functional Molecules from Wood Biomass. *Chem Catalysis* 3 (10): 100781. https://doi.org/10.1016/j.checat.2023.100781

Zhu, Y., Y. Liao, W. Lv, et al. 2020. Complementing Vanillin and Cellulose Production by Oxidation of Lignocellulose with Stirring Control. *ACS Sustainable Chemistry & Engineering* 8 (6): 2361–74. https://doi.org/10.1021/acssuschemeng.9b04837

Zuber, S. H., N. A. A. Hashikin, M. F. M. Yusof, and R. Hashim. 2022. Lignin and Soy Flour as Adhesive Materials in the Fabrication of *Rhizophora* spp. Particleboard for Medical Physics Applications. *The Journal of Adhesion* 98 (5): 429–48. https://doi.org/10.1080/00218464.2020.1839430

9 Lignin-based antimicrobials and therapeutic commodities

Narayan Singh, Vasudha Maurya, Unnati Kushavah, and Alok Patel

9.1 INTRODUCTION

Lignin, the second most abundant biopolymer on Earth after cellulose, serves as a main structural element in plant cell walls, which we see in the role of adding rigidity and protecting against microbe attack. It is a key player in the carbon cycle and makes up 15–30% of lignocellulosic biomass. Lignin used to be considered a byproduct of the pulp and paper industry, but it is now being recognized as a useful raw material in the production of bio-based and environmentally friendly products (Tang et al., 2020). At the same time, the production of lignin as a biopolymer is an excellent alternative, as it is sustainable and can be used to develop products of higher value. Unlike other biopolymers such as cellulose and starch, lignin contains different phenylpropanoid units, which not only provide a wide spectrum of biopolymeric characteristics but also imbue the biopolymer with the ability to absorb UV and possess antimicrobial and antioxidant properties. Because of these qualities, lignin is very sought-after for the creation of functional materials in a variety of industries, such as packaging, energy storage, medicines, and environmental remediation (Chandra et al., 2024; Zhou et al., 2024).

Advancements in the extraction, purification, and functionalization of lignin enhance its potential. Kraft, organosolv, and soda pulping processes produce distinct types of technical lignin, each exhibiting unique structural characteristics. Variations in material applications influence their compatibility and reactivity (Modi et al., 2024). Chemical and enzymatic modifications have been developed to address heterogeneity and enhance processability, allowing for the customization of lignin for specific applications such as hydrogels, polymer resins, and nanocomposites (Chandra et al., 2025). In terms of sustainability, lignin offers several advantages for the environment. It comes from biomass and is biodegradable, making it a powerful alternative to petroleum-based polymers. According to the principles of green chemistry, its use encourages resource efficiency and waste reduction. Additionally, lignin valorization reduces dependency on fossil fuels and advances the circular bioeconomy by converting industrial waste into valuable commodities (Jassal et al., 2022). The application of lignin in nanotechnology has increased its use even more. LNPs

DOI: 10.1201/9781003653189-9

TABLE 9.1
Lignin-based drug delivery systems with their function

S. No.	Lignin-based Molecule/ System	Drug/ Component	Function/Application	References
1	Lignin Nanoparticles (LNPs)	Doxorubicin, curcumin	Sustained release of anticancer drugs; enhanced solubility of drugs	Iravani (2020); Figueiredo et al. (2017)
2	PEGylated lignin nanoparticles	Paclitaxel	Enhanced stability and selected tumor targeting	Li et al. (2019)
3	Lignin-chitosan hybrid nanoparticles	Ciprofloxacin	Enhanced antibacterial property of ciprofloxacin antibiotic and improved wound healing	Xia et al. (2014)
4	Lignin-metal nanocomposites	AgNPs, ZnO	Improve synergistic antimicrobial and antioxidant effects of nanoparticles	Kai et al. (2016)
5	Lignin-coated magnetic nanoparticles	Methotrexate	Improved magnetic field-guided targeted drug delivery of nanoparticle	Medina et al. (2025)
6	Lignin–PLA conjugates	Hydrophobic drugs	Biodegradable matrix for sustained release	Gilca et al. (2015)
7	Lignin-drug conjugates	5-Fluorouracil, doxorubicin	pH-responsive drug release via covalent linkage	Chandra et al. (2025)
8	Lignin-curcumin complex	Curcumin	Improved solubility and antioxidant activity	Thakur et al. (2014)
9	Lignin microparticles	Ibuprofen, paracetamol	Controlled oral drug release	Chowdhury et al. (2016)
10	Lignosulfonate nanocarriers	Peptides, antioxidants	Water-soluble carriers for bioactive compounds	Kai et al. (2016)

have shown promise in drug delivery systems, UV-protective formulations, and as reinforcing agents in biocomposites due to their size, surface area, and availability of functional groups (Ulbrich & Alf, 2010; Tang et al., 2020) (Table 9.1). These advancements show how lignin can be utilized as a sustainable foundation to produce high-quality, eco-friendly goods. In conclusion, lignin's emergence as a renewable biopolymer is being propelled by both its intrinsic properties and the growing field of sustainable material science. As research continues to fully realize its promise through advanced processing and integration techniques, lignin is poised to play a significant role in the development of next-generation bio-based materials.

9.2 IMPORTANCE OF LIGNIN IN BIOMEDICAL AND PHARMACEUTICAL INDUSTRIES

Compared to cellulose and hemicellulose, lignin, one of the most plentiful renewable biopolymers on Earth, has historically been underutilized. Recent scientific advancements have reexamined the function of lignin by highlighting its diverse range of biological and pharmacological applications. It has antibacterial, antioxidant, anti-inflammatory, and drug delivery qualities due to its polyphenolic structure and variety of functional groups. One significant application for lignin is its capacity to transport medications. Because lignin-based nanoparticles are biodegradable, have low toxicity, and can encapsulate a variety of therapeutic compounds, they have demonstrated significant promise as carriers for controlled drug release. Chaudhary and Sinha (2023), for example, highlighted the conversion of industrial lignin waste into medicinal materials by creating lignin-based carriers that improve drug bioavailability and permit prolonged drug release (Figure 9.1). This strategy targets problems including traditional medications' poor solubility and dose-dependent toxicity, especially in antibacterial and cancer treatments (Chaudhary & Sinha, 2023). Antimicrobial materials are another important biomedical application for lignin. Lignin is a promising ingredient in the creation of antimicrobial coatings and wound dressings because of its inherent antibacterial qualities, which are ascribed to its phenolic components. Devu et al. (2024) described in detail the potent antibacterial activity of lignin-integrated metal oxides and sulfides in sensor applications (Figure 9.1). These composite materials are ideal for healthcare settings because they not only improve diagnostic instrument sensitivity but also prevent microbial growth (Devu et al., 2024).

Additionally, lignin plays a key role in the creation of hydrogels and biocompatible scaffolds for tissue engineering and wound healing. The application of porous biomaterials based on lignin that have superior antioxidant activity, biodegradability, and mechanical strength. These characteristics are essential for controlling oxidative stress in injured tissues and encouraging cellular development. The creation of multipurpose materials with long-lasting therapeutic effects is made possible by lignin's capacity to interact with other biopolymers (Nan et al., 2022). In conclusion, lignin's

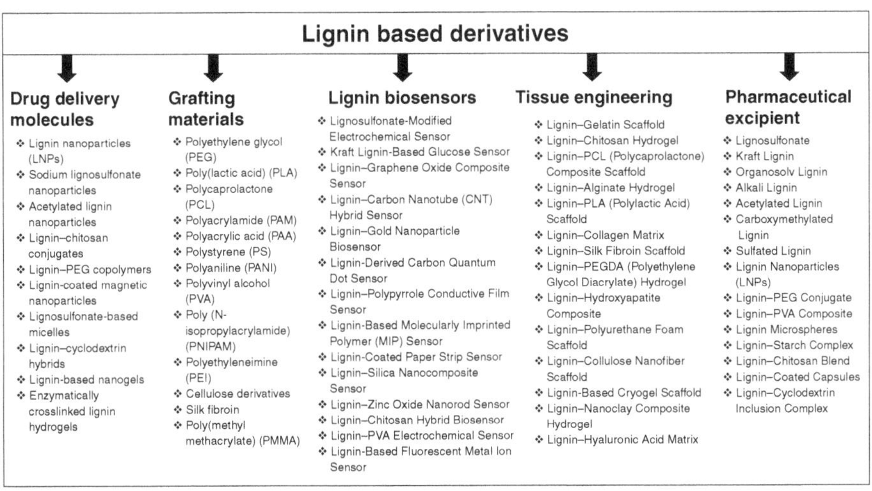

FIGURE 9.1 Lignin-based derivatives and their roles in different fields with suitable examples

diverse bioactivity, adjustable chemical structure, and environmentally friendly origin all contribute to its growing uses in biomedicine. Its transformation from an industrial waste into a useful substance for tissue engineering, medication transport, and antimicrobial protection emphasizes how important it is to solving today's biomedical and pharmaceutical problems.

Lignin, a complex aromatic biopolymer and a major constituent of lignocellulosic biomass, has historically been underutilized, primarily considered a low-value byproduct in the paper and bioethanol industries. However, because of its inherent biological qualities and structural adaptability, more and more data points to its promising potential in pharmacological and therapeutic applications (Figure 9.1). Lignin is becoming more widely acknowledged as a promising option for the creation of innovative therapeutic agents and drug delivery systems as the need for biocompatible and sustainable materials grows. The antioxidant, antibacterial, anti-inflammatory, and anticancer properties of lignin are among its most important contributions to the biomedical sector. It is suitable for wound healing, infection control, and tissue regeneration applications because of its high phenolic content, which allows for radical scavenging and interaction with microbial membranes (Ullah et al., 2022). Furthermore, lignin is a desirable ingredient for creating sophisticated biomaterials like hydrogels, cryogels, and aerogels that can be used as scaffolds for tissue engineering or as vehicles for controlled drug release due to its biodegradability and biocompatibility (Alaoui et al., 2023; Khadem et al., 2024). The transformation of lignin into nanoparticles and nanocomposites has been made easier by recent developments in material science, which have improved its physicochemical stability and allowed for precise functionalization for specific purposes (Figure 9.1). These lignin-based nanomaterials have shown enormous potential in minimizing side effects, guaranteeing controlled release profiles, and encapsulating and delivering medications (Lee et al., 2022; Billu et al., 2023; Mukheja et al., 2023). Furthermore, multifunctional composites with specific therapeutic qualities can be developed thanks to lignin's compatibility with a variety of natural and synthetic polymers. This book chapter aims to synthesize the current understanding and emerging trends surrounding lignin's application in the pharmaceutical and biomedical sectors. The chapter begins with an overview of lignin's structure, sources, and physicochemical characteristics relevant to biomedical functionality. The antimicrobial characteristics are then examined, with a focus on the mechanisms of microbial inhibition and structure-activity connections. There will be in-depth talks about the synthesis and use of lignin-derived nanomaterials, emphasizing advancements in therapeutic scaffolds, wound care, and drug delivery systems. The chapter also discusses scalability, standardization, and biocompatibility evaluations, among other practical and regulatory issues that arise when lignin-based treatments are transferred from lab research to clinical settings. Case studies will be provided to demonstrate practical uses and business initiatives utilizing the medicinal potential of lignin. The main goal is to offer a thorough reference that connects basic science and applied technologies, giving researchers, physicians, and industry participants a sophisticated grasp of how lignin can be used as a flexible and sustainable material in the pharmaceutical and biomedical fields.

9.3 LIGNIN: STRUCTURE, SOURCES, AND PROPERTIES

9.3.1 Structural complexity and heterogeneity of lignin

Lignin is a highly complex and heterogeneous biopolymer that plays a vital role in providing structural support, impermeability, and resistance to microbial attack in plant cell walls. Its structural complexity arises from the radical-mediated polymerization of three primary monolignols: p-coumaryl alcohol, coniferyl alcohol, and sinapyl alcohol, which give rise to p-hydroxyphenyl, guaiacyl, and syringyl units, respectively (Boerjan et al., 2003). The random coupling of these monolignols through various ether (*e.g.*, β-O-4) and carbon–carbon linkages results in an irregular, branched polymer without a defined repeating unit or primary sequence, making lignin highly heterogeneous. This heterogeneity varies across species, tissues, and developmental stages, as well as in response to environmental stimuli (Pesquet et al., 2019). Such complexity presents challenges for lignin depolymerization and utilization in bio-refinery applications, but also provides plants with crucial mechanical strength and defense mechanisms.

The structural complexity and heterogeneity of lignin significantly vary depending on its botanical source, such as softwoods, hardwoods, and agricultural residues. Softwood lignin is predominantly composed of guaiacyl units derived mainly from coniferyl alcohol, leading to a highly condensed and less easily degradable structure due to extensive C–C linkages. In contrast, hardwood lignin contains both guaiacyl and syringyl units, with a higher proportion of β-O-4 linkages, making it comparatively less condensed and more amenable to chemical breakdown (Li & Ragauskas, 2012). Agricultural residues, such as wheat straw or corn stover, typically contain a more diverse mix of p-hydroxyphenyl, guaiacyl, and syringyl units, contributing to even greater heterogeneity and variable reactivity (Zhen et al., 2017). These structural differences influence not only the efficiency of lignin extraction and depolymerization processes but also their potential applications in bio-based materials and chemicals.

9.3.2 Physicochemical and functional properties relevant to bioactivity

Lignin exhibits diverse physicochemical and functional properties that are highly relevant to its bioactivity and potential applications in agriculture, biomedicine, and materials science. The physicochemical properties of lignin include its complex aromatic structure, high molecular weight, and abundance of phenolic hydroxyl, methoxyl, and carboxylic groups. It is also an amorphous, hydrophobic, and thermally stable biopolymer. These functional groups are in charge of its UV-absorbing, antibacterial, and antioxidant qualities (Jiang et al., 2023). According to Yang et al. (2025), lignin's phenolic content is mostly responsible for its antioxidant action since it may scavenge reactive oxygen species (ROS) and lessen oxidative stress in biological systems. Its capacity to break down microbial cell walls and alter cellular metabolism is also linked to lignin's antibacterial qualities, which makes it a valuable natural preservative or antimicrobial agent. Lignin also demonstrates emulsifying, film-forming, and binding capabilities, which contribute to its functional roles in packaging materials, drug delivery systems, and as a biostimulant in agriculture. These properties, combined with its biocompatibility and biodegradability, underscore lignin's potential in developing sustainable bioactive products.

9.4 ANTIMICROBIAL PROPERTIES OF LIGNIN AND ITS DERIVATIVES

9.4.1 Evidence of Antimicrobial Efficacy (Bacterial, Fungal, Viral)

Lignin and its derivatives have demonstrated significant antimicrobial efficacy against a broad spectrum of bacterial, fungal, and even some viral pathogens, largely attributed to their phenolic structure and ROS modulating abilities. Recent research has shown that lignin, a natural polymer found in plant cell walls, can harm bacteria by disrupting their cell membranes and interfering with their internal functions (Spagnuolo et al., 2021). This means that lignin can effectively inhibit both types of bacteria: Gram-positive and Gram-negative. For example, studies have highlighted its powerful antibacterial effects against well-known microbes such as *P. aeruginosa*, *S. aureus*, and *E. coli* (Zhou et al., 2024). Not only does lignin show promise against bacteria, but it also combats fungi. It has been found to reduce the growth of common plant diseases, like *A. niger*, *F. oxysporum*, and *B. cinerea*, by inducing stress within the organisms and disrupting their structure (Tripodi et al., 2018). Moreover, lignin-based materials can form protective barriers that prevent fungi from spreading. Interestingly, recent studies suggest that lignin might even be effective against viruses. Certain modified forms of lignin, particularly sulfonated lignins, have demonstrated the ability to inhibit harmful viruses such as HIV and the herpes simplex virus. This happens because they can bind to viral surface proteins, blocking the viruses from entering host cells (Fukuchi et al., 2021).

9.4.2 Mechanisms of Lignin's Antimicrobial Action (*e.g.*, Oxidative Stress, Membrane Disruption)

Lignin's antimicrobial effect is mediated by several interconnected processes, the most important of which are oxidative stress induction, membrane rupture, and interference with cellular metabolism. One of the key mechanisms is the production of ROS, particularly by lignin nanoparticles or chemically changed lignin (e.g., sulfonated or oxidized forms), which causes oxidative stress in microbial cells. Elevated ROS levels damage proteins, lipids, and nucleic acids, reducing microbial viability (Zhou et al., 2024). Membrane disruption is another significant mechanism. Lignin's aromatic and amphiphilic properties allow it to interact with microbial membranes, increasing permeability or directly rupturing them. Both bacterial and fungal cells die when internal contents leak out and membrane potential is depleted (Spagnuolo et al., 2021). In addition to affecting cell activity, lignin's phenolic hydroxyl groups can connect to membrane-bound transport proteins and enzymes. Furthermore, by binding to cellular macromolecules or chelating essential metal ions, lignin impairs microbial metabolic pathways and enzymatic function, inhibiting energy production and replication. In the case of viruses, sulfonated lignin derivatives have been shown to inhibit viral attachment and penetration into host cells by binding to viral envelope glycoproteins (Fukuchi et al., 2021). Because of its multiple modes of action, lignin is a strong and broad-spectrum antimicrobial agent that is less likely to develop resistance than single-target antibiotics.

9.4.3 Influence of lignins in structural modification on antimicrobial activity

The antibacterial activity of lignin can be significantly enhanced or modified by structural alterations such as sulfonation, acetylation, oxidation, and nanoparticle formation. The molecular weight, surface charge, solubility, and availability of functional groups of lignin are all impacted by these physical or chemical changes, and these factors are crucial in controlling how lignin interacts with microbial cells. For example, sulfonation increases the lignin polymer's solubility in water and gives it a strong negative charge that enhances its ability to bind to positively charged areas on bacterial cell walls or viral glycoproteins. Sulfonated lignins have been shown to be more successful in preventing bacterial growth and viral attachment and entry (Fukuchi et al., 2021). Lignin oxidation can increase the ability to produce reactive phenolic and carbonyl groups and the antibacterial effect mediated by oxidative stress (Zhou et al., 2024). In a similar vein, acetylation alters hydrophobicity and may affect membrane interactions by lowering the number of free hydroxyl groups. The conversion of lignin into LNPs greatly increases surface area and enhances interaction with microbial cells. LNPs enable improved ROS release, improved membrane penetration, and more efficient antibacterial function delivery. Studies show that the antibacterial and antifungal capabilities of lignin nanoparticles are significantly stronger than those of their bulk counterparts (Cavallo et al., 2020). Overall, these recent investigations indicate that structural changes enhance lignin's antimicrobial effectiveness and customize its physicochemical characteristics for certain uses, making modified lignin a potential material for coatings, packaging, biocontrol agents, and biomedical formulations.

9.5 LIGNIN-DERIVED NANOMATERIALS FOR ANTIMICROBIAL APPLICATIONS

9.5.1 Lignin nanoparticles (LNPs): synthesis and functionalization

LNPs are a promising nanoformulation of lignin that boosts its surface area, reactivity, and antibacterial potential. LNPs are frequently produced via top-down techniques like ultrasonication and milling or bottom-up techniques like dialysis, anti-solvent precipitation, or self-assembly in water-organic solvent systems (Figueiredo et al., 2018). These methods can be used to manipulate important biological activity-determining parameters such as particle size, dispersion, and surface charge. LNPs can be further functionalized with cationic agents, metal nanoparticles (such as Ag and ZnO), or hydrophobic groups to improve their antibacterial activity, dispersion, or target specificity (Chai et al., 2021). Functionalization may enhance interactions with microbial membranes, promote the generation of ROS, and have synergistic effects when combined with other antimicrobial components.

9.5.2 Antimicrobial Coatings and Films Using Lignin Nanocomposites

The development of antimicrobial coatings and films has shown interest in lignin nanocomposites, specifically for use in food packaging, medical devices, and agricultural films. When LNPs are added to biopolymers like polylactic acid (PLA), chitosan, or starch matrices, they impart antimicrobial, UV-blocking, and antioxidant properties (Cavallo et al., 2020). These films effectively prevent microorganisms from growing on surfaces by releasing ROS or antibacterial chemicals at the contact point gradually. These films have demonstrated notable suppression of S. aureus and E. coli in food packaging, increasing shelf life and lowering contamination. Furthermore, lignin-based coatings are appropriate for medical textiles and wound dressings because they can reduce microbial adherence and create a moisture barrier (Li et al., 2019).

9.5.3 Synergistic Formulations with Other Antimicrobial Agents

Lignin and its nanoparticles can be used in combination with other antimicrobial agents to boost efficacy and reduce the dosage of conventional biocides. For instance, LNPs combined with silver nanoparticles (AgNPs), essential oils, or antibiotics (like ciprofloxacin) have a stronger antibacterial effect than either agent alone due to their enhanced membrane penetration and extended release (Park & Kim, 2019). This synergy results from binding interactions that enhance the secondary agent's cellular absorption as well as lignin's capacity to generate ROS. Additionally, LNPs might act as transporters, improving the stability and bioavailability of hydrophobic antimicrobial compounds in aqueous environments, thereby expanding their applications in medicine and agriculture.

9.6 THERAPEUTIC POTENTIALS BEYOND ANTIMICROBIAL ACTIVITY

9.6.1 Antioxidant, Anti-Inflammatory, and Anticancer Properties of Lignins

Because of its high phenolic hydroxyl group content, which can scavenge free radicals and prevent oxidative damage in biological systems, lignin has significant antioxidant, anti-inflammatory, and anticancer properties. In certain formulations, the antioxidant activity is on par with well-known antioxidants like tocopherol or BHT (Yang et al., 2025). Since oxidative stress is a major trigger for inflammatory pathways, this trait is crucial for reducing inflammation. It has been shown that lignin and lignin-derived nanoparticles inhibit pro-inflammatory cytokines like TNF-α, IL-6, and IL-1β by potentially changing NF-κB signaling (Rehman et al., 2021). Furthermore, lignin could be employed as a drug or as a nanocarrier to fight cancer. It has inherent cytotoxicity against a range of cancer cell lines, including those from the breast, colon, and liver, because it can induce oxidative stress and apoptosis in tumor cells while protecting healthy cells. Furthermore, according to Figueiredo et al. (2018), lignin has a synergistic cytotoxic impact that can improve the effectiveness and administration of chemotherapy medications like doxorubicin.

9.6.2 Drug delivery systems: lignin-based carriers and release profiles

Lignin's biodegradability, biocompatibility, abundance of functional groups, and capacity to form conjugates or nanoparticles make it a desirable material for drug delivery systems. Using hydrogen bonds, π–π interactions, or electrostatic forces, LNPs, lignosulfonates, or lignin hydrogels can encapsulate hydrophilic and hydrophobic medications (Figueiredo et al., 2018). Lignin-based carriers are appropriate for targeted release in the gastrointestinal tract or tumor microenvironments because they offer controlled and sustained drug release patterns that are frequently pH-responsive. LNPs loaded with anticancer medications, for instance, exhibit improved solubility, stability, and therapeutic efficiency while lowering systemic toxicity (Yang et al., 2025). The molecular structure of lignin can be changed to alter the release rate, or it can be co-formulated with polymers such as PEG, PLA, or chitosan.

9.6.3 Lignin as a scaffold material in tissue engineering

Lignin has emerged as a promising scaffold material for tissue engineering due to its mechanical strength, antioxidant properties, and compatibility with bioactive molecules. It enhances the mechanical integrity, cell adhesion, and ROS-scavenging capacity of scaffolds used in bone, cartilage, and skin tissue regeneration when added to biocomposites (such as lignin–gelatin and lignin–PLA) (Mateus et al., 2024). Additionally, lignin's phenolic groups can promote angiogenesis and cell proliferation, both of which are essential for tissue healing. Electrospun nanofibers or scaffolds based on 3D-printed lignin have demonstrated potential in bone regeneration and wound healing, particularly when paired with growth hormones or bioactive ceramics. Furthermore, the natural antibacterial and anti-inflammatory qualities of lignin aid in preventing infection and accelerating the healing of injured tissues.

9.7 APPLICATION IN MEDICAL AND CONSUMER PRODUCTS

9.7.1 Wound dressings, dental materials, and implants

Lignin-based materials have demonstrated significant promise in wound care, dental applications, and biomedical implants due to their antioxidant, antimicrobial, and anti-inflammatory properties. By scavenging ROS, lowering local inflammation, and preventing microbial infections, LNPs or lignin-containing hydrogels in wound dressings speed up healing. Additionally, lignin enhances the mechanical strength and moisture retention of dressings, thereby fostering the optimal healing environment (Icriverzi et al., 2019). In order to improve mechanical stability and reduce bacterial colonization, particularly against *Streptococcus mutans*, a primary cause of tooth decay, lignin is being added to dental cements and resin composites. The antioxidant qualities of lignin reduce oxidative stress and post-operative inflammation in oral tissues (Figueiredo et al., 2018). Lignin composites have been investigated for bone scaffolding in implants when paired with bioceramics like hydroxyapatite. These materials enhance bioactivity, reduce oxidative damage, and promote cell adhesion and proliferation.

9.7.2 Use of lignins in personal care and hygiene products

Lignin is becoming a popular ingredient in personal care and hygiene products like lotions, deodorants, sunscreens, and even sanitary items. People are drawn to it because it comes from natural sources, breaks down easily, and has some great benefits. For one, it has strong UV-absorbing properties thanks to its aromatic and phenolic structures, making it a sustainable option compared to synthetic UV filters in sunscreens (Yang et al., 2025; Spagnuolo et al., 2021). Its antioxidant and antibacterial qualities help protect the skin from environmental damage and can even extend the shelf life of products (Cavallo et al., 2020). In sanitary products like diapers and feminine hygiene items, lignin is often used in antimicrobial coatings or as a bio-based superabsorbent in the absorbent cores to help reduce odor and prevent infections (Icriverzi et al., 2019). With its low toxicity, skin-friendly nature, and ability to trap odors and microorganisms, it's an excellent choice for these uses.

9.7.3 Lignins as packaging materials with bioactive properties

Lignin integration into biodegradable packaging materials provides active protection against microbiological contamination and oxidative degradation, particularly in food and pharmaceutical packaging. Combining lignin-based films with polymers such as cellulose, starch, or PLA results in antibacterial, antioxidant, and UV-protective properties that improve food safety and product shelf life (Spagnuolo et al., 2021). These films of lignin-containing nanocomposites can either act as microbial barriers or gradually release bioactive chemicals. Furthermore, lignin is more environmentally friendly and compostable than standard petroleum-based plastics, and it improves packaging's mechanical strength and moisture resistance.

9.8 CHALLENGES AND OPPORTUNITIES

9.8.1 Biocompatibility and toxicity concerns

While lignin comes from a natural source, there are still some big issues with its toxicity and biocompatibility that need to be worked out before it can be fully used in clinical and consumer products. Generally speaking, when lignin is isolated using gentle methods, it's considered biocompatible. But technical lignins, which are produced through harsher industrial processes like kraft or organosolv methods, can lead to unpredictable biological reactions since they might still have leftover chemicals or vary in molecular structure (Figueiredo et al., 2018). Toxicology studies show that lignin's cytotoxicity depends on the dose and is influenced by things like functionalization, surface charge, and particle size. Even though LNPs usually demonstrate low toxicity in both lab settings and living organisms (Yang et al., 2025), there isn't much information available on long-term exposure or how they behave systemically. So, it's critical to focus on material purification, surface modifications, and thorough preclinical evaluations to make sure they're safe for use in biomedical and cosmetic applications.

9.8.2 Standardization and scalability of lignin-based products

One of the most major technical obstacles in lignin usage is a lack of standardization. The composition of lignin, a highly heterogeneous polymer, changes substantially depending on the extraction method (kraft, soda, organosolv, enzymatic) and botanical source. This unpredictability has an impact on biological properties, chemical reactivity, and solubility, making it difficult to generate reproducible products (Ragauskas et al., 2014; Li & Ragauskas, 2012). To keep quality in check, both in labs and in industry, we really need to establish clear extraction and purification methods and solid ways to characterize lignin (like FTIR, NMR, and GPC) (Modi et al., 2024). Yet another obstacle is scalability. Although lignin is widely available (as a byproduct of the pulp and bioethanol industries), there are still technological and financial obstacles to its conversion into high-value uses like packaging, medicines, or nanomaterials. To scale up lignin valorization, efficient, economical, and environmentally friendly processing technologies (such as green solvents, mechanochemistry, and bio-refinery integration) are required.

9.8.3 Regulatory and commercialization aspects

From a regulatory perspective, lignin-based products are still emerging, and their approval processes are not yet well-defined. For lignin-derived nanomaterials and bioactives, regulatory bodies like the FDA or EMA need thorough toxicological, pharmacokinetic, and environmental safety data, all of which are frequently absent (Ullah et al., 2022). Additionally, the classification of lignin (as an active component, polymer, or excipient) can change depending on the use, making the registration process more difficult (Mukheja et al., 2023). Commercial interest in lignin is increasing, as evidenced by the numerous startups and businesses (such as Lignin Industries, Stora Enso, and Domtar) creating lignin-based goods for medicine delivery, resins, cosmetics, and bioplastics (Jassal et al., 2022). However, a lack of regulatory frameworks, uneven supplier quality, and little consumer awareness hinder market penetration. Coordinated efforts in interdisciplinary research, public-private partnerships, and policymaking will be necessary to overcome these obstacles and establish lignin as a sustainable substance that can be used in consumer and medical markets.

9.9 FUTURE DIRECTIONS

9.9.1 Advances in lignin functionalization and hybrid systems

It is anticipated that further developments in lignin functionalization would significantly increase its bioactivity and adaptability. For improved antibacterial, antioxidant, or drug-binding qualities, emerging techniques include precision chemical alterations (such as click chemistry, enzymatic tailoring, and graft copolymerization) to introduce specific functional groups like amines, carboxyls, or sulfonates (Yang et al., 2025). Furthermore, in order to develop multifunctional nanocomposites for use in wound healing, tissue regeneration, biosensing, and smart packaging, hybrid systems combining lignin with other biopolymers (such as chitosan, cellulose, and silk fibroin) or inorganic materials (such as AgNPs, hydroxyapatite, and ZnO) are being investigated (Rehman et al., 2021). Synergistic benefits like enhanced mechanical strength, regulated medication release, and broad-spectrum antibacterial activity are made possible by these hybrid platforms.

9.9.2 Integration with Synthetic Biology and Green Chemistry

Green chemistry and synthetic biology together have the potential to revolutionize lignin value. By optimizing the monomer composition and linking patterns for downstream applications, synthetic biology allows for the engineered production of customized lignin in microorganisms or plants, lowering the need for post-processing (Pesquet et al., 2019; Lizundia et al., 2020). For instance, "designer lignins" with a lower molecular weight or particular chemical functions that are perfect for use in biomedicine or as materials could be produced by genetically altering the lignin production routes in crops. By reducing the use of hazardous chemicals and energy, green chemistry principles direct the environmentally sustainable extraction, modification, and processing of lignin (Yang et al., 2025). Enzymatic catalysis, ionic liquids, and solvent-free techniques are being developed to create high-purity lignin with little harm to the environment. It is anticipated that these multidisciplinary methods will greatly improve the sustainability, repeatability, and scalability of lignin.

9.9.3 Prospective Impact on Global Healthcare and Sustainability

Lignin's integration into biomedical and consumer products could have a transformational impact on global healthcare and sustainable development. Lignin is a bio-based, biodegradable, and non-toxic polymer that supports the objectives of decreasing plastic waste, minimizing the negative effects of medications, and increasing access to reasonably priced medical technology. Lignin-based antimicrobial coatings, wound dressings, and medication carriers may provide affordable substitutes for synthetic materials and antibiotics in low-resource environments, reducing problems such as supply chain fragility and antimicrobial resistance (AMR). Additionally, by turning industrial waste, such as that from the production of paper or biofuel, into high-value materials, lignin valorization promotes the circular bioeconomy, increasing resource efficiency and lowering carbon footprints (Ragauskas et al., 2014). Lignin has the potential to be a key component of the global advancement of green industry models and sustainable healthcare systems if these breakthroughs are backed by international regulations, public-private partnerships, and equitable technology transfer.

9.10 CONCLUSION

Lignin, a naturally abundant and structurally versatile biopolymer, has emerged as a promising material for a wide range of therapeutic and antimicrobial applications. Its intrinsic antioxidant, antimicrobial, anti-inflammatory and anticancer properties, largely attributed to its phenolic-rich structure, enable it to act both as a bioactive agent and a functional component in advanced biomaterials. The effectiveness of lignin has been further increased by structural alterations and nanoscale formulations, such as LNPs and hybrid composites, opening up uses in active packaging, wound dressings, medication delivery systems, and personal care items. Despite the tremendous advancements, issues with toxicity, standardization, biocompatibility, and regulatory approval still need to be resolved. But developments in materials science, synthetic biology, and green chemistry are opening the door to a more regulated, sustainable, and scalable usage of lignin. In light of global health issues including AMR, material sustainability, and the need for environmentally friendly biomedical solutions, lignin is ideally positioned to play a revolutionary role in upcoming therapeutic and antimicrobial technologies. Continued interdisciplinary research and strategic industrial translation will be key to unlocking lignin's full potential in building the next generation of bio-based, safe, and effective healthcare technologies.

ACKNOWLEDGMENT

The authors acknowledge the infrastructural support provided by DAV University and Luleå University of Technology administration to carry out the present work.

REFERENCES

Alaoui, C. H., G. Réthoré, P. Weiss, and A. Fatimi. 2023. Sustainable Biomass Lignin-based Hydrogels: A Review on Properties, Formulation, and Biomedical Applications. *International Journal of Molecular Sciences* 24 (17): 13493.

Billu, V. L., K. B. Syamnath, P. M. Arun, et al. 2023. Lignin-based Nanomaterials for Food and Pharmaceutical Applications: Recent Trends and Future Outlook. *The Science of the Total Environment* 881: 163316. doi:10.1016/j.scitotenv.

Boerjan, W., J. Ralph, and M. Baucher. 2003. Lignin Biosynthesis. *Annual Review of Plant Biology* 54: 519–46. doi:10.1146/annurev.arplant.54.031902.134938.

Cavallo, E., X. He, F. Luzi, et al. 2020. UV Protective, Antioxidant, Antibacterial and Compostable Polylactic Acid Composites Containing Pristine and Chemically Modified Lignin Nanoparticles. *Molecules (Basel, Switzerland)* 26 (1): 126. doi:10.3390/molecules26010126.

Chai, Y., Y. Wang, B. Li, W. Qi, R. Su, and Z. He. 2021. Microfluidic Synthesis of Lignin/Chitosan Nanoparticles for the pH-Responsive Delivery of Anticancer Drugs. *Langmuir* 37 (23): 7219–26.

Chandra, D. K., A. Kumar, and C. Mahapatra. 2025. In Silico and In Vitro Analysis of pH-Responsive SLS Delivery via CeBTC-Modified CNT Nanocarriers: Molecular Dynamic Simulation to Experimental Assessment. *Langmuir* 41 (46): 30994–33. doi: 10.1021/acs.langmuir.5c03023.

Chaudhary, M., and V. R. Sinha. 2023. Lignin-based Carriers for Drug Delivery Applications: From an Industrial Waste to a Pharmaceutical Aid. *European Polymer Journal* 195 (112206): 112206. doi:10.1016/j.eurpolymj.2023.112206.

Chowdhury, Z. Z., S. B. Abd Hamid, M. M. Rahman, and R. F. Rafique. 2016. Catalytic Activation and Application of Micro-spherical Carbon Derived from Hydrothermal Carbonization of Lignocellulosic Biomass: Statistical Analysis using Box–Behnken Design. *RSC Advance* 6 (104): 102680–94.

Devu, C., S. Sreelakshmi, R. Chandana, et al. 2024. Recent Progress in Tannin and Lignin Blended Metal Oxides and Metal Sulfides as Smart Materials for Electrochemical Sensor Applications. *Analytical Sciences: The International Journal of the Japan Society for Analytical Chemistry* 40 (6): 981–96. doi:10.1007/s44211-024-00544-4.

Figueiredo, P., K. Lintinen, J. T. Hirvonen, M. A. Kostiainen, and H. A. Santos. 2018. Properties and Chemical Modifications of Lignin: Towards Lignin-based Nanomaterials for Biomedical Applications. *Progress in Materials Science* 93: 233–69. doi:10.1016/j.pmatsci.2017.12.001.

Figueiredo, P., K. Lintinen, A. Kiriazis, V. Hynninen, et al. 2017. In Vitro Evaluation of Biodegradable Lignin-based Nanoparticles for Drug Delivery and Enhanced Antiproliferation Effect in Cancer Cells. *Biomaterials* 121: 97–108.

Fukuchi, K., T. Koshikawa, D. Asai, et al. 2021. Lignosulfonate Rapidly Inactivates Human Immunodeficiency and Herpes Simplex Viruses. *Medicines (Basel, Switzerland)* 8 (10): 56. doi:10.3390/medicines8100056.

Gilca, I. A., V. I. Popa, and C. Crestini. 2015. Obtaining Lignin Nanoparticles by Sonication. *Ultrasonics Sonochemistry* 23: 369–75.

Icriverzi, M., A. Bonciu, L. Rusen, et al. 2019. Human Mesenchymal Stem Cell Response to Lactoferrin-based Composite Coatings. *Materials* 12 (20): 3414. doi:10.3390/ma12203414.

Iravani, S. 2020. Biomedical Applications of Lignin-Based Nanoparticles. In: Shukla, A. (eds) *Nanoparticles and Their Biomedical Applications*. Springer, Singapore. https://doi.org/10.1007/978-981-15-0391-7_8

Jassal, V., C. Dou, N. Sun, S. Singh, B. Simmons, and H. Choudhary. 2022. Finding Values in Lignin: A Promising Yet Under-Utilized Component of the Lignocellulosic Biomass. *Frontiers in Chemical Engineering* 4, 1059305.

Jiang, B., H. Jiao, X. Guo, et al. 2023. Lignin-based Materials for Additive Manufacturing: Chemistry, Processing, Structures, Properties, and Applications. *Advanced Science (Weinheim, Baden-Wurttemberg, Germany)* 10 (9): e2206055. doi:10.1002/advs.202206055.

Kai, D., M. J. Tan, P. L. Chee, et al. 2016. Towards Lignin-based Functional Materials in a Sustainable World. *Green Chemistry: An International Journal and Green Chemistry Resource: GC* 18 (5): 1175–1200. doi:10.1039/c5gc02616d.

Khadem, E., M. Ghafarzadeh, M. Kharaziha, F. Sun, and X. Zhang. 2024. Lignin Derivatives-based Hydrogels for Biomedical Applications. *International Journal of Biological Macromolecules* 261: 129877.

Lee, J. H., J. S. Im, X. Jin, T. M. Kim, J. W. Choi. 2022. In Vitro and in Vivo Evaluation of Drug-Encapsulated Lignin Nanoparticles for Release Control. *ACS Sustainable Chemistry & Engineering* 10 (18): 5792–02. doi:10.1021/acssuschemeng.1c08529.

Li, J., W. J. Peng, Z. J. Fu, et al. 2019. Achieving High Electrical Conductivity and Excellent Electromagnetic Interference Shielding in Poly (Lactic Acid)/Silver Nanocomposites by Constructing Large-Area Silver Nanoplates in Polymer Matrix. *Composites Part B, Engineering* 171: 204–13. doi:10.1016/j.compositesb.2019.05.003.

Li, M. F., and A. J. Ragauskas. 2012. Chemical Composition and Structural Characterization of Lignin from Different Sources for Biofuels and Biomaterials Production. *BioResources* 7 (1): 1051–67.

Lizundia, E., I. Armentano, F. Luzi, et al. 2020. Synergic Effect of Nanolignin and Metal Oxide Nanoparticles into Poly (l-Lactide) Bionanocomposites: Material Properties, Antioxidant Activity, and Antibacterial Performance. *ACS Applied Bio Materials* 3 (8): 5263–74.

Mateus, S., L. A. Letícia, F. Barbosa, M. M. Larissa, M. D. Mattos, and V. Pereira. 2024. Curcumin-Loaded Lignin Nanoparticles: Exploring Sonication Effects on Drug Loading and Particle Properties for Future Biomedical Use. *Journal of Materials Research and Technology* 33: 5149–59.

Medina-Moreno, A., M. M. El-Hammadi, G. I. Martínez-Soler, J. G. Ramos, G. García-García, and J. L. Arias. 2025. Magnetic and pH-Responsive Magnetite/Chitosan (Core/Shell) Nanoparticles for Dual-Targeted Methotrexate Delivery in Cancer Therapy. *Drug Delivery and Translational Research* 15 (5): 1646–59. doi:10.1007/s13346-024-01701-y.

Modi, U. D., V. K. Ghodadara, S. M. Mitchla, and M. Lakdawala. 2024. Advanced Techniques in the Purification of Lignin: Challenges, Methods, and Industrial Applications. *International Journal of All Research Education and Scientific Methods* 12 (8): 1332–39.

Mukheja, Y., J. Kaur, K. Pathania, et al. 2023. Recent Advances in Pharmaceutical and Biotechnological Applications of Lignin-based Materials. *International Journal of Biological Macromolecules* 241 (124601): 124601. doi:10.1016/j.ijbiomac.2023.124601.

Nan, N., W. Hu, and J. X. Wang, 2022. Lignin-Based Porous Biomaterials for Medical and Pharmaceutical Applications. *Biomedicines* 10 (4): 747. doi:10.3390/biomedicines10040747.

Park, J., and B. Kim. 2019. An Analysis of South Korea's Energy Transition Policy with Regards to Offshore Wind Power Development. *Renewable and Sustainable Energy Reviews* 109: 71–84. doi:10.1016/j.rser.2019.04.031.

Pesquet, E., A. Wagner, and J. H. Grabber. 2019. Cell Culture Systems: Invaluable Tools to Investigate Lignin Formation and Cell Wall Properties. *Current Opinion in Biotechnology* 56: 215–22. doi:10.1016/j.copbio.2019.02.001.

Ragauskas, A. J., G. T. Beckham, M. J. Biddy, et al. 2014. Lignin Valorization: Improving Lignin Processing in the Biorefinery. *Science (New York, N.Y.)* 344 (6185): 1246843. doi:10.1126/science.1246843.

Rehman, A., Z. Peng, H. Li, et al. 2021. Genome Wide Analysis of IQD Gene Family in Diploid and Tetraploid Species of Cotton (Gossypium Spp.). *International Journal of Biological Macromolecules* 184: 1035–61. doi:10.1016/j.ijbiomac.2021.06.115.

Spagnuolo, M., M. Pagliaro, and F. Meneguzzo. 2021. Lignin: A Natural Biopolymer with Antibacterial and Antioxidant Properties. *Applied Sciences* 11 (4): 112933.

Tang, Q., Y. Qian, D. Yang, X. Qiu, Y. Qin, and M. Zhou. 2020. Lignin-based Nanoparticles: A Review on Their Preparations and Applications. *Polymers* 12 (11): 2471. doi:10.3390/polym12112471.

Thakur, V. K., M. K. Thakur, P. Raghavan, and M. R. Kessler. 2014. Progress in Green Polymer Composites from Lignin for Multifunctional Applications: A Review. *ACS Sustainable Chemistry & Engineering* 2 (5): 1072–92. doi:10.1021/sc500087z.

Tripodi, A., E. Bahadori, D. Cespi, et al. 2018. Acetonitrile from Bioethanol Ammoxidation: Process Design from the Grass-Roots and Life Cycle Analysis. *ACS Sustainable Chemistry & Engineering* 6 (4): 5441–51. doi:10.1021/acssuschemeng.8b00215.

Ulbrich, W, and A. Lamprecht. 2010. Targeted Drug-Delivery Approaches by Nanoparticulate Carriers in the Therapy of Inflammatory Diseases. *Journal of the Royal Society, Interface* 7 Suppl 1 (suppl_1): S55–66. doi:10.1098/rsif.2009.0285.focus.

Ullah, I., Z. Chen, Y. Xiea, et al. 2022. Recent Advances in Biological Activities of Lignin and Emerging Biomedical Applications: A Short Review. *International Journal of Biological Macromolecules* 208: 819–32. doi:10.1016/j.ijbiomac.2022.03.182.

Xia, T., Y. Guan, M. Yang, et al. 2014. Synthesis of Polyethylenimine Modified Fe 3 O 4 Nanoparticles with Immobilized Cu 2^{+} for Highly Efficient Proteins Adsorption. *Colloids and Surfaces. A, Physicochemical and Engineering Aspects* 443: 552–59. doi:10.1016/j.colsurfa.2013.12.026.

Yang, L., Y. Jing, L. Zhou, et al. 2025. Synthesis of Viral-like Lignin/Ag Nanoparticles with Spiky Surfaces for Antibacterial and Antioxidant Applications. *International Journal of Biological Macromolecules* 305 (Pt 1): 141034. doi:10.1016/j.ijbiomac.2025.141034.

Zhen, G., X. Lu, G. Kumar, P. Bakonyi, K. Xu, and Y. Zhao. 2017. Microbial Electrolysis Cell Platform for Simultaneous Waste Biorefinery and Clean Electrofuels Generation: Current Situation, Challenges and Future Perspectives. *Progress in Energy and Combustion Science* 63: 119–45. doi:10.1016/j.pecs.2017.07.003.

Zhou, M., O. A. Fakayode, M. Ren, H. Li, J. Liang, and C. Zhou. 2024. Green and Sustainable Extraction of Lignin by Deep Eutectic Solvent, Its Antioxidant Activity, and Applications in the Food Industry. *Critical Reviews in Food Science and Nutrition* 64 (20): 7201–19. doi: 10.1080/10408398.2023.2181762.

Zhou, Z., J. Wang, X. Xu, et al. 2024. Lignin-based Nanoparticles for Combination of Tumor Oxidative Stress Amplification and Reactive Oxygen Species Responsive Drug Release. *Bioconjugate Chemistry* 35 (8): 1207–17. doi:10.1021/acs.bioconjchem.4c00261.

10 Smart nano-biomaterials from lignin for packaging and cosmetics

Meenu Thakur, Ambika Raita, Kirti Thakur, and Suraj Kumar

10.1 INTRODUCTION

Lignocellulosic biomass holds the potential to be used for the synthesis of polymeric biomaterials, high-value chemicals, and renewable sources of fuel and thus has become increasingly popular in current research (Saleem et al., 2025). Out of all types of green biomass, lignin, a waste from the paper and pulp industry, comes in second place after cellulose in abundance. Among all types of lignin-based materials, lignin nanoparticles (LNPs) act as renewable and cost-effective resources that can be easily degraded by the fungi and bacteria. Thus, lignin-based nano-biosensors have been the critical area of research due to high surface area, real-time sensing, and less cost involved (Papa et al., 2025).

Almost 70 million tons of lignin are recovered annually through the pulp manufacturing process (Salehian et al., 2013). The percentage of lignin depends upon the type of wood used, and the range varies from 34% of lignin in the case of pine, which is softwood, to 24% in the case of eucalyptus, which is hardwood, to 11% in the case of gramineous plants (Chauhan, 2020). There are various chemical groups, such as aldehyde, methoxy, phenolic, and aliphatic hydroxy groups, owing to its nontoxic, antioxidant, and antimicrobial properties (Zhu et al., 2024). Lignin-based smart nano-biosensors are an emerging option for cosmetics and packaging (Figure 10.1) to detect specific analytes, freshness of the food, and potential contaminants (Moreno et al., 2020). In the present chapter, all the lignin-based applications in packaging and cosmetics have been discussed.

10.2 LIGNIN: A RENEWABLE AND FUNCTIONAL BIOPOLYMER

Development of biomass-based renewables is the most appealing alternative for smart materials capable of responding to the external stimuli. They not only reduce the greenhouse emissions due to the use of synthetic polymer but can also be implemented in the applications for sensing, biomedical, and smart nanomaterials (Moreno et al., 2020; Hammitt et al., 1992; Rathour et al., 2023). Lignin has been considered the second most abundant lignocellulosic waste after cellulose, which is

DOI: 10.1201/9781003653189-10

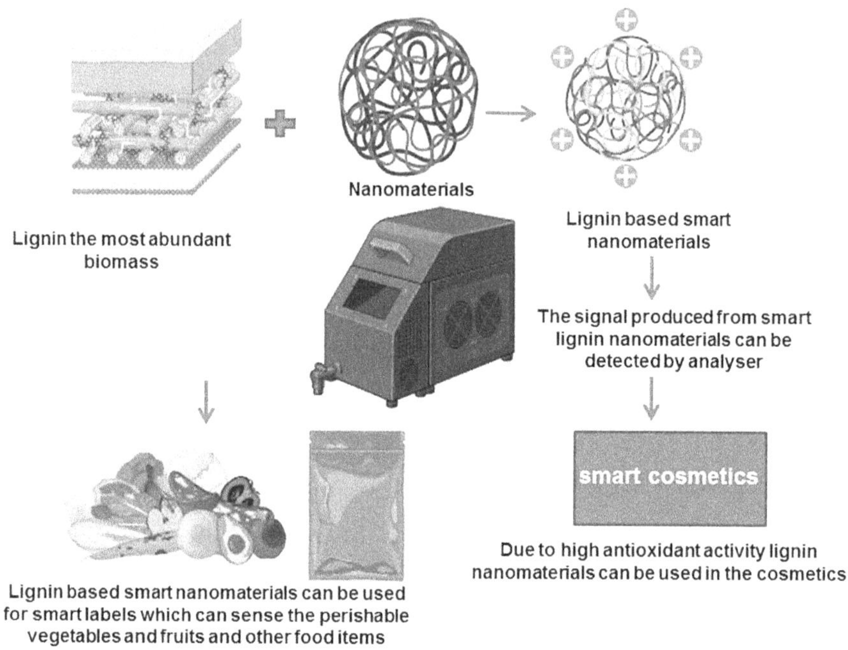

FIGURE 10.1 Use of lignin nanomaterials for smart biosensing in packaging and cosmetics

a very low-cost byproduct produced as the waste in the biochar and lignocellulosic biorefineries (Habibi et al., 2010; Lavoine et al., 2012; Rathour et al., 2025). In previous years many studies have focused on the upgrade of lignin toward advanced engineering materials (Ragauskas et al., 2014). Lignin is considered an aromatic polymer with amazing properties such as high carbon content, thermal stability, biodegradability, and antioxidant activity with antimicrobial properties depending upon the procedure of extraction, as it can change with the extraction conditions followed (Yang et al., 2018; Larrañeta et al., 2018). Generating smart biomaterials using lignin is now an emerging field with value-added applications. This new class of smart nanomaterials is able to convert the stimuli to defined outputs and even change the functional property, which requires the modification or insertion of new functional groups that can respond to the external stimuli. This property can lead to the implication of these lignin-based smart materials in sensing and bio-sensing applications, controlled encapsulation with their role in biomedical systems, and shape memory-based programmable materials. Lignin polymers, along with oligomers, dimers, and monomers, can be obtained from plant biomass and can be converted into the smart nanobiomaterial by conjugating with the nanomaterials. The major source of lignin is the pulp and paper industry. Lignin can be extracted from cellulosic fibers by conventional thermochemical treatments, resulting in the depolymerization and solubilization of the lignin. There are various forms of lignins for material applications, such as lignosulfonates (LS), kraft lignin (KL), organosolv lignin (OSL), and soda lignin (SL), and other types of lignin greatly influence the application of lignin

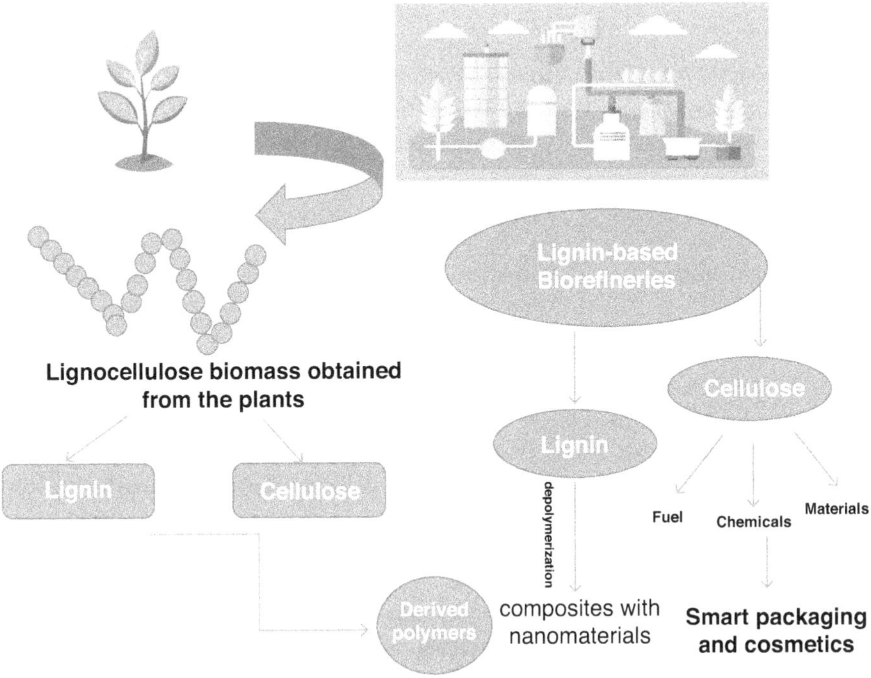

FIGURE 10.2 Lignin as a multifunctional polymer with diverse applications

biomaterials (Galletti & Antonetti, 2012). Lignin can be further modified using the chemical treatments targeting its functional groups, such as hydroxyl, methoxyl, carbonyl, or carboxyl groups, through hydroxyalkylation, esterification, and amination, and various semi-synthetic routes have been used to form application-based lignin polymers (Huang et al., 2019; Zieglowski et al., 2019). The grafting technique has been further used to modify lignin copolymers where lignin has been used as a macromonomer. One of the biggest challenges is the recalcitrant lignin resulting from the pulping process. Catalytic fermentation is one of the conceits for getting higher yield and selective lignin materials for the aromatic compounds (Li et al., 2020). Lignin-based carbon quantum dots (LCQDs) have emerged as a novel class of materials for applications in biosensing applications (Zhang et al., 2019). It works on the quenching of fluorescence emission in the presence of some analytes, and lignin nanoparticles provide the low cost and functional properties. In the present chapter lignin-based smart nanomaterials have been discussed with their applications in packaging and cosmetics (Figure 10.2).

10.3 NANOTECHNOLOGY AND SMART BIOSENSING

10.3.1 Definition of Smart Biosensing

Biosensing is a traditional concept that occurs in many life forms for their survival. For example, living organisms, from prokaryotes to complex eukaryotes, have the

TABLE 10.1
Carbon-based nanomaterials with biosensing applications

Sr. No	Carbon-based Nano- Material	Biosensor Type	Application
	Graphene quantum dot	Nano-biosensor	Use to detect ds DNA
	Carbon quantum dots	Nano-biosensor	Detection of gene associated with gastric cancer.
	Carbon nanodots	Electrochemical nanosensor	Used for detection of uric acid and creatine.

ability to sense their environment and surroundings. The advancement of nanotechnology revolutionized the concept of biosensing; these nanotechnology-enabled biosensors are called smart nano-biosensors (Banerjee et al., 2021). Smart biosensing is an advanced technique to detect biological indicators using a transducing system. The transducing system converts these biological changes into readable forms such as optical, thermal, electrical, and electrochemical signals. Biosensing is used to identify small molecules like glucose and uric acid and large biomolecules like DNA, RNA, or even whole cells. Smart materials used in biosensing are special types of materials that respond to external conditions like temperature, pH, light, electricity, or even biological substances. Smart nanomaterials such as nanoparticles (gold nanoparticles), quantum dots, and carbon-based materials like graphene are also used. These materials are more effective, as they detect target molecules more effectively. The incorporation of smart material in biosensors is a big step forward in making smart biosensors (Guo et al., 2020). Various bio-based nanomaterials in smart biosensing applications, including:

1. **Plant-based materials** like lignin, cellulose, and hemicellulose. They can detect pathogens, hormones, biomarkers, DNA, RNA, etc.
2. **Carbon-based** materials like graphene, carbon nanotubes, nanorods, etc. (Table 10.1)

Smart nanosensing enables the detection of different biomarkers of health care and environment monitoring. (Thiruvengadam et al., 2025)

10.3.2 Working principle

10.3.2.1 Optical biosensor

An optical biosensor is a device that interacts with an optical signal and detects several analytes. They provide many advantages in terms of high sensitivity, selectivity, and real-time response, which make them suitable for many applications like disease diagnosis and food quality control. The working principle of optical biosensors is that these biosensors use optical technologies like surface plasmon resonance (SPR), lossy-mode resonance (LRS), interferometry, and localized surface plasmon resonance for sensing mechanisms. When any biological analyte, like a pollutant or target molecule, interacts with a functionalized sensing surface, there is a change in the refractive index. Optical biosensors detect changes in refractive index or wavelength shift. These changes were then measured using the spectrometric method. Results are then displayed on the screen. Nanomaterials like gold nanoparticles, zinc oxide, and graphene oxide are used for enhancing binding affinity and real-time detection, as these nanomaterials have a large surface area and hence can detect smaller analytes more effectively (Soares et al., 2024).

10.3.2.2 Electrochemical biosensor

The electrochemical nano-biosensor uses a nano electrode array that is fabricated through CMOS technology (complementary metal oxide semiconductor). Each nanoelectrode enables high sensitivity, high electric field strength, and fast electron transfer. The nano electrode array is organized into four independent banks, each with a working electrode, a counter electrode, and a reference electrode. The surface of the working electrode is modified using diazonium salt electro-reduction that forms a phenylacetic acid monolayer. The sensor measures how easily the electricity flows. As the analyte binds with the sensor, it blocks the flow of electrons, making it harder for electricity to pass. It causes an increase in resistance; this is how we can measure the concentration of analyte (Cheng et al., 2024).

10.3.2.3 Piezoelectric biosensor

A piezoelectric nano-biosensor is a device that is formed by the integration of piezoelectric materials and other nanomaterials to detect biological or chemical analytes. The piezoelectric nano-biosensor operates on the principle of ultrasonic guided wave sensing using piezoelectric materials like lithium niobate-based cantilever beam array. These beams are coated with a gold layer that improves their conductivity. These gold-coated layers are also modified with gold nanoparticles and specific antigens/antibodies to enhance their sensitivity. This layer captures the target biomolecule. As the signal is applied to one beam, there is the generation of acoustic waves in the device. Each of these waves exhibits unique geometry, depending on the mass or some other event. Because of this, there is a remarkable frequency shift of beams. This shift can be measured with the help of some tools like fast Fourier transform (FFT), power spectral density (PSD), and spectrograms. Signals are then captured and analyzed (Mandal et al., 2022).

10.3.3 Importance of sensitivity, selectivity, and biocompatibility

An important property exhibited by smart nano-biosensors is their sensitivity, selectivity, and biocompatibility. Sensitivity refers to the ability of biosensors to detect mutagen even if it is present in very low concentration. To detect sensitivity, scientists have developed specific molecules (antibodies) that are attached to nanoparticles. When any target molecule (antigen) binds to these, the signal is detected, resulting in the sensitive detection of the target molecule (Yu & Irudayaraj, 2007). Cancer is the 2nd most common cause of death in today's world. Current methods like PCR and gene sequencing are costly and time-consuming, so scientists developed a new high resistivity silicon (HR-Si) conductometric biosensor that detects mutant DNA even when it is present ten times less than wild-type DNA, showing its sensitivity. (Perera et al., 2025).

Nowadays, differently shaped nanomaterials have been developed that detect very low concentrations of target molecules very precisely, increasing the sensitivity of these biosensors. e.g., gold nanorods can be used to create a special sensor called a gold nanorod molecule probe. These biosensors can detect even very low concentrations of target molecules, making the sensor very precise (Yu & Irudayaraj, 2007). The selectivity of nano-biosensors refers to the ability to distinguish targeting molecules from other molecules in the sample. This can be demonstrated when different molecules are taken in the sample; the sensor produces the strongest response to the targeting molecule, and the response to the other molecules present in the sample is much lower. Thus, a nano-biosensor shows selectivity (Mohammadnejad et al., 2023).

Nowadays nanoparticles are widely used in medicine, drug delivery (chemotherapeutics, genes, proteins, etc.), and as imaging agents because of their useful property known as biocompatibility (Sharma et al., 2023). Biocompatibility refers to the ability of the material to perform its job inside the body without causing harm to a living system. Biocompatibility in the body can be affected by immune responses and cytotoxicity. The biocompatible biosensors consist of nanomaterial that can survive inside the body and resist internal mechanical force. Biocompatibility is a critical property in nano-biosensors. These biosensors inside the body are in direct contact with cells or tissue, and biocompatibility ensures their survival inside the body without causing harm to living cells or tissue. The nanomaterials that are compatible with living systems are used to detect or monitor disease. They monitor vital signs like heart rate and blood pressure in real time. They detect how different cells or tissues respond to different drugs (Gideon et al., 2024).

10.3.4 Integration of Nanomaterials for Responsive Sensing

Integration of nanomaterials in biosensors significantly improved their performance. One of the major advantages is their improved sensitivity and selectivity. The nanomaterials used in biosensors are cost-effective, portable, and enable easy and fast responsive sensing (Razlansari et al., 2022). Modern packaging also relies on nanomaterials such as nanoparticles and nanotubes. They provide better monitoring, smart labeling, etc. (Fuertes et al., 2016). Nowadays advanced nanomaterials are used in biosensors that offer better brightness, stability, and tunability in detection systems. These include quantum dots (QDs), carbon quantum dots (CQDs), graphene quantum dots (GQDs), nanoclusters, etc. (Razlansari et al., 2022). QDs are useful nanomaterials that are often added to the biosensor, as they are stable, have large surface areas, and have excellent conductivity. QDs are usually made up of metal chalcogenides (semiconductors/nonsemiconductors); they are made biocompatible by capping with glutathione. They are often used for the detection of dengue viruses and work on real biological fluids like human saliva (Gray & Adegoke, 2024). Graphene is also a well-known carbon-based nanomaterial with a 2-dimensional honeycomb-like structure. It is a useful nanomaterial for sensing applications, as it has:

- High sensitivity, selectivity, and stability
- Good water solubility and tunable optical behavior.
- It has a large surface area and allows easy doping (Wu et al., 2025).

They are biocompatible and can be used in biomedical applications (Sadana et al., 2025). Metal nanoparticles like gold, copper, and silver are also integrated into biosensors because of their unique properties. Gold nanoparticles, because they have special optical and electrical properties, are used in nano-biosensors. Gold nanoparticles offer high sensitivity and selectivity and are ideal for biological signal detection. Also, they are chemically more stable and environmentally friendly. They are widely used in biosensors, as they detect many biomolecules like nucleotides, amino acids, etc. (Kumalasari et al., 2024)

10.4 LIGNIN-DERIVED NANOSTRUCTURES FOR BIOSENSING

10.4.1 Preparation of Lignin Nanoparticles (LNPs)

Converting lignin into lignin nanoparticles (LNP) is a smart solution, as it solves the problems related to using lignin in materials. Making lignin into tiny particles (nanoparticles) provides better control over its morphology and structure and allows the lignin nanoparticles to mix evenly with another polymer matrix. Various methods are available for the preparation of lignin nanoparticles (LNP) from lignin. Among various methods of preparation of lignin nanoparticles, ultrasonication is increasingly favored because it does not involve the use of any toxic organic solvents (Wan et al., 2025).

10.4.1.1 Ultrasonication

Lignin nanoparticles (LNP) are made commercially using a transonic homogenizer with a frequency of 20 kHz and intensity of 90%. In this method, 0.5 g of alkali lignin was mixed with 100 ml of deionized water, and the mixture was exposed to ultrasound at different power levels (200 W, 400 W, 600 W, etc.) for 1 hour. Samples were then labeled as LNP-200, LNP-400, and LNP-600. The lignin nanoparticles obtained subsequently were freeze-dried and stored for future use. As ultrasound causes heat, to avoid this, the mixture was kept cool using a water-cooling system (Wan et al., 2025).

10.4.1.2 Antisolvent precipitation

In this method, lignin was mixed with ethylene glycol under a magnetic stirrer for 24 hours, and then the mixture was vacuum filtered to remove impurities. Acidified water at pH 2.0 was added to it through a peristaltic pump, which leads to the formation of lignin nanoparticles. The nanoparticles were then washed and resuspended in acidified water (containing 0.01% (wt/v) sodium azide). The concentration of the nanoparticle suspensions was determined by gravimetry (Pedro et al., 2024).

10.4.1.3 Solvent shifting method

This is the simple method of preparation of lignin nanoparticles, which involves only the use of ethanol and water. The materials used in this method are lignin (softwood kraft lignin/SL, etc.), solvents like ethanol and deionized water, etc. In this method, 256 mg of lignin was mixed with 100 ml of ethanol and stirred overnight at 600 rpm to dissolve the lignin. The solution was then filtered to remove impurities. 150 ml of deionized water was added to the solution with the help of a separate funnel, and the mixture was stirred continuously at 600 rpm to control the dilution rate. After some time, lignin itself assembled into nanoparticles as water reduced the solubility of lignin in ethanol. The size of resulting lignin nanoparticles can be confirmed through dynamic light scattering (DLS). The size of the LNP depends on the dilution time.

(The longer the dilution time, the larger the particle size). High-yield production of lignin nanoparticle photonic glasses (Veettil et al., 2025).

10.4.2 Functionalization techniques

Lignin nanoparticles are used commercially in a wide range of applications; before using them, there is a proper requirement for their modification and functionalization. It includes modification of their surface to increase sensitivity or adding different metals (doping) to increase their electrical and thermal conductivity (Moreno et al., 2020).

10.4.2.1 Surface modification

Lignin is commercially used as an adsorbent material, but there are few limitations of it because of its adsorbent performance. This can be improved by surface modification of lignin nanoparticles. Surface modification of LNP improves mechanical stability, enhances water resistance, and increases water adsorption capacity. It can be explained by an example of surface modification of lignin nanoparticles by an amine group, i.e., by polyethyleneimine. Lignin nanoparticles were first coated with polyethyleneimine, and then glutaraldehyde was added so that polyethyleneimine cross-linked on the outer surface (Kwak et al., 2020). In this way surface modification of LNPs by polyethyleneimine takes place (Table 10.2).

10.4.2.2 Doping

Doping refers to the addition of specific atoms to lignin-derived materials to improve their electrical, chemical, and structural property. Doping with nonmetals (nitrogen, carbon, sulfur, etc.) is quite common. Nowadays researchers focus on doping with metals as metals for adjusting their optical properties, electrical behavior and for

TABLE 10.2
Enhancement of properties due to surface modifications of Lignin nanoparticles

Sr. No	Property	Before	After
1	Surface charge	Negative	Highly positive
2	Functional group	Mostly phenolic and aromatic	Amine and amide
3	Mechanical strength	Weak	Strong
4	Stability	Unstable	Highly stable

enhancing their performance (Aldakhil et al., 2024). To perform doping lignin is firstly modified by quaternization to make it reactive, then it is mixed with activators like K_2CO_3 (potassium carbonate), SiO_2 (silica), etc. Then the mixture is heated at 200°C in atmospheric nitrogen to form a porous carbon structure. Nitrogen is doped in this process, which will ultimately lead to the electrochemical performance (Feng et al., 2023).

10.4.2.3 Conjugation with biomolecules

Conjugation is the process of attachment of biomolecules like amino acids, proteins and drugs to lignin nano particles. This enhances the target ability, biocompatibility and functionality of lignin nano particles. Biomolecules with their function that they perform when conjugated with LNP are given (Table 10.3).

These conjugations have various applications such as the controlled release of cancer drugs, making lignin nanoparticles more sensitive toward pH, temperature etc. and their use in medical diagnosis and imaging. Different bonds are formed when lignin conjugates with biomolecules like ester bonds, amide bonds, etc. that provide stability to LNPs (de Lima et al., 2024)

10.4.3 Lignin composite with metal/metal oxide nanoparticles

Lignin composite with metal/metal oxide nanoparticles is an important class of biohybrid nanomaterial. It involves combining the natural functional properties of lignin with metal/metal oxide nanoparticles to form a material that has multifunctional properties. The metal /metal oxide used in the process has unique optical, magnetic, and catalytic properties. The materials that are formed as a result of this combination offer high sensitivity, selectivity, and electrical conductivity (Saegher et al., 2024). Some metal/metal oxide nanoparticles that are incorporated with lignin are given (Table 10.4) with their application.

10.4.4 Stability, responsiveness, and biocompatibility of lignin-based nanostructures

Lignin nanoparticles exhibit both hydrophilic and hydrophobic characteristics, as there is the presence of both phenolic, aliphatic, and hydroxyl groups. This contributes to the stability of LNPs. Lignin nanoparticles that are formed without chemical modification have excellent stability. They offer thermal and electrical tolerance,

TABLE 10.3
Biomolecules along with their functions upon conjugation with lignin nanoparticles

Sr. No	Biomolecule	Function
1	L. histidine	Enhance biocompatibility
2	Polyethylene glycol (PEG)	Increase solubility
3	Resveratrol (RSU)	Antioxidant, cancer therapy

TABLE 10.4
Types of metals and applications used for conjugation

Metal	Property	Application	References
Palladium	Catalytic property	Lignin depolymerization	Saegher et al. (2024)
Gold	Reducing	Sensing, imaging	Wang et al. (2020)
Silver	Antibacterial, UV blocking capability.	UV-protective clothing, smart antibacterial textile, hygiene application.	Chen et al. (2025)
Copper, nickel, iron, cobalt, etc.	Secondary metals	Enhances sensitivity, activity, etc.	Saegher et al. (2024)

broad pH, and good dispersibility. This property makes them useful tools in pharmaceutical, packaging, and environmental applications (Zhang et al., 2022). However, the stability of lignin nanoparticles largely depends upon the source from which lignin is obtained, its molecular weight, strategies used for modification, and various other factors as follows:

10.4.4.1 Chemical composition

Chemically, lignin consists of syringyl (S) and para-hydroxyphenyl units. These units are either interconnected by ether or a C-C bond. They provide structural rigidity to lignin. Their composition varies depending on the type of source (softwood, hardwood, grass, etc.).

10.4.4.2 Molecular weight

Lignin obtained from different sources has different molecular weights, which can lead to thermal instability.

10.4.4.3 Thermal instability

Thermal stability is an important factor that affects the stability of lignin nanoparticles. For example, softwood has a high glass transition temperature and is more prone to thermal degradation (Jiang et al., 2023).

10.4.4.4 Responsiveness

Which refers to the ability to change behavior depending on the surrounding environment. This property of lignin nanoparticles is very useful in cancer therapy and many other applications. During cancer therapy lignin nanoparticles release their drugs in an acidic medium (inside a cancer cell) and do not release in a normal cell.

10.4.4.5 Biocompatibility

It refers to the ability of a material to not harm a living system so that we can incorporate it into a biological system. To find the biocompatibility of LNP, researchers tested LNP on two types of mouse cells, i.e., immune cells and breast cancer cells. They used several cell viability arrays to check the toxicity of lignin nanoparticles. At very low concentrations, 90% of cells are viable, showing very good biocompatibility. At high concentrations, only 50% of immune cells survived. This study shows that LNPs are biocompatible and are safe for biological systems at low concentrations. So they can be used for drug delivery and cancer treatment at low concentrations (Imlimthan et al., 2019).

10.5 APPLICATIONS IN PACKAGING

10.5.1 Smart packaging: concept and market needs

Packaging is not just about keeping products safe anymore, but it also plays a big role in how customers experience products and how brands address environmental goals. Traditional packaging is no longer sufficient due to increasing customer experience expectations, increasing product complexity, and minimizing the carbon footprint of manufactured products. Functional packaging is required to accommodate a variety of additional consumer needs, including foods processed with fewer preservatives, products that meet increased regulatory requirements, and packaging that allows for cradle-to-grave tracking and thus may serve as a protection against lawsuits. As a result, smart packaging, packaging with built-in sensors, is gaining attraction (Figure 10.3). It helps extend shelf life, track freshness, communicate product quality, and enhance safety across food, cosmetics, pharmaceuticals, and beyond (Zubair et al., 2024).

Active packaging is a subtype of smart packaging that offers extra functionality for product preservation and extends shelf life. An active packaging system includes oxygen scavenging, moisture absorption and control, carbon dioxide and ethanol generation, and antimicrobial migrating and non-migrating systems. On the other hand, intelligent packaging focuses on communicating valuable information of the package to facilitate decision-making to achieve the benefits of enhanced product quality and safety by monitoring the condition of the product or environment using

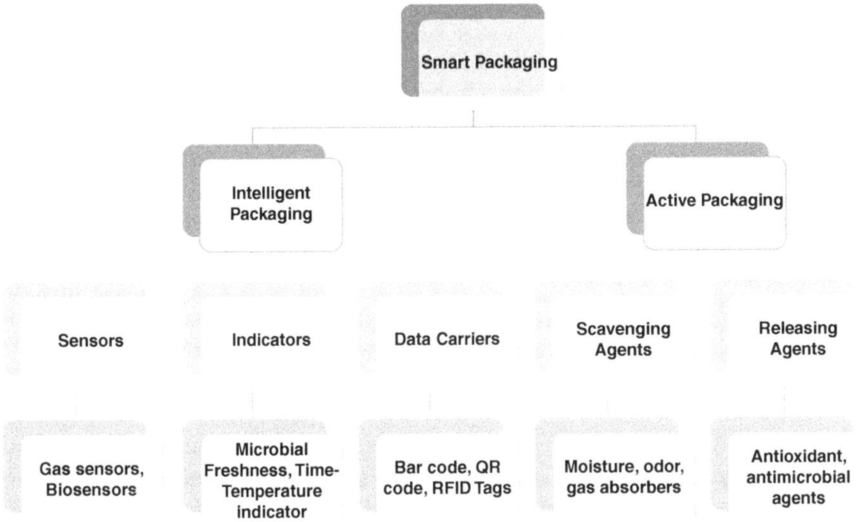

FIGURE 10.3 Classification of smart packaging using lignin nanoparticles

barcode labels, radio frequency identification tags, time-temperature indicators, gas indicators, biosensors, etc. It helps both retailers and consumers make better decisions based on real-time product condition (Siciliano et al., 2024).

10.5.2 Lignin-based nano-biosensors in packaging

Lignin is a plant-based polymer found in abundance in nature. It has excellent biodegradability and biocompatibility; lignin possesses huge potential for the production of a variety of functional and sustainable materials that can serve as replacements for plastic. It's biodegradable and safe for the environment and has a unique chemical structure that makes it more water-resistant than many other bio-based materials. Moreover, it has many amazing properties, such as UV blocking and acting as an antioxidant and antimicrobial agent for food packaging. Lignin-based bionanomaterials are now being explored for use in high-performance biosensors (Anushikha & Gaikwad, 2024). These biosensors are tiny but powerful. Their small size gives them a large surface area, which helps them react quickly and detect even the smallest changes. Thus, they are perfect for smart packaging. They can help monitor the freshness and safety of products from food to cosmetics to medicine by tracking things like spoilage, leaks, gas levels, pH, or temperature (Zubair et al., 2024).

10.5.2.1 Temperature-time indicators

Lignin-based nano-biosensors can be used to track product quality and environmental conditions. Temperature-time indicators are simple, cost-effective, and easy to use for monitoring the impact of temperature on the quality and safety of a product. TTI, also known as external markers because they are attached outside the box, tracked and analyzed a variety of food categories, including fruits, meat, milk, and marine products under chilling or freezing conditions.

10.5.2.2 Antimicrobial packaging

Antimicrobial packaging works by adding natural agents like amino acids, bacteriocins, polysaccharides, or even enzymes and fungi into the packaging material. When combined with active packaging systems, these ingredients help extend shelf life and keep the product safe by slowing down spoilage. Interestingly, when these antimicrobials are used at the nanoscale, they perform even better, offering stronger protection against harmful microbes. Some antimicrobial compounds are volatile, meaning they don't even need direct contact with the food. For example, in some types of ground beef packaging, these volatile agents can keep the meat fresh just by working in the surrounding air inside the pack (Moreno et al., 2020).

10.5.2.3 Spoilage detection

Lignin is gaining attention for its role in detecting food spoilage. When combined with color-changing dyes or metal nanoparticles, it can sense gases like ammonia and hydrogen sulfide that are released during food decomposition. These sensors visibly change color, making it easy to check freshness without opening the package. Since lignin is naturally abundant and biodegradable, it offers a sustainable alternative to synthetic materials. For instance, lignin–silver nanoparticle films have been used in meat packaging to detect spoilage through noticeable color changes (Kwon & Ko, 2022).

10.6 CASE STUDIES AND SOME EXAMPLES OF LIGNIN-BASED NANO-BIOMATERIALS

10.6.1 Types of lignin used in packaging and cosmetics

10.6.1.1 Lignin in packaging

Lignin is the second most abundant biopolymer. It is cost-effective, nontoxic, and can be used as an alternative for food packaging (Zubair et al., 2024). Different types of lignin offer different properties in packaging applications. They provide high tensile strength and have light-blocking ability in the UV–visible range. This property is essential in food packaging to prevent photooxidation of food products (Abbadessa et al., 2023). Based on type, source of lignin, and its extraction procedure, lignin has a different structure, properties, and uses (Zadeh et al., 2018). Types of lignin that can be used in packaging are: Among all the lignins listed above, alkali lignin (AL) and LS are derived from the pulping procedure in the paper industry. Both (AL and LS) show radical scavenging activity. AL is soluble in water at higher pH and has a high water absorption capacity. On the other hand, LS is highly soluble in water and has a low water absorption capacity (Zadeh et al., 2018). Another type of lignin that is used in packaging applications is KL, which has high mechanical strength and provides UV protection (Abbadessa et al., 2023). Different types of lignin used in packaging, with their source and properties, are given in Table 10.5.

TABLE 10.5
Types of lignin along with properties used for packaging

Sr. No	Type of Lignin	Source	Properties	Function	Reference
1	Alkali lignin (AL)	Kraft pulping procedure	*Solubility* soluble at high pH. *Antioxidant*—moderate *Water absorption* high	Packaging	Zadeh et al. (2018)
2	Lignosulfonate (LS)	Sulfite pulping procedure	*Solubility*- high *Antioxidant* low Water absorption—low	Packaging	Zadeh et al. (2018)
3	Kraft lignin (KF)	Kraft pulping process	*Mechanical strength*- high UV protection—improved	Packaging	Abbadessa et al. (2023)
4	Organosolv lignin (OSL)	Organic solvent pulping procedure	*Tensile strength*—high, Strong oxidant	packaging	Abbadessa et al. (2023)

10.6.1.2 Lignin in cosmetics

Different types of lignin offer antioxidant and antimicrobial properties and have the capacity to absorb UV rays, enabling protection against UV radiation. Types of lignin used in cosmetics are KL is derived from the kraft pulping procedure. Kraft pulping was developed by the Swedish inventor Carl F. Dahl. The first kraft pulp mill was built in Sweden in 1890 (Argyropoulos et al., 2023). KL has UV absorption capacity, as it has aromatic functional groups, such as a methoxy group at ortho and para positions that have electron donation capacity, which is useful in UV blocking. It also has antioxidant properties and other properties that help in the reduction of skin damage (Gagosian et al., 2022).

Another type of lignin is lignosulfonate, which is derived from the sulfite pulping procedure. They are used in the production of eco-friendly sunscreens. They are non-phototoxic and have antioxidant activity. (Lorquin et al., 2021) Different types of lignin used in cosmetics are given in Table 10.6.

KL is derived from the kraft pulping procedure. Kraft pulping was developed by the Swedish inventor Carl F. Dahl. The first kraft pulp mill was built in Sweden in 1890. (Argyropoulos et al., 2023). KL has UV absorption capacity, as it has aromatic functional groups, such as a methoxy group at ortho and para positions that have electron donation capacity, which is useful in UV blocking. It also has antioxidant properties and other properties that help in the reduction of skin damage (Gagosian et al., 2022). Another type of lignin is lignosulfonate, which is derived from the sulfite pulping procedure. They are used in the production of eco-friendly sunscreens. They are non-phototoxic and have antioxidant activity. (Lorquin et al., 2021) Different types of lignin used in cosmetics are given in Table 10.6.

10.6.1.2.1 Need for smart sensing in cosmetic formulations and skin applications

Smart sensing technologies are gaining importance in the cosmetic industry. Traditional cosmetic products function passively and often lack the ability to provide real-time feedback or adapt to individual skin conditions. Since skin is a dynamic organ—its pH, moisture, temperature, and sebum levels fluctuate throughout the day—there is a growing demand for personalized, responsive, and sustainable skincare solutions. Smart materials enable cosmetic products to detect and respond to such variations. These materials can sense skin conditions and trigger the release

TABLE 10.6
Types of lignin and properties used in cosmetics

Sr. No	Type of Lignin	Source	Properties	Reference
1	Kraft lignin	Kraft pulping procedure	Uv absorption antioxidant protection reduces skin damage	Gagosian et al. (2022)
2	Lignosulfonates	Sulfite pulping procedure	Antioxidant emulsifier and as stabilizer	Gagosian et al. (2022)
3	Technical lignin	Byproduct of the pulping process	UV absorb capacity antioxidant property	Mondal et al. (2023)

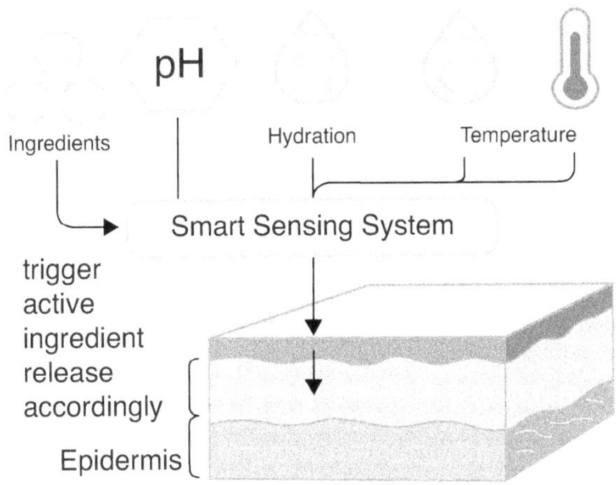

FIGURE 10.4 Representing a smart sensing system that detects different skin parameters and triggers the release of active ingredients accordingly

of active ingredients accordingly. This not only improves treatment outcomes but also minimizes unnecessary exposure to compounds, ensuring targeted delivery and better functionality (Wang et al., 2020).

Among these smart materials, biopolymer-based nanomaterials, especially lignin, show significant potential. Lignin, a natural aromatic polymer sourced from lignocellulosic biomass, exhibits intrinsic antioxidant, antimicrobial, and UV-protective properties. Its phenolic structure allows for chemical modifications that support stimuli-responsive behaviors—such as pH- or temperature-sensitive release—making it a strong candidate for smart cosmetic applications (Bains et al., 2022). Moreover, smart sensing promotes sustainability. Being biodegradable, lignin serves as a safer and eco-friendly alternative to synthetic materials, benefiting both human skin and the environment (Figure 10.4).

10.6.1.2.2 Why lignin-based biosensors

- **Natural antioxidant** – protects skin from oxidative damage
- **UV absorber** – blocks harmful UVA and UVB rays
- **Stimuli-responsive** – reacts to skin pH, moisture, or temperature
- **Dual function** – protects and senses at the same time
- **Biodegradable** – eco-friendly and sustainable choice

10.6.1.2.3 Skin pH and hydration monitoring

Healthy skin generally maintains a pH between 4.5 and 6.5, while the body's water content ranges between 50% and 70%. Deviations from these values may indicate skin disorders such as dryness or compromised barrier function. Lignin-based smart nanomaterials are being developed to monitor these parameters effectively. These can be incorporated into creams, films, or wearable patches that actively sense skin conditions (Bains et al., 2022). For pH detection, lignin nanoparticles can be functionalized with pH-sensitive dyes such as litmus or phenolphthalein or integrated with weak conductive polymers whose electrical resistance changes with pH. These features enable the creation of visibly or electronically responsive skin patches. For instance, a change in skin pH beyond the optimal range (< 4.5 or > 6.5) can cause a noticeable color shift or resistance variation, offering real-time skin condition feedback.

For hydration monitoring, hydrogels containing lignin can absorb moisture and swell proportionally, providing visual or electronic indications of skin hydration levels (Wang et al., 2020). One example is the zinc–lignin–acrylic acid hydrogel developed by Wang et al. (2020), which demonstrates remarkable stretchability (~700%), conductivity (~0.97 S/m), UV resistance, freeze tolerance, and strong skin adhesion. Additionally, noninvasive hydration monitoring techniques are emerging. Shen et al. (2023) introduced a compact near-infrared device capable of detecting moisture levels by measuring water-specific light absorption signals. Their results confirmed the device's effectiveness in real-time skin hydration assessments.

10.6.1.2.4 UV exposure indicators

UV radiation is a major cause of skin damage, leading to sunburn, premature aging, and increased cancer risk. Conventional sunscreens provide protection but lack real-time feedback on UV exposure levels. Lignin-based materials offer strong potential in the development of UV-responsive sensors due to their inherent UV-absorbing properties. For example, lignin–porphyrin polymers enhance UV–visible light absorbance and photoluminescence, making them suitable for UV detection and metal ion sensing. Lignin nanoparticles (LNPs), with their excellent UV-barrier function, have been used in materials that visibly react to UV light, making them ideal for wearable UV indicators (Moreno et al., 2022).

10.6.1.2.5 Controlled release systems

Lignin-based nanomaterials are well-suited for the controlled delivery of active agents. These materials can respond to environmental triggers such as pH or temperature, allowing for the release of therapeutic lignin's aromatic rings, which effectively block both UVA and UVB rays, making it valuable for both protection and sensing (Bains et al., 2022). According to Ariyanta et al. (2023), lignin's phenolic content contributes not only to UV absorption but also to free radical scavenging. They highlight its successful integration into sunscreens alongside inorganic filters, improving UV protection while reducing cytotoxic effects. This targeted delivery approach enhances treatment effectiveness and minimizes side effects (Prakashan et al., 2024).

Lignin nanoparticles (LNPs), owing to their small size and tunable surface properties, are efficient carriers for slow-release systems. They also contribute to product stability and performance. Techniques such as mini-emulsion and solvent exchange are used to control particle size and release kinetics (Piccinino et al., 2021). Emerging lignin-based QDs have demonstrated promising capabilities in drug delivery. For instance, curcumin has been successfully encapsulated and released over time. However, more environmentally friendly production methods are still needed (Moreno et al., 2020).

10.6.1.2.5 Integration with wearable devices or cosmetic packaging

Lignin is compatible with flexible and stretchable substrates, making it a suitable material for wearable devices and advanced cosmetic packaging. These systems can monitor parameters like pH, hydration, or UV exposure and transmit data to smartphones for real-time analysis. Lignin enhances both product protection and user interaction. Its antioxidant and UV-blocking properties offer added benefits. Lignin–PDMS composites, for example, have been used to create soft, skin-conforming sensors capable of detecting pressure and movement (Moreno et al., 2020). Lignin-based hydrogels can also respond to stimuli like touch or heat, making them practical for smart patches and interactive cosmetic containers (Piccinino et al., 2021). Table 10.7 describes the merits and demerits of lignin-based smart materials and cosmetics.

10.7 ENVIRONMENTAL AND HEALTH BENEFITS

Lignin is an aromatic biopolymer naturally found in lignocellulosic biomass. Owing to its renewable nature and biodegradability, it plays a growing role in green technologies, valued for its diverse functional properties and ecological sustainability (Moreno et al., 2020; Bains et al., 2022). Smart biopolymers such as lignin exist in multiple structural forms, including homopolymers, copolymers, and precision polymers, offering broad applicability in cosmetic and biomedical fields (Morganti et al., 2023). Lignin-based smart nano-biomaterials, derived from biomass waste, support sustainable development by substituting synthetic polymers and conforming to circular economy practices through environmentally friendly production techniques (Moreno et al., 2020; Ariyanta et al., 2023). By integrating with nanobiotechnology, these materials enable the creation of high-performance cosmeceuticals and medical applications. Thus, the advancement of lignin-based smart biopolymers contributes to improved human well-being while minimizing ecological impact.

10.7.1 Biodegradability and Eco-Safety of Lignin-Based Systems

Lignin is a naturally degradable phenolic biopolymer found in lignocellulosic residues such as *Saccharum officinarum* (sugarcane) bagasse. It breaks down in the

TABLE 10.7
Merits and demerits of lignin-based smart materials in cosmetics

Sr. No	Aspect	Merits	Demerits
1.	Smart sensing	Real-time monitoring of pH, hydration, and UV levels	Requires precise tuning and calibration
2.	Sustainability	Biodegradable and eco-friendly	Limited scalability in industrial production
3.	UV & antioxidant	Natural UV protection and free radical scavenging	Efficacy may vary depending on lignin source
4.	Controlled release	Enables targeted delivery, reduces side effects	Requires complex formulation methods

environment without producing toxic by-products, offering a safe alternative to synthetic compounds (Abraham et al., 2024). Its origin as an abundant waste product further supports the sustainable reuse of biomass (Moreno et al., 2020). Toxicological assessments confirm lignin's safety for topical use. Concentrations up to 187.5 μg/ml maintain cell viability within ISO safety guidelines. It has shown no mutagenic effects in *S. typhimurium* strains and caused no irritation during patch tests (Antunes et al., 2023). These findings indicate its suitability for cosmetic applications. Eco-friendly synthesis methods like acoustic treatment and ionic gelation produce nano-lignin without harmful reagents (Gilca et al., 2015; Morganti et al., 2023).

These nano-systems improve performance while keeping environmental risks low. They offer enhanced UV blocking, antioxidant effects, and dispersibility (Ariyanta et al., 2023). The environmental safety of lignin systems depends on factors such as particle size, chemical purity, and plant origin. With proper control, they remain nontoxic and effective in formulations for healthcare and personal care products (Morganti et al., 2023; Abraham et al., 2024). As a renewable resource, lignin contributes to circular economy goals. It minimizes industrial waste and supports greener alternatives for high-value applications (Moreno et al., 2020).

10.7.2 Reduction in Synthetic Polymer and Heavy Metal Usage

10.7.2.1 Heavy metal adsorption

Lignin's phenolic and carboxyl groups naturally bind heavy metals like Pb^{2+} and Cu^{2+}. For example, Guo et al. (2020) demonstrated effective adsorption of Pb^{2+}, Cu^{2+}, Cd^{2+}, Zn^{2+}, and Ni^{2+} following Langmuir behavior. Functionalization, such as amination, further enhances this capacity and kinetics.

10.7.2.2 Biodegradable composite materials

Incorporating KL into biodegradable polymers like PLA (polylactic acid) enhances mechanical strength, UV resistance, and antioxidant behavior. This integration reduces or replaces synthetic stabilizers and plasticizers (Esakkimuthu et al., 2022; Zhang et al., 2022).

10.7.2.3 Eco-friendly MOF systems

In lignin-based MOFs such as Zn-MOF-FA and Cu-MOF-CA, lignin-derived ligands eliminate the need for toxic organic linkers and heavy metals while supporting drug delivery and pollutant adsorption functions (Tian et al., 2025).

10.7.2.4 Circular economy impact

Valorizing industrial lignin waste (~50 million tons annually) into functional materials reduces reliance on synthetic polymers and heavy metal additives, supporting sustainable and resource-efficient material design (Moreno et al., 2020).

10.7.3 Role in Healthcare Industry

The need for eco-friendly and biocompatible materials in healthcare has driven interest in natural biopolymers. Lignin, a major byproduct of the paper and biofuel industries, is abundant, biodegradable, and rich in functional groups. These properties make it ideal for designing smart nanomaterials for biomedical use (Moreno et al., 2020). Engineered lignin systems have shown potential in targeted drug delivery, tissue repair, and skin regeneration. When combined with metals or responsive polymers, lignin-based materials exhibit antimicrobial, antioxidant, and pH- or temperature-sensitive behaviors (Tian et al., 2025). These features are valuable for applications such as cancer therapy, wound healing, and bioactive cosmetics (Morganti et al., 2023). Lignin has also been explored in wearable biosensors for noninvasive sweat analysis. Its flexibility, skin compatibility, and biodegradability make it a sustainable choice for next-generation health monitors. By replacing or enhancing synthetic materials, lignin-based biomaterials support safer, smarter, and more sustainable solutions in modern medicine.

10.8 CHALLENGES AND SCALABILITY ISSUES

10.8.1 Inconsistency in structure of lignin

Lignin is found in plants like wood or grass, and depending on how it's extracted, its structure can change. This means that every batch can behave somewhat differently, which makes it difficult to maintain consistent quality in packaging or sensor performance (Creteanu et al., 2024).

10.8.2 Needs blending with strong biopolymers

On its own, lignin is structurally fragile and doesn't form strong, flexible films. To make it usable in packaging, it usually needs to be mixed with other materials like cellulose or PLA, which adds more steps and complexity to the process (Boarino et al., 2022).

10.8.3 Low visibility

Because lignin is brown to almost black, it can block or affect the visibility of color changes in sensors—which is a big issue when you're relying on color shifts to show food spoilage (Sadeghifar & Ragauskas, 2020).

10.8.4 Accuracy of biosensor response varies

Making lignin-based sensors that can detect specific spoilage gases (like ammonia or hydrogen sulfide) without reacting to other things is tricky. This often needs chemical tweaks, which can raise costs and complicate the design (Papa et al., 2025).

10.8.5 Short shelf life

Natural materials like lignin can degrade over time, especially when exposed to air, humidity, or light. Keeping these sensors stable and working well in real-world storage conditions is still a big challenge (Boarino & Klok, 2023).

10.8.6 Not easily scalable

Making lignin-based nanoparticles consistently and in large amounts is still a work in progress. Many of the lab methods used today are hard to apply on a large industrial scale.

10.8.7 Can be expensive to produce on a large scale

Even though lignin is natural and often a byproduct, turning it into a nano-biosensor can be costly. The processing and modification steps can make it more expensive than regular plastic sensors (Nadányi et al., 2022).

10.8.8 Safety check stake time

Even though lignin is bio-based, it still needs to be approved for use in food packaging. Its safety for humans and the environment must be tested, especially with metals or other chemicals. (Sadeghifar & Ragauskas, 2020)

10.9 CONCLUSION

Lignin is the most abundant sustainable smart material for packaging and cosmetics due to its antioxidant, antimicrobial, and structural variability. Various case studies on applications of lignin nanomaterials as smart packaging and cosmetics have been discussed in the chapter, confirming their utility. But there are some major challenges or limitations, such as the structural or conformational heterogeneity and dependence of structure activity on various derivatives, that further complicate the polymer film incorporation due to the incompatibility of lignin with the matrix. However, these problems can be overcome by the structural modifications discussed in the chapter, making it more adaptable for packaging and cosmetics applications. Moreover, the mechanism of antimicrobial activity of lignin is still not clear, which can be strengthened by conducting lab-based studies on lignin and uses in packaging and cosmetics.

REFERENCES

Abbadessa, A., I. Dogaris, S. K. Farahani, et al. 2023. Layer-by-Layer Assembly of Sustainable Lignin-based Coatings for Food Packaging Applications. *Prog. Org. Coat.* 182: 107676. https://doi.org/10.1016/j.porgcoat.2023.107676

Abraham, B., H. Shakeela, L. P. Devendra, et al. 2024. Lignin Nanoparticles from Ayurvedic Industry Spent Materials: Applications in Pickering Emulsions for Curcumin and Vitamin D3 Encapsulation. *Food Chem.* 458: 140284. https://doi.org/10.1016/j.foodchem.2024.140284

Aldakhil, F., N. A. Alarfaj, S. A. Al-Tamimi, and M. F. El-Tohamy. 2024. Development of Silver-Doped Carbon Dots Sensor Derived from Lignin for Dual-Mode Fluorometric and Spectrophotometric Determination of Valsartan in a Bulk Powder and a Commercial Product. *Heliyon* 10 (23): e40848. https://doi.org/10.1016/j.heliyon.2024.e40848

Antunes, F., I. F. Mota, J. F. Fangueiro, G. Lopes, P. M. Pintado, and C. Santos. 2023. From Sugarcane to Skin: Lignin as a Multifunctional Ingredient for Cosmetic Application. *Int. J. Biol. Macromo.* 234: 123592. https://doi.org/10.1016/j.ijbiomac.2023.123592

Anushikha, and K. K. Gaikwad. 2024. Lignin as a UV Blocking, Antioxidant, and Antimicrobial Agent for Food Packaging Applications. *Biomass Convers. Biorefin.* 14 (15): 16755–67. https://doi.org/10.1007/s13399-022-03707-3

Argyropoulos, D. D. S., C. Crestini, C. Dahlstrand, et al. 2023. Kraft Lignin: A Valuable, Sustainable Resource, Opportunities and Challenges. *Chem. Sus. Chem.* 16 (23): e202300492. https://doi.org/10.1002/cssc.202300492

Ariyanta, H. A., E. B. Santoso, L. Suryanegara, et al. 2023. Recent Progress on the Development of Lignin as Future Ingredient Biobased Cosmetics. *Sustain. Chem. Pharm.* 32: 100966. https://doi.org/10.1016/j.scp.2022.100966

Bains, R., A. Kumar, A. S. Chauhan, and P. Das. 2022. Dimethyl Carbonate Solvent Assisted Efficient Conversion of Lignocellulosic Biomass to 5-Hydroxymethylfurfural and Furfural. *Renew. Energy* 197: 237–43. https://doi.org/10.1016/j.renene.2022.07.076

Banerjee, A., S. Maity, and C H. Mastrangelo. 2021. Nanostructures for Biosensing, with a Brief Overview on Cancer Detection, IoT, and the Role of Machine Learning in Smart *Biosensors. Sensors* 21 (4): 1253. https://doi.org/10.3390/s21041253

Boarino, A., and H. A. Klok. 2023. Opportunities and Challenges for Lignin Valorization in Food Packaging, Antimicrobial, and Agricultural Applications. *Biomacromolecules* 24 (3): 1065–77. https://doi.org/10.1021/acs.biomac.2c01385

Boarino, A., A. Schreier, Y. Leterrier, and H. A. Klok. 2022. Uniformly Dispersed Poly(Lactic Acid)-Grafted Lignin Nanoparticles Enhance Antioxidant Activity and UV-Barrier Properties of Poly(Lactic Acid) Packaging Films. *ACS Appl. Polym. Mater.* 4 (7): 4808–17. https://doi.org/10.1021/acsapm.2c00420

Chauhan, P. S. 2020. Lignin Nanoparticles: Eco-Friendly and Versatile Tool for New Era. *Biores. Technol. Rep.* 9: 100374. https://doi.org/10.1016/j.biteb.2019.100374

Chen, Y., S. Wang, G. Hu, F. Kong, and J. Hu. 2025. Ultraviolet Protection and Antibacterial Properties of Textile Fabric Made of Silver Nanoparticles/Alkaline Lignin/Regenerated Cellulose Fiber. *Ind. Crops Prod.* 233 (October): 121366. https://doi.org/10.1016/j.indcrop.2025.121366

Cheng, S. T., C. F. Shen, J. H. Shiau, et al. 2024. Development of an Electrochemical Biosensor with TiN Nano-Electrode Arrays for IL-6 Detection. Sens. *Actuators Rep.* 8 (December): 100204. https://doi.org/10.1016/j.snr.2024.100204

Creteanu, A., C. N. Lungu, and M. Lungu. 2024. Lignin: An Adaptable Biodegradable Polymer Used in Different Formulation Processes. *Pharmaceuticals* 17 (10): 1406. https://doi.org/10.3390/ph17101406

De, L., H. C. Pedro, V. Ribeiro, M. Renato, A. M. S. Plath, and R. Grillo. 2024. Lignocellulosic–Biomolecules Conjugated Systems: Green-Engineered Complexes Modified by Covalent Linkers. *J Mater. Chem. B* 12 (10): 2471–80. https://doi.org/10.1039/D3TB02581K. https://doi.org/10.3390/pharmaceutics13101636

Esakkimuthu, E. S., D. DeVallance, I. Pylypchuk, A. Moreno, and M. H. Sipponen. 2022. Multifunctional Lignin-Poly (Lactic Acid) Biocomposites for Packaging Applications. *Front. Bioeng. Biotechnol.* 10: 1025076. https://doi.org/10.3389/fbioe.2022.1025076

Feng, P., H. Wang, P. Huang, et al. 2023. Nitrogen-Doped Lignin-Derived Porous Carbons for Supercapacitors: Effect of Nanoporous Structure. *Chem. Eng. J.* 471 (September): 144817. https://doi.org/10.1016/j.cej.2023.144817

Fuertes, G., I. Soto, R. Carrasco, M. Vargas, J. Sabattin, and C. Lagos. 2016. Intelligent Packaging Systems: Sensors and Nanosensors to Monitor Food Quality and Safety. *J. Sensors.* 4046061: 1–8. https://doi.org/10.1155/2016/4046061

Gagosian, V. S. C., F. C. Claro, A. C. De, et al. 2022. The Potential Use of Kraft Lignins as Natural Ingredients for Cosmetics: Evaluating Their Photoprotective Activity and Skin Irritation Potential. *Int. J. Biol. Macromol.* 222 (December): 2535–44. https://doi.org/10.1016/j.ijbiomac.2022.10.037

Galletti, A. M. R., and C. Antonetti. 2012. Biomass Pretreatment: Separation of Cellulose, Hemicellulose, and Lignin – Existing Technologies and Perspectives. In *Biorefinery: From Biomass to Chemicals and Fuels*, edited by M. Aresta, A. Dibenedetto, an F. Dumeignil, 165–98. Berlin, Boston: De Gruyter. https://doi.org/10.1515/9783110260281.101

Gideon, O., H. S. Samuel, and I. A. Okino, 2024. Biocompatible Materials for Next-Generation Biosensors. *Discov. Chem.* 1 (1): 34. https://doi.org/10.1007/s44371-024-00042-6

Gilca, I., P. Andrei, I. Valentin, and C. Crestini. 2015. Obtaining Lignin Nanoparticles by Sonication. *Ultrason. Sonochem.* 23: 9–375. https://doi.org/10.1016/j.ultsonch.2014.08.021

Gray, K. I., and O. Adegoke. 2024. Electropolymerized InZnSe@PtAg Quantum Dots@molecularly Imprinted Polymer on Screen-Printed Carbon Electrodes for the Ultrasensitive Detection of NS1 Dengue Virus Protein with Smartphone-based Sensing in Saliva. *Microchem. J.* 207 (December): 111704. https://doi.org/10.1016/j.microc.2024.111704

Guo, Z., H. Liu, W. Dai, and Y. Lei, 2020. Responsive Principles and Applications of Smart Materials in Biosensing. *Smart Mater. Med.* 1: 54–65. https://doi.org/10.1016/j.smaim.2020.07.001

Habibi, Y., L. A. Lucia, and O. J. Rojas, 2010. Cellulose Nanocrystals: Chemistry, Self-Assembly, and Applications. *Chem. Rev.* 110 (6): 3479–500. https://doi.org/10.1021/cr900339w

Hammitt, J. K., R. J. Lempert, and M. E. Schlesinger. 1992. A Sequential-Decision Strategy for Abating Climate Change. *Nature* 357(6376): 315–18. https://doi.org/10.1038/357315a0

Huang, K., S. Ma, S. Wang, et al. 2019. Sustainable Valorization of Lignin with Levulinic Acid and Its Application in Polyimine Thermosets. *Green Chem.* 21 (18): 4964–70. https://doi.org/10.1039/C9GC02384D

Imlimthan, S., A. Correia, P. Figueiredo, et al. 2019. Systematic in Vitro Biocompatibility Studies of Multimodal Cellulose Nanocrystal and Lignin Nanoparticles. *J. Biomed. Mater. Res.* 108(3): 770–83. https://doi.org/10.1002/jbm.a.36856

Jiang, B., H. Jiao, X. Guo, et al. 2023. Lignin-Based Materials for Additive Manufacturing: Chemistry, Processing, Structures, Properties, and Applications. *Adv. Sci.* 10(9): 2206055. https://doi.org/10.1002/advs.202206055

Kumalasari, M. R., R. Alfanaar, and A. S. Andreani. 2024. Gold Nanoparticles (AuNPs): A Versatile Material for Biosensor Application. *Talanta Open* 9 (August): 100327. https://doi.org/10.1016/j.talo.2024.100327

Kwak, H. W., H. Lee, and K. H. Lee. 2020. Surface-Modified Spherical Lignin Particles with Superior Cr(VI) Removal Efficiency. *CMSHAF* 239 (January): 124733. https://doi.org/10.1016/j.chemosphere.2019.124733

Kwon, S., and S. Ko. 2022. Colorimetric Freshness Indicator based on Cellulose Nanocrystal–Silver Nanoparticle Composite for Intelligent Food Packaging. *Polymers* 14 (17): 3695. https://doi.org/10.3390/polym14173695

Larrañeta, E., M. Imízcoz, J. X. Toh, et al. 2018. Synthesis and Characterization of Lignin Hydrogels for Potential Applications as Drug Eluting Antimicrobial Coatings for Medical Materials. *ACS Sustain. Chem. Engg.* 6 (7): 9037–46. https://doi.org/10.1021/acssuschemeng.8b01371

Lavoine, N., I. Desloges, A. Dufresne, and J. Bras. 2012. Microfibrillated Cellulose – Its Barrier Properties and Applications in Cellulosic Materials: A Review. *Carbohydr. Polym.* 90 (2): 735–64. https://doi.org/10.1016/j.carbpol.2012.05.026

Li, H., A. Bunrit, N. Li, F. Wang. 2020. Heteroatom-Participated Lignin Cleavage to Functionalized Aromatics. *Chem. Soc. Rev.* 49 (12): 3748–63. https://doi.org/10.1039/D0CS00078G

Lorquin, F., J. Lorquin, M. Claeys-Bruno, et al. 2021. Lignosulfonate Is an Efficient SPF Booster: Application to Eco-Friendly Sunscreen Formulations. *Sustain. Chem. Pharm.* 24 (December): 100539. https://doi.org/10.1016/j.scp.2021.100539

Mandal, D., M. M. Indaleeb, A. Younan, and S. Banerjee. 2022. Piezoelectric Point-of-Care Biosensor for the Detection of SARS-COV-2 (COVID-19) Antibodies. *Sens. And Biosens. Res.* 37 (August): 100510. https://doi.org/10.1016/j.sbsr.2022.100510

Mohammadnejad, J., N. Basirhaghighi, F. Yazdian, et al. 2023. Electrochemical Nanobiosensor Based on Reduced Graphene Oxide and Gold Nanoparticles for Ultrasensitive Detection of microRNA-128. *Int. Immunopharmacol.* 117 (April): 109960. https://doi.org/10.1016/j.intimp.2023.109960

Mondal, S., A. Jatrana, S. Maan, and P. Sharma. 2023. Lignin Modification and Valorization in Medicine, Cosmetics, Environmental Remediation and Agriculture: A Review. *Environ. Chem. Lett.* 21 (4): 2171–97. https://doi.org/10.1007/s10311-023-01585-3

Moreno, A., M. H. Sipponen, and O. J. Rojas. 2020. Lignin-based Hybrid and Composite Materials for Advanced Functional Applications. *Green Chem.* 22 (3): 551–69.

Moreno, A., J. Liu, M. Morsali, and M. H. Sipponen. 2022. Chemical Modification and Functionalization of Lignin Nanoparticles. In *Manufacturing of Nanostructured Lignin Particles (MNLP)*, edited by D. Puglia, C. Santulli, and F. Sarasini, 385–431. Elsevier. https://doi.org/10.1016/B978-0-12-823702-1.00003-7

Morganti, P., G. Morganti, and M. B. Coltelli. 2023. Natural Polymers and Cosmeceuticals for a Healthy and Circular Life: The Examples of Chitin, Chitosan, and Lignin. *Cosmet.* 10 (2): 42. https://doi.org/10.3390/cosmetics10020042

Nadányi, R., A. Ház, A. Lisý, et al. 2022. Lignin Modifications, Applications, and Possible Market Prices. *Energies* 15 (18): 6520. https://doi.org/10.3390/en15186520

Papa, P., L. Giuseppina, G. Rossella, et al. 2025. Lignin-based Nanostructured Sensor for Selective Detection of Volatile Amines at Trace Levels. *Sensors* 25 (11): 3536. https://doi.org/10.3390/s25113536

Pedro, M. D. S., D. F. L. Feitosa, J. D. Filho, et al. 2024. Synthesis of Lignin Nanoparticles and Their Application in the Stabilization of Water-in-Water Pickering Emulsions: A New Technology for Valorization of Lignin from Sugarcane Bagasse. Pipitone G. eds. *Int. J. Chem. Eng.* (1): 7072901. https://doi.org/10.1155/2024/7072901

Perera, G. S., X. Huang, F. A. Bagherjeri, et al. 2025. Rapid and Selective Detection of TP53 Mutations in Cancer Using a Novel Conductometric Biosensor. *Biosens. Bioelectron.* 276 (May): 117252. https://doi.org/10.1016/j.bios.2025.117252

Piccinino, D., E. Capecchi, E. Tomaino, et al. 2021. Nano-Structured Lignin as Green Antioxidant and UV Shielding Ingredient for Sunscreen Applications. *Antioxidants* 10(2): 274. https://doi.org/10.3390/antiox10020274

Prakashan, A., T. Varghese, and J. Parameswaranpillai. 2024. Lignin-based Nanocarriers for Controlled Drug Delivery Applications: A Comprehensive Review. *Adv. Drug Deliv. Rev.* 201: 114771.

Ragauskas, A., J. Beckham, T. Gregg, et al. 2014. Lignin Valorization Improving Lignin Bioprocessing in Biorefinery. *Science* 344(6185): 1246843. https://doi.org/10.1126/science.1246843

Rathour, R. K., Behl, M., Dhashmana, et al. (2023). Non-Food Crops Derived Lignocellulose Biorefinery for Sustainable Production of Biomaterials, Biochemicals and Bioenergy: A Review on Trends and Techniques. *Ind. Crops Prod.* 204: 117220. https://doi.org/10.1016/j.indcrop.2023.117220

Rathour, R. K., Behl, M., Sakhuja, et al. (2025). Recent Updates in Biomass Pretreatment Techniques for a Sustainable Biofuel Industry: A Comprehensive Review. *Waste Biomass Valoriz.* 16: 3833–58. https://doi.org/10.1007/s12649-025-03016-6

Razlansari, M., F. Ulucan-Karnak, M. Kahrizi, et al. 2022. Nanobiosensors for Detection of Opioids: A Review of Latest Advancements. *Eur. J. Pharm. and Biopharm.* 179 (October): 79–94. https://doi.org/10.1016/j.ejpb.2022.08.017

Sadana, S., N. Rajamohan, R. Manivasagan, et al. 2025. Graphene-based Materials for Photocatalytic and Environmental Sensing Applications. *Results Engg.* 27 (September): 105725. https://doi.org/10.1016/j.rineng.2025.105725

Sadeghifar, H., and A. Ragauskas. 2020. Lignin as a UV Light Blocker—A Review. *Polymers* 12 (5): 1134. https://doi.org/10.3390/polym12051134

Saegher, T. D., B. Atanasova, P. Vermeir, Geem, et al. 2024. Enhancing the Performance of Heterogeneous Palladium based Catalysts in the Mild Reductive Depolymerization of Soda Lignin through Addition of a Non-Noble Metal and Tuning of the Preparation Strategy. *RSC Sustain.* 2 (5): 1551–67. https://doi.org/10.1039/D4SU00054D

Saleem, A., W. Lu, S. Haiqing, et al. 2025. Cutting-Edge Innovations in Lignin-Based Nanoparticles: A Review of Synthesis Techniques, Characterization, and Diverse Applications. *Int. J. Biol. Macromol.* 307 (May): 142123. https://doi.org/10.1016/j.ijbiomac.2025.142123

Salehian, P., K. Keikhosro, Z. Hamid, and J. Azam. 2013. Improvement of Biogas Production from Pine Wood by Alkali Pretreatment. *Fuel* 106 (April): 484–89. https://doi.org/10.1016/j.fuel.2012.12.092

Sharma, V., M. L. Tsai, P. Nargotra, et al. 2023. Journey of Lignin from a Roadblock to Bridge for Lignocellulose Biorefineries: A Comprehensive Review. *STOTN.* 861(160560). https://doi.org/10.1016/j.scitotenv.2022.160560

Shen, L., X. Wang, and Z. Zhang. 2023. Near-Infrared Spectroscopy for Non-Invasive Skin Hydration Assessment. *Sens. Actuat. B-Chem.* 389: 133776.

Siciliano, S., C. G. Lopresto, and F. Lamonaca, 2024. From Traditional Packaging to Smart Bio-Packaging for Food Safety: A Review. *Euro. Mediterr. J Env. Int.* 9 (4): 1971–86. https://doi.org/10.1007/s41207-024-00627-8

Soares, S., S. Maria, K. Ragini, et al. 2024. The Role of Smart Optical Biosensors and Devices on Predictive Analytics for the Future of Aquaculture Systems. *Opt. Laser Technol.* 177 (October): 111049. https://doi.org/10.1016/j.optlastec.2024.111049

Thiruvengadam, M., H. Y. Chi, H. J. Choi, et al. 2025. Sustainable and Smart Nano-Biosensors: Integrated Solutions for Healthcare, Environmental Monitoring, Agriculture, and Food Safety. *Ind. Crops Prod.* 233 (October): 121337. https://doi.org/10.1016/j.indcrop.2025.121337

Tian, Z., L. Guanyan, C. Xiangmeng, et al. 2025. Innovative Lignin-based MOFs and COFs for Biomedicine, Energy Storage, and Environmental Remediation. *ACHM.* 8. https://doi.org/10.1007/s42114-024-01201-0

Veettil, U. T., A. J. Huertas-Alonso, T. S. Plivelic, H. M. Sipponen. 2025. High-Yield Production of Lignin Nanoparticle Photonic Glasses. *Green Chem.* 27 (7): 2130–37. https://doi.org/10.1039/D4GC05797J

Wan, Z., H. Zhang, M. Niu, W. Zhang, Y. Guo, and H. Li. 2025. Preparation of Lignin Nanoparticles by Ultrasonication and Its Incorporation in DLP 3D Printing UV-Curable Resin as Bio-Filler. *Ind. Crops Prod.* 224 (February): 120394. https://doi.org/10.1016/j.indcrop.2024.120394

Wang, B., G. Yang, J. Chen, and G. Fang. 2020. Green Synthesis and Characterization of Gold Nanoparticles Using Lignin Nanoparticles. *Nanomaterials* 10 (9): 1869. https://doi.org/10.3390/nano10091869

Wu, R., Y. Cao, Z. Chen, and J. J. Zhu, 2025. Fluorescent Graphene Quantum Dots: Properties Regulation, Sensing Applications, and Future Prospects. *Adv. Energy Sens. Mater.* 4 (3): 100140. https://doi.org/10.1016/j.asems.2025.100140

Yang, W., E. Fortunati, D. Gao, et al. 2018. Valorization of Acid Isolated High Yield Lignin Nanoparticles as Innovative Antioxidant/Antimicrobial Organic Materials. *ACS Adv. Sens. Energy Mater.* 6: 3502–3514. https://doi.org/10.1021/acssuschemeng.7b03782

Yu, C., and Irudayaraj, J. 2007. Quantitative Evaluation of Sensitivity and Selectivity of Multiplex Nano SPR Biosensor Assays. *Biophys. J.* 93 (10): 3684–92. https://doi.org/10.1529/biophysj.107.110064

Zadeh, E. M., S. F. O'Keefe, and Y. T. Kim, 2018. Utilization of Lignin in Biopolymeric Packaging Films. *ACS Omega* 3 (7): 7388–98. https://doi.org/10.1021/acsomega.7b01341

Zhang, B., Y. Liu, M. Ren, et al. 2019. Sustainable Synthesis of Bright Green Fluorescent Nitrogen-Doped Carbon Quantum Dots from Alkali Lignin. *Chem Sus Chem* 12 (18): 4202–10. https://doi.org/10.1002/cssc.201901693

Zhang, Z., C. B. Marín, M. Lefebvre, C. Lefebvre, V. Terrasson, and E. Guénin. 2022. The Preparation of Stable Spherical Alkali Lignin Nanoparticles with Great Thermal Stability and No Cytotoxicity. *Int. J. Biol. Macromol.* 222 (December): 1830–39. https://doi.org/10.1016/j.ijbiomac.2022.09.272

Zhu, Y., L. Hang, Z. Qing-Sheng, Z. Bing. 2024. Effect of DES Lignin Incorporation on Physicochemical, Antioxidant and Antimicrobial Properties of Carboxymethyl Cellulose-Based Films. *Int. J. Biol. Macromol.* 263 (April): 130294. https://doi.org/10.1016/j.ijbiomac.2024.130294

Zieglowski, M., S. Trosien, J. Rohrer, et al. 2019. Reactivity of Isocyanate-Functionalized Lignins: A Key Factor for the Preparation of Lignin-based Polyurethanes. *Front. Chem.* 7 (August): 562. https://doi.org/10.3389/fchem.2019.00562

Zubair, M., Z. Rauf, S. Fatima, and A. Ullah. 2024. Lignin-Derived Bionanocomposites as Functional Food Packaging Materials. *Sustain. Food Technol.* 2 (4): 945–66. https://doi.org/10.1039/D4FB00105B

11 Environmental and economic considerations of lignin valorization

Sa-ad Abubakari, Hussein Alhassan, Anas Abdulwahab Lawal, Mark Dakurah, Saleem Ullah, and Irshad Ahmad

11.1 INTRODUCTION

Among the biopolymers we have in nature, lignin is the second most abundant after cellulose, accounting for 20–30% of lignocellulosic biomass (LCB) dry weight. Its structure features include a complex aromatic heteropolymer composed of three phenylpropanoid units: p-coumaryl alcohol, guaiacyl alcohol, and syringyl alcohol, all of which provide rigidity, water transport, and pathogen resistance in plants (Jiju et al., 2025). In total, almost 200 billion tons of LCB are produced annually, with lignin representing a major underutilized fraction (Lobato-Rodríguez et al., 2023). Some of the sources of lignin include hardwoods, softwoods, and grasses, with industrial lignin derived mainly from kraft pulping, lignosulfonate processes, or upcoming green extraction technologies such as organosolv and deep eutectic solvents (DES). Traditional use has largely been restricted to combustion of energy, which represents a low-value application compared to its chemical potential. Biorefineries, contrary to petroleum refineries, aim to separate the components of biomass into fuels, chemicals, and materials, thereby improving circular bioeconomy objectives. In this context, lignin plays a double role. On one side, it is often regarded as a recalcitrant byproduct limiting enzymatic hydrolysis of cellulose. On the other hand, lignin valorization is progressively recognized as a key to improving the economic feasibility and environmental sustainability of biorefineries. Its pleasant nature makes it a unique renewable feedstock for high-value products such as resins, adhesives, carbon fibers, bio-based plastics, and smart biomaterials (Lobato-Rodríguez et al., 2023; Zhao et al., 2024). Recent advances have proven that innovative pretreatment strategies, such as the use of DES, can selectively separate lignin while preserving functional groups important for downstream valorization. Some of the advantages of this treatment strategy, DES, include offering biodegradability, lower cost, and recyclability advantages as compared to ionic liquids, making them promising candidates for large-scale application (Lobato-Rodríguez et al., 2023). By recovering high-quality lignin streams, biorefineries can move from energy-focused to multiproduct systems, thereby increasing profitability and aligning with the United Nations Sustainable Development Goals.

DOI: 10.1201/9781003653189-11

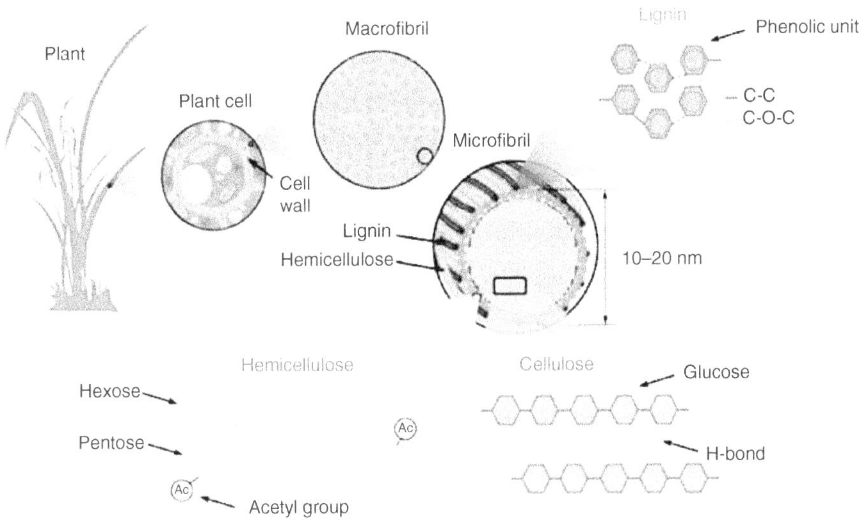

FIGURE 11.1 Schematic structure of lignocellulose components Brienza et al. (2024)

Lignin is no longer seen and regarded as a waste product but as a foundation of future biorefinery models. Revealing its potential through sustainable extraction and valorization technologies is central to advancing green industries, reducing fossil dependence, and achieving an integrated bioeconomy (Figure 11.1).

11.1.1 Importance of Lignin Valorization in the Context of Green Industries

Lignin has lately attracted rising interest for its potential role in empowering green industries through the production of bio-based chemicals, fuels, and advanced materials. In the context of the circular economy and bioeconomy, lignin valorization is particularly significant because it emphasizes waste minimization, renewable resource use, and cascading material flows. Through lignin valorization, it also enhances the economic viability of lignocellulosic biorefineries by altering lignin into higher-value products like vanillin, phenolic resins, carbon fibers, and adhesives, and thus, bioplastics industries can reduce dependence on fossil-based chemicals and open new revenue streams (Kylili et al., 2023). Unlike pyrolysis, which often suffers from poor selectivity and wide product distributions, recent innovations in electrocatalytic valorization further expand lignin's potential, and go through processes enable the targeted cleavage of lignin's β-O-4 linkages under mild conditions, improving efficiency and selectivity (Ebrahimpourboura et al., 2024).

Moreover, it has been reported from the studies of life cycle assessment (LCA) that lignin-derived products can achieve substantial reductions in greenhouse gas (GHG) emissions and fossil fuel depletion compared to conventional or traditional processes, although environmental trade-offs remain depending on the valorization pathway (Kylili et al., 2023). For example, lignin-based carbon fibers and resins can reduce energy intensity and carbon emissions compared to petrochemical counterparts, while valorization into soil amendments provides clear environmental and economic advantages (Kylili et al., 2023).

According to research carried out, a techno-economic analysis (TEA) of EC lignin valorization showed that large-scale biorefineries could generate nearly 50 tons per day of unique chemicals, with production costs estimated at around $16.86/kg. Most importantly, life cycle modeling indicates that integrating EC valorization units into ethanol biorefineries can reduce overall CO_2 emissions by 10–46%, depending on the allocation method employed (Ebrahimpourboura et al., 2024). All these successes of lignin valorization suggest that, when coupled with advanced catalytic approaches, it can reinforce the competitiveness and environmental performance of green industries. All these findings also underscore lignin's influence not only on energy efficiency but also on broader sustainability targets, and therefore, lignin valorization holds great promise for advancing green industries by transforming an underutilized byproduct into a valuable resource stream (Tables 11.1 and 11.2).

11.1.2 Objectives and Scope of the Chapter

The objective of this chapter is to critically examine the environmental and economic magnitudes of lignin valorization as a critical component of sustainable biorefineries and green industries. Specifically, the chapter's aim is to

1. Outline lignin's structural and functional role in biomass processing.
2. Evaluate the environmental impacts of lignin valorization processes through LCA, emission profiles, and circular economy principles, and
3. Analyze the economic feasibility, market potential, and policy drivers that shape the commercialization of lignin-derived products.

TABLE 11.1
Representative examples of HTL of lignin

Lignin Source	Concentration of Lignin (wt%)	Temperature (°C)	Pressure (Mpa)	Oil Yield (%)	Monophenol Yield (%)
Lignoboost Backhammer Sweden	5.5	330	13	62	18
Black liquor Backhammer Sweden	5	380	30	80	5
Indulin AT	10	374	22	58–79	Not reported
Black liquor of beech, poplar, oak and chestnut	19.4	280	7	67	5

Source: Ali et al. (2024)

TABLE 11.2
Technical advantages and disadvantages with different applications

Technology	Advantages	Disadvantages
Lime kiln fuel	• Easy to operate	• Low value • Off gas needs purification
Gasification	• No co-additives required. • Platform chemical • Different options • Market size	• Storage challenges if shut down • Energy consumption
HTL	• No co-additives required • Generate phenols • Market size	• Storage challenges if shut down • Low concentration of feed • Energy consumption • Requires fatty acids
Thermosets	• Highly desired materials	• Requires > 50 wt% other components • Mechanical properties are reduced • Market size
Thermoplasts	• Highly desired materials	• Requires > 50 wt% other components • Mechanical properties are poor • Recyclability is challenging
Biomedical devices	• High value • Natural component • Inherent properties	• Low TRL level • Heterogenous nature

Source: Ali et al. (2024).

The scope of the chapter extends to both the scientific and industrial perspectives. On the environmental side, it addresses the role of lignin valorization in reducing GHG emissions, minimizing waste, and mitigating ecotoxicological risks. On the economic side, it considers TEAs, cost–benefit trade-offs, market opportunities, and barriers to scale-up. The chapter also integrates real-world case studies to highlight successful industrial applications and distill lessons for future innovation. By combining technical insights with socio-economic considerations, the chapter provides a holistic understanding of how lignin valorization contributes to climate goals, circular bioeconomy strategies, and industrial competitiveness. This dual focus positions lignin not as a byproduct, but as a cornerstone for advancing sustainable value chains and guiding stakeholders from researchers to policymakers toward practical and scalable solutions.

11.2 ENVIRONMENTAL IMPACTS OF LIGNIN VALORIZATION

Lignin is more than just a byproduct. Its value is gaining attention due to the environmental benefits it offers. Converting it into fuels, chemicals, or materials and not burning it gives the opportunity to reduce emissions, minimize waste, and create circular value chains. These, however, depend on how it is processed and used. Hence, it is necessary to understand its environmental impact (Bajwa et al., 2019).

11.2.1 Life Cycle Assessment (LCA) of lignin-based products

Life cycle assessment examines the effects that a product has at each stage of production, all the way to disposal. When choosing which developing lignin applications to pursue for future commercialization, it is important to make sure that the approaches have the least negative effects on the environment (Bajwa et al., 2019). Depending on the available data and situation, several hierarchical approaches have been used by researchers for LCA studies; these are subdivision, system expansion by substitution, and allocation by mass, energy, or economic value. The allocation technique used in the multi-output process that produces the lignin output has a significant impact on the lignin's long-term climatic impact in any application. Potential rates and market fluctuations, the surrounding system (i.e., which products will be replaced by lignin and other system co-products), what appears as the primary product, the system's driver, or any changes made to the system—such as identifying the rationale behind lignin extraction from the pulp mill are all important factors in allocation. Because emerging technologies with multi-output processes are sensitive to temporal constraints, all of these factors should be considered in LCAs (Hermansson et al., 2020). Producing chemicals like catechol or adipic acid from lignin has yielded smaller carbon footprints and has used less fossil fuel when compared to petroleum-based methods. Corona et al. (2018) studied the synthesis of bio-based adipic acid using lignin obtained from the manufacturing of bioethanol, as opposed to the traditional petrochemical production process. Because a biorefinery side-stream is used as feedstock material and nitrous oxide emissions are avoided, the LCA forecasts a lower total environmental effect for the bio-based pathway. Compared to traditional adipic acid, bio-based adipic acid is expected to reduce GHG emissions by 62–78%, with an estimated 4.87 kg CO_2 eq. per kg AA. Several studies have investigated the possible effects on the overall environmental performance of pulp and paper factories as well as biorefineries of converting lignin wastes into higher-value products, such as fuel, as opposed to their traditional (conventional) use for energy recovery (Hermansson et al., 2020). Secchi et al. (2019) also investigated the effect of partially removing biorefinery lignin byproduct for market sale and concluded that the facility's environmental performance may be enhanced by using natural gas, a fuel with lower carbon intensity than lignin, for heat and power generation. However, there was no use for the extracted lignin byproduct within their system boundary. It was also shown that the life cycle carbon emissions of ethanol can be reduced to −82 g CO_2e/MJ of ethanol by using 100% second-generation biorefinery lignin as a composite filler. Furthermore, burning this byproduct lignin on-site or off-site with coal to produce electricity will produce 40 g CO_2e/MJ of ethanol, whereas applying it as a land amendment produces –2 g CO_2e/MJ of ethanol (Pourhashem et al., 2013). The lignin substitutes, energy inputs, and processing techniques are only a few of the things that affect LCA results. Therefore, even though lignin has a lot of potential, careful consideration is necessary to make sure that we are choosing the best environmental options.

11.2.2 Carbon footprint and greenhouse gas emissions

The ability to replace materials and chemicals derived from petroleum is one of its main benefits. For example, lignin-based composites, polyols, and adhesives can lessen dependency on fossil feedstocks, frequently resulting in large GHG savings (Tuck et al., 2012). Furthermore, by turning a waste stream into high-value materials, valorizing lignin instead of burning it for electricity enhances biorefinery sustainability and complies with waste minimization and emission reduction targets (Trezza et al., 2025). Since it makes it easier to incorporate renewable carbon into material and chemical cycles, lignin valorization contributes significantly to the advancement of circular bioeconomy models in addition to its immediate climatic advantages (Kemala et al., 2025). The carbon footprint of one kilogram of Kraft lignin might range from 4 kg of CO_2 eq to a negative impact, depending on the allocation strategy selected. Also, the sensitivity to process parameters and system modeling is shown by LCA studies that compare lignin-derived adipic acid with conventional petroleum-derived alternatives, reporting GHG savings varying from 90% (i.e., a large reduction) to +100% (i.e., doubling the emissions) (Moretti et al., 2021). The various GHG emissions profiles of lignin valorization processes are mostly determined by the allocation mechanism, system limits, and the chosen route of valorization. Net emissions are greatly impacted by energy-intensive extraction or conversion processes as well as the energy's carbon intensity. Such effects can be mitigated through the use of greener pretreatment technologies such as DES, enzymatic methods, or organosolv processes, which help reduce the energy intensity of lignin conversion. Life cycle studies provide numerical evidence of these benefits. Bernier et al. (2013) estimated the cradle-to-gate impact of producing 1 kg of Kraft lignin at 0.6 kg CO_2-eq, a figure that already incorporates a biogenic carbon credit of 2.3 kg CO_2-eq per kg of lignin. This credit reflects the assumption that biogenic carbon remains stored for more than a century, thereby reducing the apparent footprint (Mersmann et al., 2025). However, system boundaries and allocation choices can strongly influence such outcomes. For instance, Secchi et al. (2019) point out that diverting lignin into higher-value applications rather than combustion necessitates sourcing replacement energy often from natural gas or cogeneration systems, which may offset some of the expected climate benefits. Achieving genuine carbon neutrality or even a net negative balance requires not only careful process design but also the consistent use of low-carbon or renewable energy sources throughout the valorization pathway, alongside strategies to minimize emissions from conversion steps.

11.2.3 Waste minimization and circular economy

A vital part of the circular bioeconomy, lignin valorization offers prospects to conserve resources, lessen environmental impacts, and create sustainable substitutes for fossil-based products (Trezza et al., 2025). It is often described as a simple step toward making biorefineries more circular, but its role is in fact more nuanced. The pulp and paper industry generates lignin in huge quantities, and for decades, most of it has been burned for energy. While that practice does provide some value, it also confines lignin to low-grade uses and leaves higher-value opportunities untapped. Converting lignin into products such as bioplastics, adhesives, or functional composites provides an alternative pathway—one that treats lignin as a renewable feedstock rather than a waste stream (Trezza et al., 2025). Such a shift reduces reliance on landfilling and combustion and helps keep carbon cycling within the biorefinery (Gupta et al., 2025).

From a circular economy perspective, embedding renewable carbon in materials reduces the need for virgin resources and nudges the system closer to zero waste (Dessie et al., 2023). At the same time, the real sustainability gains are not easy to pin down. LCA is the main tool we have, and it can capture both the benefits like lower GHG emissions and the drawbacks, such as toxicity or eutrophication. The problem is that only a handful of thorough studies exist so far (Kylili et al., 2023), which leaves industry and policymakers with more questions than answers when it comes to deciding how best to use lignin. On the technological side, new approaches such as "lignin-first" biorefineries, which stabilize reactive intermediates at the fractionation stage—show promise in improving processability and opening up more efficient valorization routes (Brienza et al., 2024). In the sustainable bioeconomy, lignin has the potential to change a multitude of industries by providing renewable alternatives to chemicals and materials sourced from fossil fuels. By lowering carbon emissions, cutting waste, and encouraging resource efficiency, lignin valorization helps reach global sustainability objectives, including climate neutrality and the circular economy. It is projected that the use of lignin will grow as market demand and technological developments spur innovation, enhancing its standing as a vital resource for a more robust and sustainable future (Trezza et al., 2025). Yet, this remains easier said than done. The variability of lignin, its structural recalcitrance, and the high energy demands of several conversion processes all hamper industrial integration (Sethupathy et al., 2022). In practice, then, lignin valorization must be viewed as both a significant potential and a continual difficulty. Before lignin can fully contribute to waste minimization and the circular bioeconomy, infrastructure must be expanded, processing technologies refined, and knowledge gaps in environmental evaluation closed (Figure 11.2).

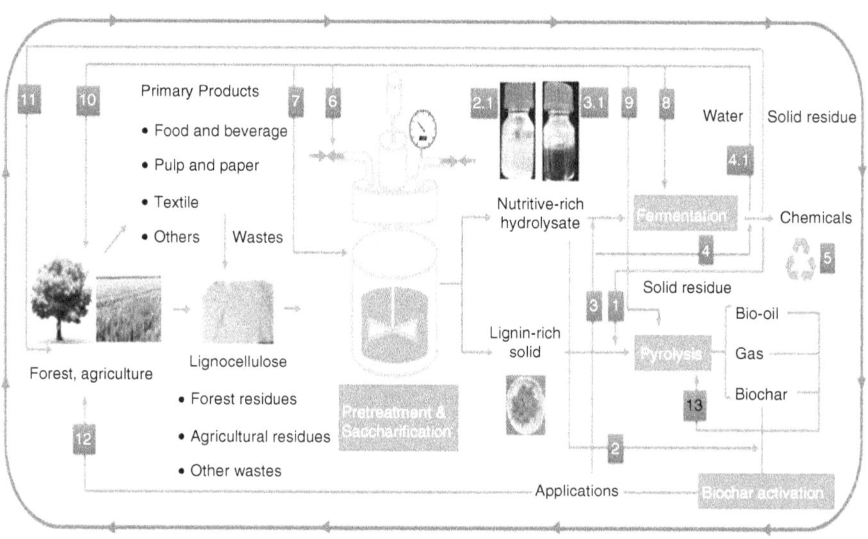

FIGURE 11.2 Process framework for full utilization of biomass, zero waste, and circular production in lignocellulose valorization through an integrated biorefinery approach Dessie et al. (2023)

11.2.4 Ecotoxicological considerations

Despite being bio-based, lignin-derived chemicals are not inherently harmless, and their ecotoxicological behavior must be carefully assessed. For example, toxicity tests have revealed that alkali lignin can affect aquatic creatures. In one study, fungal decolorization of 0.2% alkali lignin reduced—but did not completely eliminate—toxicity in *Vibrio fischeri* bioluminescence tests, demonstrating that treatment processes influence, but do not guarantee, environmental safety (Rybczyńska-Tkaczyk & Korniłłowicz-Kowalska, 2017). Also, lignin's phenolic content and molecular complexity can generate risks once released into the environment. For example, aromatic derivatives such as guaiacol, which occur naturally in the atmosphere, have been identified as environmentally relevant toxicants, reminding us that downstream effects on both air and water systems must be carefully assessed (Pflieger & Kroflič, 2017). In soil, lignin's resistance to breakdown raises its own concerns. Composting studies indicate that waste products rich in lignin can put ecological strain on microbial populations by changing their habitats, which may have long-term effects for soil health and stability (Huang et al., 2025). These findings emphasize the need for conducting targeted ecotoxicity studies rather than presuming safety solely because lignin is a renewable substance. On the other hand, lignin-derived carbon materials (LDCMs) are increasingly being explored as eco-friendly adsorbents for water and soil remediation. Thanks to their large surface area, tunable porosity, and diverse functional groups, these materials can capture pollutants such as dyes, pharmaceuticals, and heavy metals while fitting neatly into circular economy strategies. Lignin-based porous carbons have demonstrated impressive performance—for example, achieving 98% photodegradation of methylene blue under UV with hydrotropic lignin nanoparticles, as well as removing contaminants like PFOA (perfluorooctanoic acid) and ibuprofen through photocatalytic or adsorption mechanisms (Nath et al., 2025). In a similar vein, lignin-derived activated carbons and multifunctional carbon materials have shown strong adsorption of organic dyes, heavy metals (e.g., Cu, Zn, Pb), and even toxic gases such as SO_2 and H_2S (Dong et al., 2025). Taken together, these results underscore lignin's dual role: it can be developed into a sustainable tool for remediation, but its complex environmental interactions also make thorough ecotoxicological evaluation essential before large-scale application.

11.3 ECONOMIC ASPECTS OF LIGNIN VALORIZATION

11.3.1 Market potential and value chains

Lignin, an abundant byproduct of pulp and other lignocellulosic processes, is one of the most underutilized biomass components with significant potential for producing bio-based chemicals, fuels, and materials that can lessen dependence on fossil resources. Lignin valorization supports global sustainability goals like SDG 12 (responsible consumption) and SDG 13 (climate action) by fostering circular economy models and low-carbon industries. Emerging applications include bio-based composites, carbon fibers, adhesives, coatings, and nanomaterials such as lignin nanoparticles, which improve polymer matrices with antioxidant, antimicrobial, and UV-protective properties (Boarino & Klok, 2023). Additionally, LDCMs are increasingly used in energy storage devices, activated carbon, and water purification, offering diverse market opportunities (Jiju et al., 2025). Global production of kraft lignin (KL) alone is estimated at around 70 million tons annually, with extraction rates of 20–40% depending on mill configuration and feedstock (Argyropoulos et al., 2023). Finland has demonstrated the potential of lignin valorization to underpin national bioeconomy strategies, with targets to double the bioeconomy's contribution to GDP by 2035 (Österberg et al., 2024). Commercially emerging examples include Stora Enso, which markets lignin-based anode materials (Lignode®) for sodium-ion and lithium-ion batteries (Espinoza-Acosta et al., 2018). UPM BioPiva® lignin for the replacement of fossil phenol in phenolic resins. Lignosulfonates (LS) also represent a consolidated market segment with an estimated 900 kt/y production in 2023, used in dispersants, binders, and feed additives. Market prices vary significantly by lignin type and purity, ranging from €220–650/t for KL, €230–650/t for LS, and €1,100–1,250/t for organosolv lignin (OL) (Correa-Guillen et al., 2025). The value chain is strongly influenced by supply constraints, purity issues, and end-use potential. EU regulations are increasingly pushing industries to valorize biomass side streams, reinforcing lignin's role in circular bioeconomy strategies (Correa-Guillen et al., 2025).

11.3.2 Cost–benefit analysis

Techno-economic assessments (TEAs) consistently show that while lignin valorization raises production costs relative to combustion for energy, the added value often offsets this (Klein et al., 2023). The "lignin-first" biorefinery concept, prioritizing lignin valorization ahead of other biomass fractions, has been found to deliver competitive minimum selling prices (MSPs) relative to fossil alternatives (Moretti et al., 2022). Applications explored include adipic acid production via fermentation, hydrogenolysis for phenolic monomers, and lignin oils for maritime fuel (Moretti et al., 2022). Hydrolysis lignin (HL) has been assessed for superactive carbon (SAC), carbon dots, and bioethanol co-production, with results suggesting economic viability but requiring scale-up. Similarly, KL-derived activated carbon was shown to be competitive depending on lignin price and yield, while hydrogen production from KL remained less feasible due to cost and yield constraints (Correa-Guillen et al., 2025). For OL, TEAs demonstrated profitability for activating carbon, carbon fibers, and aromatics, particularly under lignin-first integration with pulp and sugar industries. However, production of lignin monomers and oligomers remains unprofitable at current scales (Gan et al., 2022). Importantly, combined techno-economic and life cycle assessment (TEA-LCA) studies reveal a nuanced trade-off: lignin valorization can reduce carbon emissions while remaining economically competitive (Arts et al., 2023). Yet, some low-emission pathways remain unprofitable (Brienza et al., 2024), highlighting the challenge of aligning environmental and economic objectives.

11.3.3 Investment and commercialization challenges

The transition from laboratory findings to industrial deployment requires substantial capital and operational investments. High capital expenditures, catalyst costs, and energy demands make processes expensive and limit scalability (Ali et al., 2024). Regulatory compliance, byproduct recovery inefficiencies, and the lack of reliable, uncontaminated feedstocks, particularly lignin sourced from the pulp and paper industries containing impurities, further constrain industrial adoption (Zhu et al., 2023). Furthermore, commercial lignin extraction capacity is unevenly distributed, with notable closures (e.g., Stora Enso's Sunila mill at 50,000 t/y) but also new demo plants (e.g., Metsä Group, Södra), with scale-up remaining as a bottleneck due to high upfront costs, limited infrastructure, and uncertainties in raw material supply (Correa-Guillen et al., 2025). Also, operational costs are heavily influenced by lignin purity, ash content, and variability in feedstock. For example, high ash limits LS applications in high-value markets, while color and odor restrict KL (Sethupathy et al., 2022; Covinich et al., 2024). In many TEAs, feedstock price and yield accounted for the majority of operating costs, directly affecting market competitiveness. These issues slow the transition of lignin valorization from concept to industrial adoption.

11.3.4 Policy and incentives frameworks

Government initiatives and bioeconomy policies are critical enablers. The EU has mandated valorization of biomass side streams and supports lignin projects through frameworks such as the Bio-based Industries Joint Undertaking (BBI-JU, now CBE-JU), which funds circular bio-based applications, including lignin coatings (e.g., LIGNICOAT) (Correa-Guillen et al., 2025). National strategies, such as Finland's Bioeconomy Strategy, envision valorizing 20–40% of available KL into diverse applications, potentially adding €670–1,500 million in economic value (Österberg et al., 2024). Incentives such as subsidies, carbon credits, and green certifications can further enhance lignin's competitiveness by reducing risks for early adopters. In the U.S., the Department of Energy's Bioenergy Technologies Office (BETO) provides grants and funding for lignin valorization research, encouraging innovation in biorefineries and high-value lignin-derived products (Jiju et al., 2025). Overall, lignin valorization sits at the intersection of technological innovation, market demand, and supportive policy. Its economic feasibility depends not only on process optimization and scale-up but also on coordinated efforts between academia, industry, and governments to overcome barriers and align economic incentives with sustainability goals.

11.4 CASE STUDIES AND INDUSTRIAL EXAMPLES

11.4.1 Real-world examples of lignin valorization in industry

A number of industrial projects have shown how lignin valorization is being executed at scale, connecting environmental benefits with economic viability. The world's largest KL extraction capacity, which has been in operation since 2015, is at Stora Enso's Sunila Mill in Finland. It has a yearly lignin volume of 50,000 tons, and the mill has developed Lignode®. This Lignode® is a lignin-based hard carbon developed as a replacement for fossil graphite in lithium-ion battery anodes. This invention not only minimizes dependence on mined graphite but also improves battery performance via faster charging and improved cold-temperature resilience, placing lignin as a planned material for sustainable electrification (Stora, 2022). Moving to North America, the LignoForce™ system has also been successfully produced by West Fraser's Hinton mill in Alberta, Canada, manufacturing 30 tons per day of high-quality lignin. The lignin obtained is used as a renewable replacement in fuel applications, resins, and antioxidants. The plant reveals robust replicability potential across pulp mills due to its relatively low capital cost and compatibility with present infrastructure, offering both environmental and economic benefits by debottlenecking recovery boilers while producing higher-value products (Kropat et al., 2021). Another remarkable example emanates from the Netherlands, where lignin has been used in asphalt production. Lignin obtained from companies such as Avantium and Vertoro has discovered significant environmental gains from an LCA of lignin-based asphalts using biorefinery. Substituting bitumen with lignin in roadways minimizes climate change influences by 35–70% for top-layer asphalts and 25–50% for base layers compared to conventional options (Moretti et al., 2022). Together, these exemplary cases indicate that lignin valorization is not just theoretical but is already being implemented in energy storage, construction, and material industries, which also contributes to emission minimization targets in infrastructure sectors.

11.4.2 Success stories and lessons learned

Numerous current techno-economic and life cycle studies highlight equally the successes and the key lessons learned in lignin valorization. One notable success is the establishment of the reductive catalytic fractionation (RCF) as a lignin-first strategy. RCF permits selective depolymerization of lignin while protecting carbohydrate fractions for ethanol production. A techno-economic study indicates that crude RCF oil can be produced at around $1.13/kg, but high capital costs, especially from high-pressure reactors and solvent recovery, remain a major challenge (Bartling et al., 2021). The important lesson here is that reducing solvent loading and reactor pressure, or using other solvents like ethanol or ethylene glycol, is likely to minimize both economic costs and environmental problems while improving scalability. Another success story stems from electrocatalytic (EC) lignin valorization. Researchers have shown that selective cleavage of β-O-4 linkages under mild conditions can be achieved using hydrogen atom transfer (HAT) mediators. A techno-economic and LCA analysis of integrating EC into biorefineries has also shown great potential for chemical production of approximately 50 tons per day, with up to 46% minimization in CO_2 production depending on allocation methods (Ebrahimpourboura et al., 2024). Lessons from this study stress that catalyst cost, stability, and product pricing powerfully affect profitability, meaning further innovation in low-cost, durable mediators is of importance. In this final success story, it has to do with LCAs of lignin-derived products that include phenolic resins, vanillin, and carbon fibers that provide broader insights. While lignin replacement can significantly cut fossil fuel use and minimize GHG production, studies also expose trade-offs in categories like ecotoxicity and human toxicity due to chemical inputs in processing (Kylili et al., 2023). The lesson learned here is that complete LCAs are needed to enhance environmental gains in climate performance. Collectively, these examples reveal that lignin valorization is rolling from concept to practice, with concrete environmental and economic benefits.

11.5 CHALLENGES AND FUTURE PERSPECTIVES

11.5.1 Technical, regulatory, and market challenges

Lignin valorization holds great potential, but its advancement is slowed by various obstacles. To mention a few, they include technical challenges, regulatory hurdles, and market barriers. The inherent complexity of lignin results in its technical challenges. In fact, it is heterogeneous and has a recalcitrant structure that makes its depolymerization difficult, leading to small yields and a broad mixture of products with limited downstream usability. Conventional fractionation techniques, like kraft or sulfite pulping, immensely change native lignin, thereby reducing its suitability for high-value applications (Dixon et al., 2024). Moreover, the strong carbon–carbon links and high bond dissociation energies in lignin functional groups upsurge energy and catalyst requirements, further elevating processing costs (Sethupathy et al., 2022). Additional technical barriers lie in lignin isolation and standardization, as batch-to-batch variability in lignin molecular weight, purity, and functional groups confounds consistent product development. Regulatory challenges, as one of the key factors that hold lignin potential back add another layer of complexity. Stringent approval processes that require rigorous toxicity, safety, and environmental impact studies for several lignin-derived products, especially those targeting the pharmaceutical, food, or biomedical markets, pose a significant barrier to lignin valorization (Sethupathy et al., 2022). Also, the absence of coordinated international standards for lignin-derived chemicals and materials confines global trade and drastically reduces commercialization. This is particularly relevant when lignin-based hydrogels, surfactants, or biocomposites are considered for consumer markets, where regulatory agencies require comprehensive proof of safety and reproducibility. Market challenges are equally important in lignin valorization. Currently, most industrial lignin is burned for energy because it remains the most economically viable option. Competing with low-cost petroleum-based products is tough, as fluctuating oil prices undercut the competitiveness of lignin-derived alternatives (Sethupathy et al., 2022). Moreover, the market for high-value lignin-based materials, for example, carbon fibers, resins, or advanced polymers, is still developing, and there are deficiencies in mature value chains. Other constraints, such as infrastructure limitations, high transportation costs of biomass residues, and insufficient investment in pilot-to-industrial-scale facilities, further worsen the situation (Dixon et al., 2024). Lignin valorization, therefore, faces these entangled challenges: the technical difficulty of overcoming lignin's structural recalcitrance, stringent regulatory frameworks governing product safety, and market barriers related to cost competitiveness and underdeveloped value chains. Addressing these challenges will demand integrated advances in plant biology, process engineering, supportive policies, and innovative business models to unlock lignin's full possibilities as a renewable feedstock in the circular bioeconomy.

11.5.2 Research gaps and innovation opportunities

Notwithstanding the enormous potential of lignin as a renewable feedstock for high-value chemicals, fuels, and materials, many research gaps hinder its large-scale utilization. One of the main challenges is lignin's complex and heterogeneous structure, which makes its depolymerization and conversion highly variable across plant sources and processing methods. These differences limit the development of standardized valorization routes and hinder commercial-scale adoption. In addition, most existing processes are energy-intensive, depend on harsh chemicals, or exhibit low selectivity, leading to inefficiencies and undesired by-products (Ali et al., 2024). Microbial and enzymatic processes work under mild conditions but face challenges of slow reaction rates, low tolerance to lignin-derived aromatics, and limited scalability for industrial deployment. Engineered microbes, such as *Ralstonia eutropha* H16, show promise in changing lignin-derived aromatics into bioplastics, but their native metabolic pathways are often imperfect or inefficient without extensive genetic modification (Zhao et al., 2024). Another gap lies in the limited efficiency and scalability of catalytic and biological depolymerization methods. While catalytic approaches like the zeolites and sulfonated carbon catalysts expand product yields and selectivity, issues of high catalyst cost, deactivation, and recovery remain unsettled (Ali et al., 2024). Also, not having a full understanding of lignin's molecular structure and function impedes its progress. Mass spectrometry, FTIR, and NMR are some of the examples of the advanced characterization techniques that have upgraded structural elucidation, yet other strong analytical frameworks are required to control targeted pretreatment and conversion strategies (Ali et al., 2024). Inadequate comprehensive LCAs for upcoming lignin-based products, creating doubt about their real environmental significance compared to fossil-derived changes. For the innovation side, current progressions in hub metabolite-based autoregulation systems in engineered microbes enhanced dynamic control of gene expression in response to lignin-derived substrates, thereby making it efficient and minimizing the metabolic burden on cells (Zhao et al., 2024). Several opportunities are developing. Electrochemical depolymerization is renowned for gaining sight for its capacity to function under mild conditions, integrate renewable electricity, and provide fine-tuned selectivity toward desirable products. Also, merging DES with electrochemical methods gives greener, more selective pathways for lignin conversion (Ali et al., 2024). In the situation of material applications, innovations in bioplastic creation from lignin signify a critical opportunity to address global plastic pollution. Engineering microbial strains with improved aromatic demethylation and aldehyde detoxification pathways has already proven better-quality yields of biodegradable polyhydroxybutyrate (PHB) from lignin-derived aromatics (Zhao et al., 2024). These strategies indicate the potential of integrating synthetic biology, catalysis, and process engineering to create new product portfolios. Hence, the crucial research gaps spin around scalability barriers, lignin's structural complexity, process inefficiencies, and catalyst restrictions. As for the innovation opportunities, it vastly embodies electrochemical valorization, advanced catalytic systems and comprehensive LCAs to validate sustainability claims, and synthetic biology for microbial upgrading. Solving these stringent situations, especially these gaps, will help fast-track lignin's revolution from an underutilized byproduct into a cornerstone of green industries.

11.5.3 Future Outlook for Sustainable Lignin Valorization

The future of sustainable lignin valorization needs interdisciplinary collaboration between the combination of green chemistry, advanced materials science, and circular bioeconomy principles. While lignin has long been tipped as an industrial byproduct, recent inventions are repositioning it as a strategic feedstock for next-generation bio-based industries. For example, LDCMs are evolving as scalable replacements to key raw materials in batteries and supercapacitors, whereas agricultural and cosmetic applications display their changeability and growing industrial value (Trezza et al., 2025). Crucial to this outlook is the development of lignin into advanced material applications. Etherification, polymer grafting, and esterification are some examples of functionalization strategies that enable lignin to be tailored for use in carbon fibers, self-healing elastomers, 3D printing resins, and bioplastics. Such adjustments solve the inherent issues like high polarity and poor melt processability, making the way for lignin-based composites with advanced outcomes and sustainability (Shorey et al., 2024). Looking forward, industrial adoption will rely not only on technical advances but also on a holistic biorefinery combination. Studies underscore that valorization strategies must go in line with the core principles of the circular bioeconomy, making sure that lignin-manufactured products originally minimize carbon footprints and do not move environmental challenges to other impact categories. Current analyses indicate that lignin-based polymers, composites, and biofuels can help achieve the global sustainability goals when combined with LCAs and techno-economic evaluations that guide scale-up decisions (Jiju et al., 2025). The future of lignin valorization is moving beyond energy recovery to high-value, application-driven solutions. By harnessing innovations in catalytic conversion, functionalization, and biorefinery combination, lignin can move from an underutilized residue into a cornerstone of sustainable material innovation. Ensuring this transition will need not only technical breakthroughs but also strong policy support and market incentives that ensure industrial confidence in lignin-based technologies. To reveal lignin's full potential, stakeholders must implement an integrated approach. Low-energy conversion technologies, scalable, comprehensive LCAs, and advanced catalysts are some of the measures researchers should focus on to ensure genuine sustainability. Industry players are also encouraged to diversify lignin applications into high-value markets such as energy storage, bioplastics, and healthcare while building consistent supply chains and pilot-to-industrial-scale infrastructure. Policymakers and regulators should harmonize standards for lignin-derived materials, streamline approval processes, and create incentives (e.g., carbon credits, green financing) that improve competitiveness against fossil alternatives. Lignin's versatility can be influenced by investors and entrepreneurs to advance business models that go in line with circular bioeconomy goals, targeting sectors like construction, packaging, and renewable energy.

11.6 CONCLUSION

Lignin, though historically reported as underutilized, has increasingly been recognized as a foundation of sustainable biorefineries. Some of the key findings are: Electrocatalytic valorization (EC), DES, and catalytic approaches like RCF are some examples of innovative extraction techniques that produce and ensure higher yields, selectivity, and lower environmental problems. It has been proven by LCAs that lignin-based products like adhesives, carbon fibers, vanillin, bioplastics, and asphalt binders can minimize GHG production and fossil fuel use as compared to petrochemical counterparts. Lignin-modified asphalts in the Netherlands, Stora Enso's Lignode® for batteries, and West Fraser's LignoForce™ are some of the industrial examples that prove that lignin valorization is already commercially feasible in certain areas. Yet, some challenges still persist that include high processing costs, lignin's structural heterogeneity, stringent regulatory hurdles, lack of standardization, and competition with low-cost petroleum-based products that hamper large-scale adoption.

ACKNOWLEDGMENT

The authors would like to acknowledge the King Fahd University of Petroleum and Minerals (KFUPM), Dhahran 31261, Saudi Arabia, for the conduction of this study.

REFERENCES

Ali, S., A. Rani, M. A. Dar, et al. 2024. Recent advances in characterization and valorization of lignin and its value-added products: Challenges and future perspectives. *Biomass* (Switzerland) 4(3): 947–77. https://doi.org/10.3390/biomass4030053

Argyropoulos, D. D. S., C. Crestini, C. Dahlstrand, et al. 2023. Kraft lignin: A valuable, sustainable resource, opportunities and challenges. *Chem Sus Chem* 16 (23): 1–34. e202300492.https://doi.org/10.1002/cssc.202300492

Arts, W., K. V. Aelst, E. Cooreman, J. V. Aelst, S. Van den Bosch, B. F. Sels. 2023. Stepping away from purified solvents in reductive catalytic fractionation: A step forward towards a disruptive wood biorefinery process. *Energy & Environmental Science* 16 (6): 2518–39.

Bajwa, D. S., G. Pourhashem, A. H. Ullah, S. G. Bajwa. 2019. A concise review of current lignin production, applications, products and their environmental impact. *Industrial Crops and Products* 139: 111526. https://doi.org/10.1016/j.indcrop.2019.111526

Bartling, A. W., M. L. Stone, R. J. Hanes, et al. 2021. Techno-economic analysis and life cycle assessment of a biorefinery utilizing reductive catalytic fractionation. *Energy and Environmental Science* 14 (8): 4147–68. https://doi.org/10.1039/d1ee01642c

Bernier, E., C. Lavigne, P. Y. Robidoux. 2013. Life cycle assessment of kraft lignin for polymer applications. *The International Journal of Life Cycle Assessment* 18 (2): 520–28. https://doi.org/10.1007/s11367-012-0503-y

Boarino, A., H. A. Klok. 2023. Opportunities and challenges for lignin valorization in food packaging, antimicrobial, and agricultural applications. *Biomacromolecules* 24 (3): 1065–77.https://doi.org/10.1021/acs.biomac.2c01385

Brienza, F., D. Cannella, D. Montesdeoca, I. Cybulska, D. P. Debecker. 2024. A guide to lignin valorization in biorefineries: Traditional, recent, and forthcoming approaches to convert raw lignocellulose into valuable materials and chemicals. *RSC Sustainability* 2 (1): 37–90. https://doi.org/10.1039/D3SU00140G

Corona, A., M. J. Biddy, D. R. Vardon, M. Birkved, M. Z. Hauschild, G. T. Beckham. 2018. Life cycle assessment of adipic acid production from lignin. *Green Chemistry* 20 (16): 3857–66. https://doi.org/10.1039/C8GC00868J

Correa-Guillen, E., K. A. Henn, M. Österberg, L. Dessbesell. 2025. Lignin's role in the beginning of the end of the fossil resources era: A panorama of lignin supply, economic and market potential. *Current Opinion in Green and Sustainable Chemistry* 54: 101038.

Covinich, L. G., M. C. Area. 2024. Trends and limitations of lignin as a starting material. *BioResources* 19 (1): 1–5.

Dessie, W., X. Luo, F. He, Y. Liao, G. J. Duns, Z. Qin. 2023. Lignin valorization: A crucial step towards full utilization of biomass, zero waste and circular bioeconomy. *Biocatalysis and Agricultural Biotechnology* 51: 102777. https://doi.org/10.1016/j.bcab.2023.102777

Dixon, R. A., A. Puente-Urbina, G. T. Beckham, Y. Román-Leshkov. 2024. Enabling lignin valorization through integrated advances in plant biology and biorefining. *Annual Review of Plant Biology* 75 (1): 239–63. https://doi.org/10.1146/annurev-arplant-062923-022602

Dong, X., Y. Zhang, S. Shao, H. Li, X. Yan. 2025. Application of lignin-derived carbon materials in adsorption and separation. *Separations* 12 (4): 88. https://doi.org/10.3390/separations12040088

Ebrahimpourboura, Z., M. Mosalpuri, C. Yang, et al. 2024. Comparative techno-economic and life cycle assessment of electrocatalytic processes for lignin valorization. *Green Chemistry* 26 (22): 11303–15. https://doi.org/10.1039/d4gc01963f

Espinoza-Acosta, J. L., P. I. Torres-Chávez, J. L. Olmedo-Martínez, A. Vega-Rios, S. Flores-Gallardo, E. A. Zaragoza-Contreras. 2018. Lignin in storage and renewable energy applications: A review. *Journal of Energy Chemistry* 27 (5): 1422–38.

Gan, M. J., Y. Q. Niu, X. J. Qu, C. H. Zhou. 2022. Lignin to value-added chemicals and advanced materials: Extraction, degradation, and functionalization. *Green Chemistry* 24 (20): 7705–50.

Gupta, N., W. P. T. Wijenayake, D. Roy, et al. 2025. Smart recycling and sustainable lignocellulosic waste management. In *Value Addition and Utilization of Lignocellulosic Biomass,* edited by G. Mukherjee and S. Dhiman, 221–50. Springer Nature Singapore. https://doi.org/10.1007/978-981-96-2786-8_9

Hermansson, F., M. Janssen, M. Svanström. 2020. Allocation in life cycle assessment of lignin. *The International Journal of Life Cycle Assessment* 25 (8): 1620–32. https://doi.org/10.1007/s11367-020-01770-4

Huang, F., Y. Lu, Z. Li, C. Zhang, X. Qiao, J. Wu. 2025. The impact of ecological pressure caused by toxic lignin metabolites on microbial function transformation during composting. *International Journal of Biological Macromolecules* 307: 142021. https://doi.org/10.1016/j.ijbiomac.2025.142021

Jiju, P. S., A. K. Patel, N. S. Shruthy, et al. 2025. Sustainability through lignin valorization: Recent innovations and applications driving industrial transformation. *Bioresources and Bioprocessing* 12 (1), 1–23. https://doi.org/10.1186/s40643-025-00929-x

Kemala, P., A. H. Prianto, A. Martino, et al. 2025. Valorization of lignin as an active agent in sustainable agriculture: A review. *Journal of Industrial and Engineering Chemistry* 150: 101–15. https://doi.org/10.1016/j.jiec.2025.03.001

Klein, I., B. Liu, S. Prager, S. Zhao. 2023. SPERLU Selective Process for Efficient Removal of Lignin and Upgrading (Final Technical Report) (No. DOE-SPERO-8503-4). *Spero Renewables*, LLC, Gaviota, CA.

Kropat, M., M. Liao, H. Park, K. S. Salem, S. Johnson, D. S. Argyropoulos. 2021. A perspective of lignin processing and utilization technologies for composites and plastics with emphasis on technical and market trends. *BioResources* 16 (1): 2084–115.

Kylili, A., M. Koutinas, P. Z. Georgali, P. A. Fokaides. 2023. Lignin valorisation: Life Cycle Assessment (LCA) considerations for enabling circular bioeconomy. *International Journal of Sustainable Energy* 42 (1): 1008–27. https://doi.org/10.1080/14786451.2023.2250869

Lobato-Rodríguez, Á., B. Gullón, A. Romaní, P. Ferreira-Santos, G. Garrote, P. G. Del-Río. 2023. Recent advances in biorefineries based on lignin extraction using deep eutectic solvents: A review. *Bioresource Technology* 388 (August), 1–12. https://doi.org/10.1016/j.biortech.2023.129744

Mersmann, L., C. Dias, J. Pires, A. L. Fernando. 2025. The green shift in lignin recovery: Sustainability and impact of pretreatment methods. In *Advances in lignin research*, 192–209. Springer. https://doi.org/10.1007/978-3-031-94484-0_16

Moretti, C., B. Corona, R. Hoefnagels, I. Vural-Gürsel, R. Gosselink, M. Junginger. 2021. Review of life cycle assessments of lignin and derived products: Lessons learned. *Science of The Total Environment* 770: 144656. https://doi.org/10.1016/j.scitotenv.2020.144656

Moretti, C., R. Hoefnagels, M. van Veen, et al. 2022. Using lignin from local biorefineries for asphalts: LCA case study for the Netherlands. *Journal of Cleaner Production* 343: 131063. https://doi.org/10.1016/j.jclepro.2022.131063

Nath, S., A. Naha, K, Saikia, C. P. Choudhury, V. Venkatramanan. 2025. Degradation of organic pollutants using lignin-derived carbon materials as a sustainable approach to environmental remediation. *Biotechnology for Sustainable Materials* 2 (1): 11. https://doi.org/10.1186/s44316-025-00036-z

Österberg, M., M. Karjalainen, J. Lintunen, et al. 2024. From timber to medicine value added for the forest sector through broadening the product Portfolio. https://urn.fi/URN:ISBN:978-952-65456-3-9

Pflieger, M., A. Kroflič. 2017. Acute toxicity of emerging atmospheric pollutants from wood lignin due to biomass burning. *Journal of Hazardous Materials* 338: 132–39. https://doi.org/10.1016/j.jhazmat.2017.05.023

Pourhashem, G., P. R. Adler, A. J. McAloon, S. Spatari. 2013. Cost and greenhouse gas emission tradeoffs of alternative uses of lignin for second generation ethanol. *Environmental Research Letters* 8 (2): 025021. https://doi.org/10.1088/1748-9326/8/2/025021

Rybczyńska-Tkaczyk, K., T. Korniłłowicz-Kowalska. 2017. Biotransformation and ecotoxicity evaluation of alkali lignin in optimized cultures of microscopic fungi. *International Biodeterioration & Biodegradation* 117: 131–40. https://doi.org/10.1016/j.ibiod.2016.12.011

Secchi, M., V. Castellani, M. Orlandi, E. Collina. 2019. Use of lignin side-streams from biorefineries as fuel or co-product? Life cycle analysis of bio-ethanol and pulp production processes. *BioResources* 14 (2): 4832–65. https://doi.org/10.15376/biores.14.2.4832-4865

Sethupathy, S., G. M. Morales, L. Gao, et al. 2022. Lignin valorization: Status, challenges and opportunities. *Bioresource Technology* 347: 126696. https://doi.org/10.1016/j.biortech.2022.126696

Shorey, R., A. Salaghi, P. Fatehi, T. H. Mekonnen. 2024. Valorization of lignin for advanced material applications: A review. *RSC Sustainability* 2 (4): 804–31. https://doi.org/10.1039/d3su00401e

Stora, E. O. 2022. Lignode ® by Stora Enso. https://f.hubspotusercontent40.net/hubfs/4056792/Biomaterials/SE_Lignode_whitepaper.pdf?__hstc=254639703.00fe250699623a5900c425a0ebc8ed6d.1665941239734.1665941239734.1665941239734.1&__hssc=254639703.1.1665941239734&__hsfp=1973820561&hsCtaTracking=8feb8e38

Trezza, A., L. Mahboob, A. Visibelli, M. Geminiani, A. Santucci. 2025. Lignin waste valorization in the bioeconomy era: Toward sustainable innovation and climate resilience. *Applied Sciences* (Switzerland) 15 (14): 1–21. https://doi.org/10.3390/app15148038

Tuck, C. O., E. Pérez, I. T. Horváth, R. A. Sheldon, M. Poliakoff. 2012. Valorization of biomass: Deriving more value from waste. *Science* 337 (6095): 695–99. https://doi.org/10.1126/science.1218930

Zhao, Y., L. Xue, Z. Huang, et al. 2024. Lignin valorization to bioplastics with an aromatic hub metabolite-based autoregulation system. *Nature Communications* 15 (1): 1–17. https://doi.org/10.1038/s41467-024-53609-3

Zhu, D., B. Yang, H. Wang, M. Shahnawaz 2023. Lignin valorization: Recent trends and future perspective. *Frontiers in Bioengineering and Biotechnology* 11: 1190128.

12 Challenges and future perspective of lignin biorefinery

Rohit Kumar Singh, Parul Bali, Aditya Sharma, Pushpak Dahiya, Ranju Kumari Rathour, and Ravi Kant Bhatia

12.1 INTRODUCTION

Plant-derived lignocellulosic biomass, which includes the phenolic polymer lignin and the two carbohydrate polymers cellulose and hemicellulose, is the most abundant and sustainable energy on Earth. Although lignin biomass offers high recalcitrance, lignocellulosic biomass offers a substitute for fossil fuels in the manufacturing of second-generation biofuels and other bio-based compounds (Okolie et al., 2021). Lignocellulosic biorefineries are essential for advancing energy and environmental sustainability as well as for empowering the emerging bioeconomy (Asghar et al., 2022). According to recent consensus, the simultaneous breakdown and use of the three main components of lignocellulosic biomass (LCB)—cellulose, hemicelluloses, and lignin—is crucial for a sustainable biorefinery (Questell-Santiago et al., 2020). Lignin, which is the second most prevalent terrestrial polymer after cellulose, holds enormous promise as a sustainable source for materials, chemicals, and fuels (Liu et al., 2018; Rathour et al., 2023). In conventional biorefineries, lignin was long regarded as a low-value waste product generated during the conversion of lignocellulosic biomass into biofuels (Kumari & Singh, 2018). However, once extracted, lignin has gained attention as a rich natural reservoir of diverse chemicals and high-value products (Banu et al., 2021). Consequently, large quantities of lignin released by the pulp and paper industry and traditional biomass refineries have highlighted both the urgent need and significant opportunity to convert this polymer into value-added materials (Velvizhi et al., 2021). Although several physical and chemical techniques have been developed to extract useful molecules from lignin, these methods face limitations. Therefore, microorganisms capable of modifying or degrading lignin are essential to access embedded polysaccharides and release valuable intermediates (Silva et al., 2021).

Recent advances show that microbial bioconversion of lignin holds great potential, as certain microorganisms have evolved specialized metabolic pathways to

DOI: 10.1201/9781003653189-12

transform a wide range of aromatic compounds via a "biological funnel" into commercially useful products (Davis & Moon, 2020). Among them, lignolytic microbes such as *Pseudomonas putida* possess robust enzymatic systems to degrade lignin oligomers and aromatics, channeling them into the production of polyhydroxyalkanoates (PHAs) intracellular biopolymers that store carbon and energy (Ma et al., 2020). Importantly, PHAs represent a sustainable alternative to petrochemical plastics, as they are biodegradable polyesters that can be used to manufacture bioplastics, thus addressing global environmental concerns (Salvachúa et al., 2020). Integrating lignin-to-PHA conversion into biorefineries could therefore enhance both profitability and sustainability (Wu et al., 2020).

The lignin bioconversion process generally involves several steps: pretreatment, depolymerization into small aromatic molecules, microbial breakdown into central metabolites or intermediates, and ultimately the biosynthesis of value-added compounds (Radhika et al., 2021). However, lignin's complex chemical structure poses challenges, including inefficient fractionation, limited processability, and low yields of target products. Harsh pretreatments often exacerbate these issues, generating condensed lignin structures that are more resistant to microbial degradation (Schutyser et al., 2018). To address these limitations, the concept of a "lignin-first" biorefinery has been proposed, wherein lignin valorization is prioritized during biorefinery design. This approach focuses on modifying lignin's chemistry early in processing to produce less condensed, more processible fractions (Abu-Omar et al., 2020). Implementing this strategy requires not only novel fractionation techniques but also significant adjustments to pretreatment operations, which could influence carbohydrate processing for fuel production. The overarching aim is to optimize lignin utilization while minimizing disruptions to existing biofuel pathways. Overcoming these technical challenges requires improving lignin solubility and reactivity through advanced processing technologies. A systematic evaluation of lignin processibility under current pretreatments, identification of favorable chemical structures for enhanced conversion, and strategies to balance lignin valorization with carbohydrate-based biofuel production are critical (Chauhan et al., 2025). Additionally, the economic feasibility of lignin valorization must be carefully assessed, particularly in terms of overall biorefinery output and capital costs. This chapter explores the role of lignin in modern biorefineries, strategies for its valorization, current challenges, and future perspectives in building sustainable lignin-based bioprocesses.

12.2 CHALLENGES IN LIGNIN BIOREFINERY

Because of its structural complexity, heterogeneity, and recalcitrance, lignin is difficult to depolymerize and selectively convert into valuable compounds, posing considerable hurdles to the lignin biorefinery. Product homogeneity and process standardization are made more difficult by the variation in lignin concentration across various biomass sources (Rathour et al., 2025). Thermochemical and catalytic processes used today frequently have large energy requirements, harsh reaction conditions, and poor selectivity. They are discussed below and also depicted in Figure 12.1. Furthermore, it is still technically and financially difficult to separate and purify intermediates produced from lignin. Large-scale implementation is

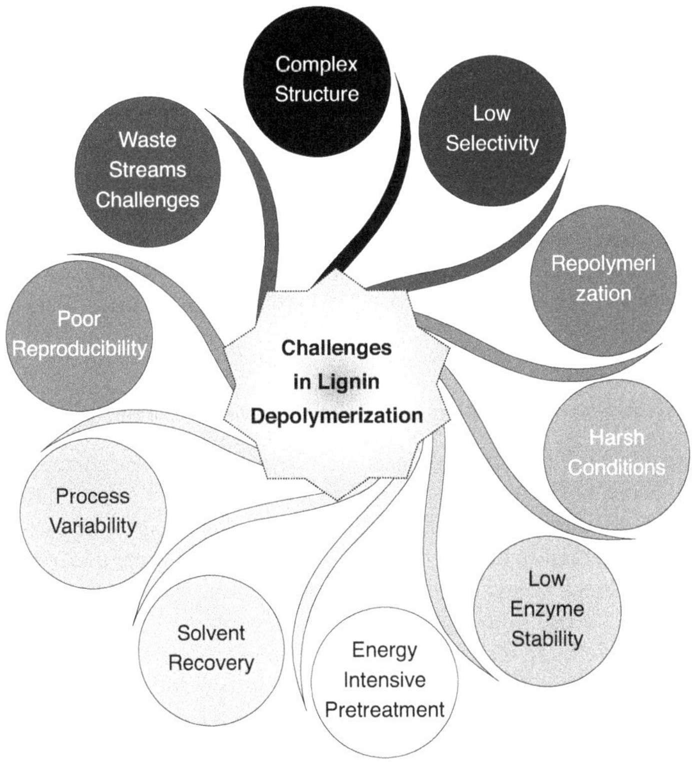

FIGURE 12.1 Challenges in lignin depolymerization

further limited by a lack of effective biocatalytic pathways, high operating costs, and a lack of knowledge on lignin reactivity, which calls for integrated, sustainable, and financially feasible biorefinery strategies.

12.2.1 Structural heterogeneity and recalcitrance

Lignin represents a critical bottleneck in biorefinery applications owing to its inherent structural heterogeneity and recalcitrance. Unlike cellulose and hemicellulose, which exhibit relatively uniform and predictable architectures, lignin is a highly irregular, polyphenolic macromolecule composed of guaiacyl (G), syringyl (S), and p-hydroxyphenyl (H) units interconnected through more than ten distinct ether and carbon–carbon linkages. The distribution of these linkages varies with plant species, developmental stage, and extraction method, giving rise to substantial structural diversity (Beckham et al., 2016; Xie et al., 2016). The inherent heterogeneity of lignin gives rise to complex product streams during depolymerization, complicating the production of high-purity lignin-derived chemicals and creating significant challenges for downstream processing. Structurally, lignin imparts rigidity and hydrophobicity to the plant cell wall by embedding within and covalently linking to cellulose and hemicellulose, resulting in a highly recalcitrant lignin–carbohydrate complex (Rathour et al., 2023) This integration not only reinforces plant cell walls but also hampers enzymatic hydrolysis of biomass through two primary mechanisms: (i) restricting access to polysaccharides via physical shielding and lignin–carbohydrate associations and (ii) sequestering hydrolytic enzymes in non-productive adsorption, thereby reducing their catalytic efficiency (Li et al., 2016). In addition, depolymerization often induces undesirable condensation reactions that generate chemically intractable residues, further diminishing the efficiency of lignin valorization (Ali et al., 2024). Taken together, these properties highlight lignin's paradoxical role—serving as a vital structural component in plants while simultaneously presenting one of the most formidable barriers to achieving cost-effective and sustainable lignin-based biorefineries.

12.2.2 Technical bottlenecks in lignin depolymerization

Efficient depolymerization of lignin into well-defined, value-added aromatic monomers remains one of the most formidable challenges in lignin valorization. This difficulty arises from the complex and irregular distribution of ether and carbon–carbon linkages within the lignin matrix, which resist uniform cleavage under conventional processing conditions (Rana et al., 2019). Condensation and repolymerization reactions occurring during both fractionation and depolymerization lower monomer yields and generate more recalcitrant residues. To overcome this, stabilization strategies are required; however, they frequently introduce additional complexity and cost (Li et al., 2018; Liu et al., 2020). Thermal depolymerization, while conceptually straightforward, is constrained by poor selectivity, high energy demands, and extensive repolymerization, which generate heterogeneous mixtures of tar, char, and low concentrations of monomers. These factors complicate separation and purification, limiting its application largely to pilot-scale operations. Pyrolysis, one of the most widely studied thermal approaches, yields gases, volatile liquids, and phenolic monomers under low-oxygen conditions (Bhatia et al., 2020). However, it faces several challenges: (i) broad and variable product distributions strongly influenced by lignin source and pretreatment, (ii) difficulty in controlling reaction pathways since bond cleavage occurs across a wide temperature range, and (iii) severe coke formation, particularly from eugenol at ~450°C and methoxy groups in guaiacol above 550°C, which reduces monomer yields and deactivates catalysts. Although harsher conditions and catalytic additives can enhance depolymerization and improve selectivity, they also intensify repolymerization and increase overall process complexity and cost (Pei et al., 2024; Ali et al., 2024). Catalytic depolymerization provides a more selective route for bond cleavage and the generation of high-value products, but its industrial adoption remains limited due to the need for harsh reaction conditions, expensive catalysts, catalyst deactivation, and difficulties in recovery and recycling. Biological depolymerization, mainly mediated by microbial enzymes, offers environmentally benign pathways but suffers from low reaction rates, sensitivity to environmental fluctuations, and challenges in scale-up and process control (Bhatt et al., 2018). Alternative chemical strategies also show potential yet face critical limitations. Oxidative depolymerization achieves efficient bond cleavage but generates significant byproducts, requires large oxidant inputs, and raises environmental concerns (Bhatia et al., 2020). In contrast, reductive depolymerization is relatively selective but depends on costly reducing agents and complex operating conditions, with limited opportunities for reagent recovery and reuse. Taken together, these technical bottlenecks reveal fundamental barriers to efficiency, selectivity, cost, and scalability in lignin valorization. Overcoming them requires integrated strategies that not only promote selective bond cleavage but also stabilize reactive intermediates, thereby enabling sustainable and economically viable lignin-based biorefineries (Ali et al., 2024).

12.2.3 Limited efficiency of catalysts and enzymes

Despite considerable advances in enzymatic lignin depolymerization, large-scale deployment of lignin-degrading enzymes remains constrained by their low stability, narrow substrate specificity, and vulnerability to harsh industrial conditions. For example, peroxidases are readily inactivated by excess hydrogen peroxide, while lignin peroxidase activity is strongly influenced by substrate size and solubility. The inherent insolubility of lignin often necessitates the use of organic solvents, which can denature enzymes and compromise catalytic efficiency (Rathour et al., 2025). Although overexpression strategies and protein engineering have improved yields of certain enzymes, many still exhibit minimal or no catalytic activity, and in some cases, the target proteins are not expressed at all. Industrial adoption of fungal enzyme systems is further impeded by incomplete genomic information and persistent challenges in protein expression and genetic modification. Another major limitation is the strong dependence of key ligninolytic enzymes, such as laccases and peroxidases, on redox mediators. However, the underlying enzyme–mediator interactions are not yet systematically understood. While mediators like hydroxyaryl aldehydes (HAA) and catechol can enhance redox efficiency, their widespread use is restricted by high costs, difficulties in recycling, potential toxicity, and inconsistent performance across different lignin feedstocks. Moreover, stereoselectivity in substrate oxidation adds another layer of complexity, as distinct enzyme–mediator systems often act selectively on specific stereoisomers of lignin linkages, producing variable outcomes (Bhatia et al., 2020). Collectively, these challenges highlight the urgent need for the development of more robust, solvent-tolerant enzymes, alongside integrated catalytic approaches that combine enzymatic and chemical strategies. Such innovations will be critical to achieving efficient, scalable, and sustainable lignin valorization (Chan et al., 2020).

12.2.4 High processing costs and energy demands

Lignin biorefineries continue to face critical barriers related to high processing costs and energy demands, which undermine both economic and environmental sustainability. Among the most expensive and energy-intensive steps are fractionation and pretreatment processes such as solvolysis and hydrolysis, which require high pressures, elevated temperatures, and costly solvents (Rathour et al., 2023). These operations not only necessitate specialized solvent recovery systems and high-pressure reactors but also drive up both capital and operating expenditures. Additional purification steps further elevate energy consumption, while variability across biomass sources adds process heterogeneity and economic uncertainty. Collectively, pretreatment and fractionation remain the primary cost bottlenecks in lignin valorization (Bhatia et al., 2020). A major contributor to energy intensity is solvent recycling, typically achieved via distillation, which can consume up to 73% of the initial biomass energy content and represent nearly 35% of total operating expenses. Such separation processes demand continuous energy input and are therefore among the most significant obstacles to lignin utilization. High-pressure reactor systems used in reductive catalytic fractionation (RCF) further escalate capital costs due to stringent requirements for maintenance, safety infrastructure, and specialized construction materials (Bartling et al., 2021; Obydenkova et al., 2022; Liu et al., 2021). Moreover, supporting facilities—including wastewater treatment units, lignin drying operations, and combined heat and power plants—add to infrastructure costs while increasing greenhouse gas emissions (Shafiei-Alavijeh et al., 2024). In addition to thermal and chemical barriers, the recalcitrance of lignin imposes severe biocatalytic limitations. Owing to its complex and heterogeneous structure, lignin is highly resistant to enzymatic attack, often requiring large enzyme dosages to achieve meaningful depolymerization. However, high enzyme consumption significantly raises raw material costs, while enzyme production, extraction, and recycling demand additional energy inputs. This creates a cost–energy feedback loop that further constrains the viability of enzymatic depolymerization. As a result, enzyme inefficiency remains one of the most persistent obstacles to economically feasible lignin valorization (Ullah et al., 2022).

12.2.5 Scale-up and Industrial Integration Issues

Similar to petroleum refineries, biorefineries operate on large-scale biomass processing but face significant challenges during scale-up. Successful scale-up requires efficient coordination between "cell factories" (biological production systems) and "steel factories" (manufacturing, recovery, and purification units operating under strict regulatory frameworks). Seamless integration of these two domains is vital to drive product and process innovation in lignin-based biorefineries (Chandel et al., 2018). Due to the inherent complexity and multiple interfaces in biomass processing, results obtained at laboratory or pilot scales are often difficult to reproduce at commercial scales handling several tons of biomass. Therefore, developing accurate process flow sheet, robust process modeling, advanced simulation, cost sensitivity analysis, life cycle assessment, forecasting, and risk assessment are essential components of scale-up strategies. With global demand for bio-based products increasing at 3–4% annually, efficient scaling of biorefineries has become a pressing need (Junqueira et al., 2016). However, lignin's complex and heterogeneous structure introduces additional uncertainty. Its variable chemical composition often leads to uneven product streams, making depolymerization and conversion difficult to standardize at an industrial scale. Variability in feedstock source and isolation methods further exacerbates batch-to-batch inconsistencies, complicating process control and downstream applications (Calvo-Flores & Martín-Martínez, 2022). Industrial integration of lignin valorization into existing biorefinery operations also presents major obstacles.

While co-location with pulp and paper industries can create synergies in resource utilization, differences in feedstock characteristics, operating conditions, and product specifications hinder seamless integration (Singh et al., 2021). Achieving a balance between carbohydrate valorization and lignin valorization remains particularly challenging, as prioritizing one stream often compromises the yield and efficiency of the other, reducing overall profitability and sustainability. These technical constraints are compounded by external challenges, including irregular biomass supply, seasonal variations, and fluctuating market demand for lignin-derived products. Such uncertainties further impact supply chain management and threaten economic viability (Liu et al., 2021; Özdenkçi et al., 2016).

12.2.6 Environmental and sustainability concerns

Environmental and sustainability considerations represent one of the most pressing challenges in lignin-based biorefineries. Many pretreatment and fractionation processes are highly energy- and chemical-intensive, resulting in significant environmental footprints (Islam et al., 2019). While lignin valorization offers a promising pathway toward sustainable biofuels and bioproducts, its environmental credentials are not without challenges. The escalating combustion of fossil fuels has driven greenhouse gas emissions, global warming, and related environmental crises, underscoring the urgency of sustainable energy alternatives. Lignin biorefineries provide an opportunity to address this by utilizing lignocellulosic biomass for the production of biofuels and biochemicals, thereby reducing dependence on finite fossil resources and contributing to lower net carbon emissions. However, achieving true environmental sustainability in these systems remains complex. Large-scale bioconversion must not only minimize energy inputs and chemical use but also mitigate the release of secondary pollutants such as NOx during fuel utilization. Equally important is the need to balance dual objectives: advancing energy security while safeguarding environmental quality. Integrating lignin-derived fuels into existing fuel infrastructures requires careful design to avoid negative impacts on air quality, ecosystems, and public health. Moreover, the transformation of lignin into high-value chemicals and renewable fuels—although environmentally favorable in principle—demands innovations that are consistent with circular economy principles and aligned with global climate and sustainability goals (Sharma et al., 2022).

12.3 ECONOMIC AND MARKET BARRIERS

Economic and market barriers significantly hinder the commercialization of lignin biorefineries. High capital and operational expenses, coupled with inefficient depolymerization and product recovery methods, make lignin valorization economically uncompetitive compared to petrochemical processes. The limited and fragmented market demand for lignin-derived chemicals and materials further restricts profitability. Inconsistent lignin quality from different biomass sources affects standardization and scalability. Additionally, inadequate policy incentives, lack of investor confidence, and insufficient integration with existing biorefineries slow market development. Overcoming these challenges requires cost reduction through technological innovation, value-chain integration, policy support, and the creation of high-value markets for lignin-based products.

12.3.1 High production costs

The economic feasibility of lignin biorefineries is constrained by a combination of technological and cost-related barriers that limit their competitiveness against well-established petrochemical processes. Pretreatment methods such as organosolv, kraft, and sulfite pulping are indispensable for separating lignin from cellulose and hemicellulose (Abolore et al., 2025). However, these processes are highly energy- and chemical-intensive, requiring specialized infrastructure that significantly elevates capital and operational expenditures (Wesseling et al., 2017). The severity of pretreatment—defined by factors such as temperature, pressure, and solvent intensity—not only impacts efficiency but can also alter lignin's chemical structure, sometimes increasing its resistance to degradation and complicating subsequent valorization steps (Chaudhary et al., 2023).This reliance on harsh pretreatment arises from lignin's intrinsic structural heterogeneity and recalcitrance, both of which have been identified as central technological challenges in recent techno-economic analyses. Depolymerization of lignin into value-added aromatic monomers and polymers further compounds costs, as effective strategies often depend on expensive metal catalysts or advanced solvent systems that demand precise operational control. Although enzymatic depolymerization presents a more sustainable alternative, it typically involves complex, low-efficiency systems that remain prohibitively expensive at scale, with most progress confined to laboratory or pilot settings (Tong et al., 2023; Liu et al., 2021). Downstream processing adds yet another layer of economic burden. The transformation of lignin into commercially valuable products requires energy-intensive purification and separation steps. Furthermore, variability in lignin composition shaped by its biomass source and extraction method leads to complex and inconsistent product streams, necessitating costly quality control measures to isolate desired compounds (Pérez-Almada et al., 2023). This variability reduces yield consistency and increases production costs, making lignin-derived materials difficult to position competitively against petroleum-based counterparts. As emphasized in several techno-economic and environmental assessments, such complexities remain major obstacles to the large-scale commercialization of lignin biorefineries (Calvo-Flores & Martin-Martinez, 2022).

12.3.2 Lack of established value chains

Unlike conventional biomass or fossil feedstocks, lignin currently lacks a well-established value chain to enable its large-scale utilization. The infrastructure required for efficient collection, fractionation, storage, and transportation of lignin-derived products remains underdeveloped. In the pulp and paper industry, most lignin is still burned on-site for heat and power generation, restricting its external availability for conversion into fuels, chemicals, or advanced materials. This practice not only limits lignin's economic potential but also perpetuates fragmented and underutilized supply chains (Rath et al., 2024; Bajwa et al., 2019). The absence of integrated supplier networks, standardized handling protocols, and dedicated distribution channels further complicates market development. Without cohesive logistics and established buyer–seller linkages, the costs of lignin handling and processing rise significantly, discouraging large-scale adoption (Wei et al., 2024). This logistical gap creates uncertainty for both industries and investors, reducing confidence in lignin-based ventures. Consequently, the lack of a robust and coordinated value chain remains one of the most significant barriers to scaling up lignin biorefineries and achieving commercial competitiveness with petroleum-derived alternatives (Dessbesell et al., 2020).

12.3.3 Market competition

Lignin-derived fuels, chemicals, and polymers face substantial competition from petroleum-based products, which dominate global markets due to their mature technologies and well-established supply chains (Mahajan et al., 2025). Over decades, fossil fuel industries have optimized production processes, achieved large economies of scale, and developed highly efficient value chains, enabling them to supply products at consistently lower costs. In contrast, lignin-based products are still emerging, with relatively high production costs and limited infrastructure constraining their market penetration (Bhuiyan & Siddique, 2025). This competitive disadvantage is particularly evident in sectors such as plastics, adhesives, resins, and fuel additives, where fossil-derived materials meet high performance standards, offer proven reliability, and are readily available at a large scale (Houssini et al., 2025). As a result, lignin-based alternatives often remain confined to lower-value applications or niche specialty markets. Without substantial technological advancements, cost reductions, and supportive policy frameworks, lignin-derived products will continue to face significant barriers in competing with petroleum-based counterparts, limiting the commercial viability and growth potential of lignin biorefineries (Rath et al., 2024).

12.3.4 Uncertain demand

The market for value-added products derived from lignin remains in its infancy and is highly fragmented (Li et al., 2024). Unlike fossil-based products, which serve established and mature markets, lignin-based compounds lack well-defined applications and standardized demand structures (Yang et al., 2021). Market interest is often influenced by external factors such as government policies, subsidies, sustainability incentives, and advances in research that enhance lignin processing and product performance. Consumer preferences for sustainable materials also affect demand; however, these trends are currently inconsistent and insufficient to drive widespread adoption of lignin-based products (Hall, 2024). Limited awareness among industries and consumers regarding the unique functionalities of lignin-derived products, such as antioxidant, antimicrobial, and UV-blocking properties, further constrains market development. These features could enable high-value applications in pharmaceuticals, cosmetics, food packaging, and specialty chemicals (Priyadarshi et al., 2024). Yet, the absence of market education, pilot-scale demonstrations, and product standardization means that many of these opportunities remain underexploited. This

TABLE 12.1
Summary of economic and market barriers in lignin biorefinery

Barrier	Key Challenges	Implications
High production costs	High pretreatment costs, expensive catalysts and enzymes, and complicated purification.	Alternative energy often suffers higher expenses compared to fossil fuels.
Lack of established value chains	Inadequate collection, transport, and supplier networks.	Fragmented logistics lead to inefficiencies and underutilization.
Market competition	Lower costs of petroleum products with established supply chains.	Currently, lignin is used mostly in niche markets, restricting its potential.
Uncertain demand	Market volatility and limited awareness of lignin's potential.	Investors hesitate due to underdeveloped applications and funding challenges.
Economies of scale	Small-scale pilot plants and inconsistent feedstocks.	Production costs remain high, making competition tough.
Product standardization issues	Diverse lignin characteristics, lack of global standards.	Quality variability creates barriers to adoption.
Limited high-value applications	Many initiatives remain in pilot or demonstration phases.	There are few available products, and reliance on fossil fuels persists.
Investment risks	High capital expenditure, lengthy payback periods, and market policy uncertainty.	Securing funding and scaling projects is difficult.
Policy and incentive gaps	Limited subsidies or mandates; fragmented regulatory frameworks.	Alternative energy struggles against biofuels and fossil fuels.
Tech–commercialization gap	Lab-scale innovations struggle to scale up.	Innovative research fails to being implemented in working environments.

unpredictability discourages significant capital investments, delays the establishment of long-term supply agreements, and reduces the confidence of businesses seeking to commercialize lignin-based technologies. Consequently, demand uncertainty represents a major barrier to the stable growth and industrial adoption of lignin biorefineries, limiting their potential to scale and achieve commercial competitiveness in the emerging bio-based economy (Kong et al., 2022). Table 12.1 summarizes the market and economic barriers and key challenges in lignin biorefinery.

12.3.5 Economies of Scale

Most lignin biorefinery initiatives are currently limited to pilot or small-scale commercial operations, which prevents them from realizing the cost advantages associated with large-scale production (Salaghi et al., 2021). Operating at a smaller scale results in higher per-unit production costs, reduced bargaining power with raw material suppliers, and limited influence when negotiating contracts with potential buyers. In contrast, fossil fuel refineries and petrochemical plants benefit from decades of scaling, optimized supply chains, and large throughput capacities, enabling them to achieve significantly lower costs and consistent product quality (Velvizhi et al., 2021). The pathway to scaling up lignin biorefineries is further complicated by the inherent variability of lignin as a feedstock. Its composition and structure vary depending on the biomass source, pretreatment method, and extraction process, demanding site-specific optimization for each production facility (Liu et al., 2024). This heterogeneity introduces uncertainty in product yields, quality, and reproducibility, making standardization and large-scale process design more challenging. Additionally, the high capital investments required for building large biorefinery plants are often hindered by uncertain markets and fluctuating policies. As a result, the failure to achieve economies of scale continues to delay the competitiveness of lignin-derived products compared to fossil-based alternatives (Makepa & Chihobo, 2024; Lin & Weng, 2019).

12.3.6 Product standardization issues

A major barrier to the commercialization of lignin-based products is the lack of product standardization. Lignin is a highly heterogeneous biopolymer, and its structural and physicochemical properties vary significantly depending on the plant species and the extraction method employed (Li et al., 2024). Differences in molecular weight distribution, solubility, purity, and functional group composition make it difficult to maintain consistent quality across production sites. For instance, kraft lignin from pulping processes exhibits higher sulfur content and distinct reactivity compared to sulfur-free organosolv lignin, leading to variability in downstream performance (Liu et al., 2022). This inconsistency poses significant challenges for industries requiring uniform raw materials to meet strict performance standards, such as those producing plastics, resins, adhesives, and carbon-based composites (Dona & Smith, 2025). The lack of globally recognized specifications, classification systems, and quality benchmarks further limits the adoption of lignin-derived products. Without harmonized standards, manufacturers face difficulties in scaling production, certifying product performance, and establishing trust with end users (Zhao et al., 2025). Addressing these standardization issues will be critical for integrating lignin-based products into mainstream markets. Developing international guidelines, certification frameworks, and collaborative industry initiatives can help ensure consistent quality, facilitate regulatory compliance, and enable wider commercialization of lignin-derived fuels, chemicals, and materials.

12.3.7 Limited high-value applications

Although lignin is the most abundant source of renewable aromatics on Earth, its industrial use remains limited. Most lignin is still burned as a low-value fuel in pulp and paper mills to generate heat and electricity. While this practice provides immediate energy recovery, it fails to tap into lignin's true potential as a feedstock for value-added products (Abolore et al., 2025). Significant research efforts have focused on developing high-value applications for lignin, such as carbon fibers, functional polymers, specialty adhesives, bioplastic precursors, and advanced composites. These products could offer considerable economic benefits and contribute to the transition toward a circular bioeconomy. However, progress toward large-scale commercialization has been slow (Bajwa et al., 2019).

Several technical challenges hinder the widespread adoption of lignin-derived applications, including their heterogeneity, inconsistent quality, high processing costs, and lack of supply chains, making them less competitive than fossil alternatives. Additionally, market uncertainties, the absence of standardized specifications, and weak end-user demand limit investment in these technologies, keeping most high-value applications in lab or pilot stages, with few reaching early commercialization (Jeffri et al., 2024; Vasile & Baican, 2023). Bridging the gap between research and market-scale deployment will require not just technological innovation but also coordinated efforts in policy support, industrial partnerships, and targeted investments (Ullah et al., 2022).

12.3.8 Investment risks

The valorization of lignin into fuels, chemicals, and materials is a capital-intensive endeavor, requiring substantial upfront investment in research and development, specialized infrastructure, and advanced processing equipment. Unlike fossil-based industries, which benefit from decades of established infrastructure and financing mechanisms, lignin biorefineries face longer payback periods and greater uncertainty in achieving profitability (Figueirêdo et al., 2022). These financial challenges are compounded by the sector's limited technological maturity, as many promising processes remain at pilot or demonstration scale, leaving their commercial viability uncertain (Singh et al., 2021; Ragauskas et al., 2014). Regulatory uncertainty and fluctuating policy support further exacerbate financial risks. Incentives such as subsidies, carbon pricing, and bioeconomy targets are often inconsistent across regions, creating an unstable investment environment (Ali et al., 2023). Market unpredictability—driven by variable demand, lack of standardized lignin-based products, and competition from cost-effective petrochemical alternatives—further discourages long-term capital commitments (Ali et al., 2024). For startups and smaller firms, these factors make it challenging to secure financing, attract industrial partners, and scale operations. Consequently, high investment risk remains one of the most significant barriers to the commercialization and widespread adoption of lignin-based technologies, limiting both industrial growth and the broader transition toward sustainable bio-based economies (Kang et al., 2025; Aditya & Aditya, 2024).

12.3.9 Policy and incentive gaps

Despite its potential as a renewable and sustainable feedstock, lignin has not received the same level of policy support as other bio-based sectors, such as conventional biofuels, bioenergy, or starch- and sugar-derived biochemicals. Existing bioeconomy frameworks, renewable energy directives, and circular economy strategies tend to prioritize feedstocks that are easier to process and integrate into current infrastructures (Sulis et al., 2025). Lignin, by contrast, is often overlooked due to its structural complexity, heterogeneous nature, and the perception that it is more suitable for low-value energy recovery than for high-value applications (Jeffri et al., 2024).

Where supportive measures do exist, they are often fragmented and narrowly focused on waste minimization or greenhouse gas reduction, rather than fostering lignin's potential as a versatile platform for bio-based chemicals, polymers, and advanced materials (Hu et al., 2023). The absence of dedicated policies—including targeted subsidies, tax incentives, carbon credits, or market mandates—places lignin-derived products at a structural disadvantage when competing with petroleum-based alternatives. Inconsistent policy signals across regions further discourage long-term investments and hinder the development of stable markets (Qadir et al., 2021). Addressing these gaps through comprehensive and coherent policy frameworks, coupled with incentive structures, will be critical to accelerate the commercialization of lignin-based products. Such measures can unlock the sector's full potential, ensuring that lignin contributes meaningfully to the transition toward a sustainable, bio-based economy (Ali et al., 2024; Hudson et al., 2019).

12.3.10 Technology-commercialization gap

Recently, significant scientific progress has been made in advanced lignin valorization pathways, including catalytic depolymerization, biocatalytic and microbial conversion, and innovative lignin-based product formulations (Jiju et al., 2025). These approaches show strong potential for producing renewable fuels, chemicals, and functional materials. However, their implementation remains largely limited to laboratory or pilot-scale studies, with the transition to full-scale industrial adoption facing multiple challenges that collectively create a substantial gap between technology and commercialization (Sundaram & Muniyandi, 2025). Effective scaling of lignin technologies requires robust process engineering, standardized supply chains, and validated product performance under industrial conditions. Nevertheless, lignin's inherent heterogeneity complicates process design, reduces reproducibility, and makes optimization beyond the laboratory difficult (Yan et al., 2025). High scale-up costs, coupled with risks of inconsistent yields and uncertain market acceptance, further discourage industrial stakeholders from committing to large-scale deployment. As a result, many promising lignin-based innovations remain at the proof-of-concept stage, while the commercial landscape continues to be dominated by low-value applications such as energy combustion (Palmié et al., 2023). Bridging this gap will require coordinated efforts to enhance technology transfer, strengthen academia–industry collaboration, and establish clear frameworks for scaling pilot processes to commercial operations. Success in these areas will depend on sustained support from both public and private investment, as well as policies that facilitate innovation adoption and market integration (Ahmed et al., 2022).

12.4 FUTURE PERSPECTIVES IN LIGNIN BIOREFINERY

12.4.1 Advances in catalytic and biological lignin valorization

Lignin is often regarded as a waste product due to its negligible economic value outside of its use as a heat source. In the pulping industry, lignin content has been a target for genetic modification for decades, as it is inversely correlated with the kappa number and fodder digestibility of biomass. However, strategies such as gene silencing or spontaneous mutations that sharply and non-selectively reduce lignin levels have frequently failed. For a biorefinery to be economically viable, all biomass streams—including lignin—must be converted into high-value products. Controlling the up- or down-regulation of genes encoding lignin-biosynthetic enzymes is fundamental to generating lignin building blocks with desired properties (Liu et al., 2022). After biosynthesis, lignin monomers are transported to the plant's secondary cell walls and undergo radical polymerization, mediated by enzymes such as laccases and peroxidases. While the exact roles of these enzymes in final polymerization are not fully understood, different laccase isoforms have been shown to facilitate the radical reactions necessary for lignin assembly. Silencing specific laccases leads to significant changes in lignification, indicating their involvement. Further research is needed to identify the peroxidases (or combinations thereof) that contribute to lignin formation. Mapping all genes involved in enzyme production for lignin monomer assembly can expand possibilities for systemic manipulation, potentially allowing the creation of lignin polymers with novel structures using unconventional monomers (Ullah et al., 2022). For chemical synthesis of low-molecular-weight compounds or polymers, lignin must be as pure as possible. Impurities can reduce yields, compromise product quality, and lead to unwanted byproducts. For example, residual sugar moieties can significantly decrease the value of technical lignins. Efficient methods exist for separating lignin and sugars, typically based on differences in solubility, such as acid/base extraction (Ali et al., 2024). However, the fractionation method can modify chemical bonds and functional groups in lignin, impacting reactivity and bioconversion efficiency by cleaving labile C–O bonds while generating more recalcitrant C–C linkages. Several thermochemical and chemical strategies have been explored to valorize lignin into fuels and value-added compounds, including catalytic pyrolysis, oxidation, and hydrotreatment (hydrogenolysis and deoxygenation) (Shao et al., 2017). Pyrolysis, commonly used to generate heterogeneous mixtures of aromatic compounds (C6–C22), often produces significant char and heavy oil, particularly from softwood kraft lignin. To prevent repolymerization and self-condensation, hydrodeoxygenation—a costly and time-intensive step—is required to obtain biofuel, followed by hydroprocessing of the partially deoxygenated product (Xu et al., 2012). Among these methods, hydrotreatment stands out for achieving high lignin conversion, improved selectivity, and reduced coke formation. Despite the promise of thermochemical and catalytic methods, challenges such as high costs, complex product compositions, low quality, high energy demand, and the need for organic waste management continue to hinder commercial implementation (Huang et al., 2022). Biological approaches, such as lignin-consolidated bioprocessing, offer complementary strategies. For example, combinatorial pretreatment of maize stover has been shown to enhance lignin reactivity and increase lipid synthesis, demonstrating the potential of integrated biological and chemical approaches to valorize lignin (Xie et al., 2015).

12.4.2 Genetic engineering of plants for tailored lignin

In recent decades, there has been increasing attention on the valorization of industrial lignin to produce high-value bioproducts, largely due to its commercial availability in substantial quantities. Several techno-economic assessments indicate that the success of future lignocellulosic biorefineries will depend significantly on effective lignin utilization (Schutyser et al., 2018). Recent research on lignin valorization has focused on producing fuels, fatty acids, muconate, biopolymers, aromatic building blocks, and advanced materials such as lignin-based hydrogels and aerogels for tissue engineering, drug encapsulation, and lignin-based nanoparticles and nanocomposites (Sharma et al., 2020). Genetic modification of plants' lignin biosynthesis pathways offers the potential to enhance lignin valorization by increasing lignin removal, reducing lignin content, or altering lignin composition. For example, delignification of biomass from transgenic poplar trees with reduced cinnamyl alcohol dehydrogenase (CAD) activity requires lower alkali concentrations during the kraft pulping process compared to wild-type biomass. Similarly, organosolv methods can be employed to extract lignin from transgenic poplar plants with either high or low syringyl contents (Simões et al., 2020). It is anticipated that biomass from genetically modified plants will not require major adjustments to conventional lignin extraction techniques. However, further industrial-scale research is needed to confirm the feasibility of using both traditional and modern extraction methods on such biomass. By leveraging advanced genetic engineering, plants can be tailored to possess traits advantageous for specific applications, including increased biomass yield for biofuel production, enhanced raw materials for manufacturing industries, and improved feedstocks for pulp and paper production. These strategies not only enhance the sustainability and environmental benefits of using renewable plant resources but also increase the economic value of plant-based products.

12.4.3 Integration with Circular Bioeconomy and Biorefineries

Beyond its conventional use as a low-grade fuel, lignin represents a valuable renewable resource with considerable potential for value creation within the framework of circular bioeconomy and biorefineries. As the most abundant aromatic biopolymer on Earth, lignin can be strategically integrated into biorefinery systems to enable the sustainable and resource-efficient production of high-value chemicals, materials, and energy (Rathour et al., 2023). Incorporating lignin valorization pathways into circular bioeconomy strategies promotes waste minimization, strengthens the financial viability of biomass processing sectors, and facilitates the development of new value chains that support sustainable development and climate objectives. By employing advanced fractionation and conversion techniques that address lignin's structural complexity, integrated biorefineries are increasingly focusing on the simultaneous deconstruction of lignocellulosic biomass for maximal valorization (Liu et al., 2021). Nevertheless, significant technological and economic barriers limit the widespread use of lignin-derived products in a circular bioeconomy. Lignin's inherent chemical heterogeneity and recalcitrance impede efficient fractionation, processing, and the production of homogeneous products. High operating and capital costs remain major obstacles, as scalable and cost-effective technologies for reliable lignin valorization are still under development, often rendering bio-based applications less competitive than conventional fossil-based alternatives (Rathour et al., 2025). Moreover, an optimal pretreatment strategy for the simultaneous valorization of all biomass fractions has yet to be established, and achieving this may require substantial redesign and investment in biorefinery infrastructure to effectively utilize lignin as a primary feedstock (Abbati de Assis et al., 2018).

12.4.4 Development of high-value products (biopolymers, biofuels, fine chemicals)

Lignin is a highly promising renewable resource, and as the global demand for lignocellulosic biomass for chemical and energy production has grown, the amount of lignin available as a byproduct or waste has also increased (Vasile & Baican, 2023). The primary goal of lignin valorization has been the production of fine chemicals and high-performance bio-based products. The market for lignin-derived products spans carbon fibers, eugenol, jet fuel, vanillin, catechol, propyl guaiacol, BTX, polyhydroxyalkanoates (PHA), polyurethane (PUR) foams as polyols, and colloidal lignin particles (Zhu et al., 2023). The specific applications depend largely on the type and purity of the lignin. High-purity lignin is used to produce carbon fibers, vanillin, phenolic derivatives, and materials for pharmaceutical and medical applications, whereas low-purity lignin is primarily utilized for energy, bitumen, and biofuel production. As lignin purity increases, its potential applications expand further to include BTX aromatics, phenols, cement additives, activated carbon, and phenolic resins. For lignin-based polymers to meaningfully replace current petrochemical infrastructure, either supportive policies must accelerate the transition from petroleum, or lignocellulosic biomass-derived polymers must demonstrate superior performance or long-term socioeconomic benefits (Shu et al., 2021). Achieving cost-competitive petrochemical alternatives from lignin requires substantial advancements in separation, purification, and polymerization chemistry. Current research focuses on producing lignin nanoparticles and nanofibers, hydrogels, green composites, and other multicomponent functional materials, whose complex compositions can enhance performance through synergistic interactions (Sethupathya et al., 2022). The use of lignin monomers and formulations suitable for 3D printing is also promising; however, stereolithographic printing is currently limited by the lack of material diversity. Expanding lignin use in 3D printing could reduce environmental impact and broaden application areas (Agrawal et al., 2022). Bulk lignin enhances the functionality of composite materials, lowers costs, and increases renewability. Nonetheless, producing high-performance, cost-effective lignin-based products remains challenging. While their lifecycle and economic performance may not yet match conventional petrochemical plastics, lignin-based materials are better aligned with green design principles and sustainability objectives (Garlapati et al., 2020).

12.4.5 Digitalization, AI, and process optimization

Lignocellulosic biomass (LCB) can be transformed into a wide array of value-added products, including chemicals, enzymes, and biofuels. To maximize yields during process development, advanced modeling techniques such as artificial neural networks (ANN) and adaptive neuro-fuzzy inference systems (ANFIS) have been increasingly employed (Chauhan et al., 2025; Pradhan et al., 2022). The conversion processes for LCB are typically complex, nonlinear, and unstable, making it challenging to derive precise physical models that accurately predict process performance. Empirical modeling approaches offer a practical solution by generating accurate and statistically significant results without requiring detailed knowledge of the underlying reactions and transformations. Empirical models can be categorized as conventional or non-conventional. A widely used conventional technique is response surface methodology (RSM), which has been extensively applied to model and optimize various LCB valorization processes. However, RSM assumes a polynomial relationship between input variables and outputs, which can limit its accuracy for highly nonlinear systems. To address these limitations, researchers have increasingly turned to artificial intelligence-based, non-traditional empirical modeling techniques such as fuzzy logic, ANNs, and ANFIS (Chauhan et al., 2025). In an ANN, the basic computational unit is a neuron, which receives multiple inputs, aggregates them, applies a nonlinear transformation, and generates an output (Lofgren et al., 2022). These AI-driven models provide enhanced flexibility and predictive capability, making them particularly suitable for optimizing complex LCB valorization processes.

12.5 SUSTAINABILITY AND LIFE CYCLE ASSESSMENT (LCA)

Life cycle assessment (LCA) and sustainability are essential for assessing the economic and environmental feasibility of lignin biorefineries. From the procurement of biomass to the usage of lignin-derived products, LCA offers a thorough framework for evaluating energy inputs, greenhouse gas emissions, resource consumption, and waste generation throughout the whole value chain (Tschulkow et al., 2024). By incorporating sustainability principles, carbon neutrality is improved, the circular bioeconomy is encouraged, and reliance on fossil fuels is decreased. Data variability, methodological standardization, and process scalability are among the difficulties. However, lignin biorefineries can be sustainable solutions for future bio-based economies by using LCA-driven optimization to inform industrial adoption, policy, and technological design.

12.5.1 Environmental impacts of lignin utilization

Lignin utilization offers significant environmental advantages, but its impact depends strongly on the chosen application and processing pathway. Valorizing lignin as a sustainable feedstock for materials and chemicals often yields a substantially lower environmental footprint compared to fossil-based alternatives (Ragauskas et al., 2014). For instance, producing bio-based adipic acid from lignin can reduce greenhouse gas (GHG) emissions by 62–78%, reaching approximately 4.87 kg CO_2-eq per kg, compared to conventional petrochemical routes (Corona et al., 2018). Similarly, lignin-derived carbon fibers exhibit a carbon footprint of around 1.50 kg CO_2-eq per kg, representing a 32% reduction relative to glass fibers and an order-of-magnitude improvement over fossil-based carbon fibers (Janssen et al., 2019). Emerging electrocatalytic lignin conversion techniques, particularly those employing hydrogen atom transfer mediators, can lower total CO_2 emissions by 10–46% in integrated biorefineries, according to LCA and techno-economic analyses (Ebrahimpourboura et al., 2024). Comparative studies indicate that many lignin-based products outperform their fossil-based or energy-focused counterparts in terms of climate impact, especially when allocation methods are carefully applied (Beaucamp et al., 2022). Beyond reducing GHG emissions, lignin utilization supports circular bioeconomy objectives by transforming an abundant side-stream into high-value products such as adhesives, composites, bioplastics, and biochar, thereby reducing waste and dependence on virgin feedstocks (Mateo et al., 2025). In agriculture, lignin-based biodegradable mulches offer climate benefits by sequestering carbon in soil while avoiding the microplastic pollution associated with conventional mulches (Creteanu et al., 2024). Similarly, lignin-derived slow-release fertilizers can reduce nutrient loss and eutrophication, promoting more sustainable farming practices (Boarino & Klok, 2023). However, environmental drawbacks and uncertainties remain. LCA results are sensitive to methodological choices, including system boundaries, allocation methods, and treatment of biogenic carbon, which can complicate comparisons across studies (Hermansson et al., 2020). For example, kraft lignin's climate impact varies depending on whether it is considered a primary product or a co-product. Consequential LCA also suggests that using kraft lignin for energy may be less beneficial than conventional recovery boilers, while chemical uses, such as substituting for bitumen or phenol, show improved environmental performance (Marson et al., 2023). The energy and material inputs for lignin valorization are also critical. For bio-based adipic acid, sodium hydroxide usage and heating demand are major environmental hotspots (Corona et al., 2018). In ALPHA fractionation processes, solvent selection significantly affects GHG emissions and energy consumption; for example, ethanol performs better environmentally and economically than acetic acid. Advanced "lignin-first" biorefineries, while promising, may initially generate higher GHGs due to gaseous byproducts, though optimized product streams and energy recovery strategies can halve emissions, particularly when transport logistics are minimized (Tschulkow et al., 2024). Overall, the environmental impact of lignin utilization is highly context-dependent. When integrated effectively into biorefineries and combined with optimized processing strategies, lignin-derived products can deliver substantial GHG reductions and promote circular economy principles. Nevertheless, their environmental performance depends on robust LCA methodology, careful management of energy and chemical inputs, and the specific end-use application. Addressing these challenges is essential for mainstreaming lignin as a sustainable bio-based resource.

12.5.2 Carbon Neutrality and Climate Change Mitigation

Within the biorefinery context, particularly in lignin valorization, the careful application of LCA frameworks is essential to identify emission hotspots, accurately track biogenic carbon flows, and avoid greenwashing or double counting (Bach, 2023). Lignin-derived materials and processes contribute to climate mitigation by substituting fossil-based products with renewable, bio-based alternatives. For example, lignin-based carbon fibers exhibit a substantially lower climate impact (~1.50 kg CO_2-eq per kg), reducing emissions by 32% relative to glass fibers and performing an order of magnitude better than fossil-based carbon fibers (Janssen et al., 2019).

Similarly, PEG-modified lignin incorporated into nylon masterbatches has been shown to reduce greenhouse gas emissions by 14–36%, depending on lignin content, demonstrating concrete mitigation at the material substitution level (Fuchigami et al., 2025). Bio-based composites and carbon-storing panels offer additional potential. One life cycle study found that lignin-plastic composite panels can achieve net carbon sequestration over extended horizons—reaching carbon neutrality in as little as 54 years, and potentially within 20 years if renewable electricity and recycled components are integrated (Kappagantula et al., 2025). These findings highlight the critical influence of product lifespan, end-of-life management, and energy sourcing on mitigation potential. However, the diversity of LCA outcomes for lignin-derived products, from strongly negative (climate-positive) to substantially positive carbon footprints, underscores the importance of process efficiency, end-of-life considerations, and accurate biogenic carbon accounting (Sudheshwar et al., 2024). Bio-based materials may appear climate-neutral on average, but worst-case emissions can exceed those of conventional polymers, particularly when disposal leads to landfill-induced methane emissions or when system boundaries are inconsistently defined. Achieving true carbon neutrality requires integrated strategies: substituting renewable materials for fossil-based ones, optimizing production and energy inputs, and employing robust LCA methodologies that transparently consider all life cycle stages. For lignin-based biorefineries and products, leveraging carbon storage and fossil displacement potential offers significant promise but only if assessments maintain high integrity, applications are long-lived, renewable energy is integrated, and intentional end-of-life strategies are implemented. These approaches are essential to align lignin utilization with broader climate mitigation and net-zero objectives.

12.6 CONCLUSION

Lignin biorefineries represent both a formidable challenge and a promising frontier in the transition toward a sustainable bioeconomy. As the most abundant renewable aromatic polymer, lignin possesses immense potential for the production of high-value chemicals, fuels, and biomaterials. However, its complex and heterogeneous structure, coupled with limitations in current extraction and depolymerization techniques, continues to impede efficient and scalable utilization. Key technical bottlenecks, including catalyst deactivation, low product selectivity, energy-intensive processing, and inconsistent lignin quality, remain significant obstacles. Additionally, limited industrial demonstrations, underdeveloped product markets, and uncertainties regarding economic viability constrain lignin's integration into commercial biorefineries. Despite these challenges, the outlook for lignin valorization is highly encouraging. Advances in green chemistry, synthetic biology, metabolic engineering, and nanotechnology are creating novel pathways for selective lignin conversion and the development of tailored products. The integration of lignin valorization with carbohydrate-based processes can enhance overall biomass utilization and improve the economic resilience of biorefineries. Digital tools such as machine learning, process simulation, and LCA further enable the design of sustainable and cost-effective lignin conversion strategies. Policy support, industrial partnerships, and the standardization of bio-based products will be crucial for scaling up these technologies.

With ongoing innovation, strategic collaboration, and sustainability-driven policies, lignin can transition from a low-value byproduct to a cornerstone of circular, climate-friendly economies. Harnessing its full potential will not only improve the profitability and efficiency of biorefineries but also make a significant contribution to achieving carbon neutrality and global climate change mitigation goals.

ACKNOWLEDGMENTS

The authors would like to acknowledge the University Grant Commission (UGC) and Council of Scientific and Industrial Research (CSIR) for providing financial assistance to Mr. Aditya Sharma [NTA reference no. 231620036265] and Ms. Pushpak Dahiya [CSIR File no. 09/0237(15947)/2022-EMR 1], respectively.

REFERENCES

Abbati de Assis, C., L. G. Greca, M. Ago, M. Y. Balakshin, H. Jameel, R. Gonzalez, and O. J. Rojas. 2018. "Techno-Economic Assessment, Scalability, and Applications of Aerosol Lignin Micro- and Nanoparticles." *ACS Sustainable Chemistry & Engineering* 6 (9): 11853–68.

Abolore, R. S., S. Jaiswal, and A. K. Jaiswal. 2025. "A Comprehensive Review on Sustainable Lignin Extraction Techniques, Modifications, and Emerging Applications." *Industrial Crops and Products* 235: 121696. https://doi.org/10.1016/j.indcrop.2025.121696.

Abu-Omar, M. M., K. Barta, G. T. Beckham, J. S. Luterbacher, J. Ralph, R. Rinaldi, Y. Román-Leshkov, J. S. M. Samec, B. F. Sels, and F. Wang. 2020. "Guidelines for Performing Lignin-First Biorefining." *Energy & Environmental Science* 14 (1): 262–92. https://doi.org/10.1039/d0ee02870c.

Aditya, A., and A. Aditya. 2024, July 4. "Challenges and Opportunities for Small and Medium-Sized Enterprises (SMEs)." Ande Aditya. https://andeaditya.com/challenges-and-opportunities-for-small-and-medium-sized-enterprises-smes/.

Agrawal, R., A. Kumar, S. Singh, and K. Sharma. 2022. "Recent Advances and Future Perspectives of Lignin Biopolymers." *Journal of Polymer Research* 29: 222.

Ahmed, F., M. T. Fattani, S. R. Ali, and R. N. Enam. 2022. "Strengthening the Bridge between Academic and the Industry through the Academia-Industry Collaboration Plan Design Model." *Frontiers in Psychology* 13. https://doi.org/10.3389/fpsyg.2022.875940.

Ali, K., H. Hongbing, C. Y. Liew, and D. Jianguo. 2023. "Governance Perspective and the Effect of Economic Policy Uncertainty on Financial Stability: Evidence from Developed and Developing Economies." *Economic Change and Restructuring* 56 (3): 1971–2002. https://doi.org/10.1007/s10644-023-09497-6.

Ali, S., A. Rani, M. A. Dar, M. M. Qaisrani, M. Noman, K. Yoganathan, and D. Zhu. 2024. "Recent Advances in Characterization and Valorization of Lignin and Its Value-Added Products: Challenges and Future Perspectives." *Biomass* 4 (3): 947–77. https://doi.org/10.3390/biomass4030053.

Asghar, A., H. N. Sairash, Z. Baqar, A. Sumrin, and M. Bilal. 2022. "Current Challenges of Biomass Refinery and Prospects of Emerging Technologies for Sustainable Bioproducts and Bioeconomy." *Biofuels, Bioproducts and Biorefining* 16 (6): 1478–94. https://doi.org/10.1002/bbb.2403.

Bach, V. 2023. "Life Cycle Assessment in the Context of Decarbonization and Carbon Neutrality." *International Journal of Life Cycle Assessment* 28: 741–5. https://doi.org/10.1007/s11367-023-02190-w.

Bajwa, D., G. Pourhashem, A. Ullah, and S. Bajwa. 2019. "A Concise Review of Current Lignin Production, Applications, Products and Their Environmental Impact." *Industrial Crops and Products* 139: 111526. https://doi.org/10.1016/j.indcrop.2019.111526.

Banu, J. R., N. Preethi, S. Kavitha, V. K. Tyagi, M. Gunasekaran, O. P. Karthikeyan, and G. Kumar. 2021. "Lignocellulosic Biomass Based Biorefinery: A Successful Platform towards Circular Bioeconomy." *Fuel* 302: 121086. https://doi.org/10.1016/j.fuel.2021.121086.

Bartling, A., M. Stone, R. Hanes, A. Bhatt, Y. Zhang, M. Biddy, R. Davis, J. Kruger, N. Thornburg, J. Luterbacher, and R. Rinaldi. 2021. "Techno-Economic Analysis and Life Cycle Assessment of a Biorefinery Utilizing Reductive Catalytic Fractionation." *Energy & Environmental Science* 14: 4147–68. https://doi.org/10.1039/d1ee01642c.

Beaucamp, A., M. Muddasar, I. S. Amiinu, M. M. Leite, M. Culebras, K. Latha, and M. N. Collins. 2022. "Lignin for Energy Applications–State of the Art, Life Cycle, Technoeconomic Analysis and Future Trends." *Green Chemistry* 24 (21): 8193–226.

Beckham, G., C. Johnson, E. Karp, and D. Vardon. 2016. "Opportunities and Challenges in Biological Lignin Valorization." *Current Opinion in Biotechnology* 42: 40–53. https://doi.org/10.1016/j.copbio.2016.02.030.

Bhatia, S. K., S. P. Jagtap, A. A. Bedekar, R. K. Bhatia, A. K. Patel, D. Pant, J. R. Banu, C. V. Rao, Y. J. Kim, and Y. H. Yang. 2020. "Recent Advancement in Lignocellulosic Biomass Pretreatment and Detoxification: Effect of Key Factors, Technological Improvements and Challenges." *Bioresource Technology*. https://doi.org/10.1016/j.biortech.2019.122724.

Bhatt, A. K., R. K. Bhatia, S. Thakur, N. Rana, V. Sharma, and R. K. Rathour. 2018. "Fuel from Waste: A Review on Scientific Solution for Waste Management and Environment Conservation." In: *Prospects of Alternative Transportation Fuels*. pp.205–33.

Bhuiyan, M. M. H., and Z. Siddique. 2025. "Hydrogen as an Alternative Fuel: A Comprehensive Review of Challenges and Opportunities in Production, Storage, and Transportation." *International Journal of Hydrogen Energy* 102: 1026–44. https://doi.org/10.1016/j.ijhydene.2025.01.033.

Boarino, A., and H. A. Klok. 2023. "Opportunities and Challenges for Lignin Valorization in Food Packaging, Antimicrobial, and Agricultural Applications." *Biomacromolecules* 24 (3): 1065–77.

Calvo-Flores, F., and F. Martín-Martínez. 2022. "Biorefineries: Achievements and Challenges for a Bio-Based Economy." *Frontiers in Chemistry* 10. https://doi.org/10.3389/fchem.2022.973417.

Chan, J. C., M. Paice, and X. Zhang. 2020. "Enzymatic Oxidation of Lignin: Challenges and Barriers toward Practical Applications." *ChemCatChem* 12 (2): 401–25. https://doi.org/10.1002/cctc.201901480.

Chandel, A. K., V. K. Garlapati, A. K. Singh, F. A. Antunes, and S. S. da Silva. 2018. "The Path Forward for Lignocellulose Biorefineries: Bottlenecks, Solutions, and Perspective on Commercialization." *Bioresource Technology* 264: 370–81. https://doi.org/10.1016/j.biortech.2018.06.004.

Chaudhary, G., N. Chaudhary, S. Saini, Y. Gupta, V. Vivekanand, and A. Panghal. 2023. "Assessment of Pretreatment Strategies for Valorization of Lignocellulosic Biomass: Path Forwarding towards Lignocellulosic Biorefinery." *Waste and Biomass Valorization* 15 (1): 1–36. https://doi.org/10.1007/s12649-023-02219-z.

Chauhan, S., P. Solanki, C. Putatunda, A. Walia, A. Keprate, A. K. Bhatt, V. K. Thakur, and R. K. Bhatia. 2025. "Recent Advancements in Biomass to Bioenergy Management and Carbon Capture through Artificial Intelligence Integrated Technologies to Achieve Carbon Neutrality." *Sustainable Energy Technologies and Assessments* 73: 104123. https://doi.org/10.1016/j.seta.2024.104123.

Corona, A., M. J. Biddy, D. R. Vardon, M. Z. Birkved, H. M. Hauschild, and G. T. Beckham. 2018. "Life Cycle Assessment of Adipic Acid Production from Lignin." *Green Chemistry* 20 (16): 3857–66.

Creteanu, A., C. N. Lungu, and M. Lungu. 2024. "Lignin: An Adaptable Biodegradable Polymer Used in Different Formulation Processes." *Pharmaceuticals* 17 (10): 1406.

Davis, K., and T. S. Moon. 2020. "Tailoring Microbes to Upgrade Lignin." *Current Opinion in Chemical Biology* 59: 23–29. https://doi.org/10.1016/j.cbpa.2020.04.001.

Dessbesell, L., M. Paleologou, M. Leitch, R. Pulkki, and C. Xu. 2020. "Global Lignin Supply Overview and Kraft Lignin Potential as an Alternative for Petroleum-Based Polymers." *Renewable and Sustainable Energy Reviews* 123: 109768. https://doi.org/10.1016/j.rser.2020.109768.

Dona, N. L. K., and R. C. Smith. 2025. "Tailoring Polymer Properties through Lignin Addition: A Recent Perspective on Lignin-Derived Polymer Modifications." *Molecules* 30 (11): 2455. https://doi.org/10.3390/molecules30112455.

Ebrahimpourboura, Z., M. Mosalpuri, C. Yang, A. Ponukumati, C. Stephenson, M. Foston, and M. M. Wright. 2024. "Comparative Techno-Economic and Life Cycle Assessment of Electrocatalytic Processes for Lignin Valorization." *Green Chemistry* 26 (22): 11303–15.

Figueirêdo, M. B., I. Hita, P. J. Deuss, R. H. Venderbosch, and H. J. Heeres. 2022. "Pyrolytic Lignin: A Promising Biorefinery Feedstock for the Production of Fuels and Valuable Chemicals." *Green Chemistry* 24 (12): 4680–702. https://doi.org/10.1039/d2gc00302c.

Fuchigami, Y., Y. Ohashi, E. Takata, and T. Yamada. 2025. "Life Cycle Inventory of Polyethylene Glycol-Modified Lignin and Greenhouse Gas Emission Reduction by Replacing Existing Resins." *Journal of Wood Science* 71 (1): 34.

Garlapati, V. K., A. K. Chandel, S. P. J. Kumar, S. Sharma, S. Sevda, A. P. Ingle, and D. Pant. 2020. "Circular Economy Aspects of Lignin: Towards a Lignocellulose Biorefinery." *Renewable and Sustainable Energy Reviews* 130: 109977.

Hall, M. 2024, September 25. "Factors That Cause Market Fluctuation." *Investopedia*. https://www.investopedia.com/ask/answers/100314/what-are-key-factors-cause-market-go-and-down.asp.

Hermansson, F., M. Janssen, and M. Svanström. 2020. "Allocation in Life Cycle Assessment of Lignin." *International Journal of Life Cycle Assessment* 25: 1620–32. https://doi.org/10.1007/s11367-020-01770-4.

Houssini, K., J. Li, and Q. Tan. 2025. "Complexities of the Global Plastics Supply Chain Revealed in a Trade-Linked Material Flow Analysis." *Communications Earth & Environment* 6 (1). https://doi.org/10.1038/s43247-025-02169-5.

Hu, R., J. Zhan, Y. Zhao, X. Xu, G. Luo, J. Fan, J. H. Clark, and S. Zhang. 2023. "Bio-Based Platform Chemicals Synthesized from Lignin Biorefinery." *Green Chemistry* 25 (22): 8970–9000. https://doi.org/10.1039/d3gc02927a.

Huang, C., X. Jiang, X. Shen, J. Hu, W. Tang, X. Wu, A. Ragauskas, H. Jameel, X. Meng, and Q. L. Yong. 2022. "Lignin-Enzyme Interaction: A Roadblock for Efficient Enzymatic Hydrolysis of Lignocellulosics." *Renewable and Sustainable Energy Reviews* 154: 111822.

Hudson, B., D. Hunter, and S. Peckham. 2019. "Policy Failure and the Policy-Implementation Gap: Can Policy Support Programs Help?" *Policy Design and Practice* 2 (1): 1–14. https://doi.org/10.1080/25741292.2018.1540378.

Islam, M., H. Wang, S. Rehman, C. Dong, H. Hsu, C. Lin, and S. Leu. 2019. "Sustainability Metrics of Pretreatment Processes in a Waste-Derived Lignocellulosic Biomass Biorefinery." *Bioresource Technology* 122558. https://doi.org/10.1016/j.biortech.2019.122558.

Janssen, M., E. Gustafsson, L. Echardt, J. Wallinder, and J. Wolf. 2019. "Life Cycle Assessment of Lignin-Based Carbon Fibres." In *14th Conference on Sustainable Development of Energy, Water and Environment Systems (SDEWES)*, 1–6 October 2019, Dubrovnik.

Jeffri, N. I., N. F. M. Rawi, M. H. M. Kassim, and C. K. Abdullah. 2024. "Unlocking the Potential: Evolving Role of Technical Lignin in Diverse Applications and Overcoming Challenges." *International Journal of Biological Macromolecules* 274: 133506. https://doi.org/10.1016/j.ijbiomac.2024.133506.

Jiju, P. S., A. K. Patel, N. S. Shruthy, S. Shalu, C. Dong, and R. R. Singhania. 2025. "Sustainability through Lignin Valorization: Recent Innovations and Applications Driving Industrial Transformation." *Bioresources and Bioprocessing* 12 (1). https://doi.org/10.1186/s40643-025-00929-x.

Junqueira, T. L., O. Cavalett, and A. Bonomi. 2016. "The Virtual Sugarcane Biorefinery—A Simulation Tool to Support Public Policies Formulation in Bioenergy." *Industrial Biotechnology* 12 (1). https://doi.org/10.1089/ind.2015.0015.

Kang, F., H. Qi, Z. Huang, and J. Hu. 2025. "New Horizons in Lignin-First Strategies for Upgrading Lignocellulose." Green Carbon. https://doi.org/10.1016/j.greenca.2024.11.004.

Kappagantula, K. S., Y. Jiang, F. Pierobon, M. R. E. Rabby, J. Ramos, Y. Ni, et al., and D. J. Heldebrant. 2025. "Carbon Dioxide-Negative Composite Materials: An Economically Viable Solution for CO_2 Sequestration." *Green Chemistry* 27 (9): 2392–403.

Kong, Q., R. Li, Z. Wang, and D. Peng. 2022. "Economic Policy Uncertainty and Firm Investment Decisions: Dilemma or Opportunity?" *International Review of Financial Analysis* 83: 102301. https://doi.org/10.1016/j.irfa.2022.102301.

Kumari, D., and R. Singh. 2018. "Pretreatment of Lignocellulosic Wastes for Biofuel Production: A Critical Review." *Renewable and Sustainable Energy Reviews* 90: 877–91. https://doi.org/10.1016/j.rser.2018.03.111

Li, J., D. Guan, S. Xia, Y. Fan, K. Zhao, Z. Zhao, and A. Zheng. 2024. "Recent Advances, Challenges, and Opportunities in Lignin Valorization for Value-Added Chemicals, Biofuels, and Polymeric Materials." *Energy Conversion and Management* 322: 119123. https://doi.org/10.1016/j.enconman.2024.119123.

Li, M., Y. Pu, and A. J. Ragauskas. 2016. "Current Understanding of the Correlation of Lignin Structure with Biomass Recalcitrance." *Frontiers in Chemistry* 4: 45. https://doi.org/10.3389/fchem.2016.00045.

Li, N., Y. Li, C. Yoo, X. Yang, X. Lin, J. Ralph, and X. Pan. 2018. "An Uncondensed Lignin Depolymerized in the Solid State and Isolated from Lignocellulosic Biomass: A Mechanistic Study." *Green Chemistry* 20: 4224–35. https://doi.org/10.1039/C8GC00953H.

Lin, S., and Y. Weng. 2019. "Market Size, Productivity and Product Quality Regarding Firm Heterogeneity." *Economic Research-Ekonomska Istraživanja* 32 (1): 2924–41. https://doi.org/10.1080/1331677x.2019.1653781.

Liu, X., F. Bouxin, J. Fan, V. Budarin, C. Hu, and J. Clark. 2020. "Recent Advances in the Catalytic Depolymerization of Lignin towards Phenolic Chemicals: A Review." *ChemSusChem* 13: 4296–317. https://doi.org/10.1002/cssc.202001213.

Liu, Z. H., B. Z. Li, J. S. Yuan, and Y. J. Yuan. 2022. "Creative Biological Lignin Conversion Routes toward Lignin Valorization." *Trends in Biotechnology* 40 (12): 1550–66.

Liu, Z., N. Hao, S. Shinde, Y. Pu, X. Kang, A. J. Ragauskas, and J. S. Yuan. 2018. "Defining Lignin Nanoparticle Properties through Tailored Lignin Reactivity by Sequential Organosolv Fragmentation Approach (SOFA)." *Green Chemistry* 21 (2): 245–60. https://doi.org/10.1039/c8gc03290d.

Liu, Z., N. Hao, Y. Wang, C. Dou, F. Lin, R. Shen, R. Bura, D. Hodge, B. Dale, A. Ragauskas, B. Yang, and J. Yuan. 2021. "Transforming Biorefinery Designs with 'Plug-in Processes of Lignin' to Enable Economic Waste Valorization." *Nature Communications* 12. https://doi.org/10.1038/s41467-021-23920-4.

Lofgren, J., D. Tarasov, T. Koitto, P. Rinke, M. Balakshin, and M. Todorovic. 2022. "Machine Learning Optimization of Lignin Properties in Green Biorefineries." *ACS Sustainable Chemistry & Engineering* 10 (29): 9469–79.

Ma, H., Y. Zhao, W. Huang, L. Zhang, F. Wu, J. Ye, and G. Chen. 2020. "Rational Flux-Tuning of *Halomonas Bluephagenesis* for Co-Production of Bioplastic PHB and Ectoine." *Nature Communications* 11 (1). https://doi.org/10.1038/s41467-020-17223-3.

Mahajan, J. S., E. R. Gottlieb, J. M. Kim, and T. H. Epps. 2025. "Toward Sustainable Materials: From Lignocellulosic Biomass to High-Performance Polymers." *Accounts of Materials Research* 6 (3): 316–26. https://doi.org/10.1021/accountsmr.4c00359.

Makepa, D. C., and C. H. Chihobo. 2024. "Barriers to Commercial Deployment of Biorefineries: A Multi-Faceted Review of Obstacles across the Innovation Chain." *Heliyon* 10 (12): e32649. https://doi.org/10.1016/j.heliyon.2024.e32649.

Marson, A., J. S. Samec, and A. Manzardo. 2023. "Consequential Life Cycle Assessment of Kraft Lignin Recovery with Chemical Recycling." *Science of the Total Environment* 882: 163660.

Mateo, S., G. Fabbrizi, and A. J. Moya. 2025. "Lignin from Plant-Based Agro-Industrial Biowastes: From Extraction to Sustainable Applications." *Polymers* 17 (7): 952.

Obydenkova, S., P. Kouris, D. Smeulders, M. Boot, and Y. Van Der Meer. 2022. "Evaluation of Environmental and Economic Hotspots and Value Creation in Multi-Product Lignocellulosic Biorefinery." *Biomass & Bioenergy*. https://doi.org/10.1016/j.biombioe.2022.106394.

Okolie, J. A., A. Mukherjee, S. Nanda, A. K. Dalai, and J. A. Kozinski. 2021. "Next-Generation Biofuels and Platform Biochemicals from Lignocellulosic Biomass." *International Journal of Energy Research* 45 (10): 14145–69. https://doi.org/10.1002/er.6697.

Özdenkçi, K., C. Blasio, H. Muddassar, K. Melin, P. Oinas, J. Koskinen, G. Sarwar, and M. Järvinen. 2016. "A Novel Biorefinery Integration Concept for Lignocellulosic Biomass." *Energy Conversion and Management* 149: 974–87. https://doi.org/10.1016/j.enconman.2017.04.034.

Palmié, M., V. Parida, A. Mader, and J. Wincent. 2023. "Clarifying the Scaling Concept: A Review, Definition, and Measure of Scaling Performance and an Elaborate Agenda for Future Research." *Journal of Business Research* 158: 113630. https://doi.org/10.1016/j.jbusres.2022.113630.

Pei, Z., X. Liu, J. Chen, H. Wang, and H. Li. 2024. "Research Progress on Lignin Depolymerization Strategies: A Review." *Polymers* 16. https://doi.org/10.3390/polym16172388.

Pérez-Almada, D., Á. Galán-Martín, M. D. Mar Contreras, and E. Castro. 2023. "Integrated Techno-Economic and Environmental Assessment of Biorefineries: Review and Future Research Directions." *Sustainable Energy & Fuels* 7 (17): 4031–50. https://doi.org/10.1039/d3se00405h.

Pradhan, D., S. Jaiswal, and A. K. Jaiswal. 2022. "Artificial Neural Networks in Valorization Process Modeling of Lignocellulosic Biomass." *Biofuels, Bioproducts and Biorefining* 16 (6): 1849–68.

Priyadarshi, R., T. Ghosh, S. D. Purohit, V. Prasannavenkadesan, and J. Rhim. 2024. "Lignin as a Sustainable and Functional Material for Active Food Packaging Applications: A Review." *Journal of Cleaner Production* 469: 143151. https://doi.org/10.1016/j.jclepro.2024.143151.

Qadir, S. A., H. Al-Motairi, F. Tahir, and L. Al-Fagih. 2021. "Incentives and Strategies for Financing the Renewable Energy Transition: A Review." *Energy Reports* 7: 3590–606. https://doi.org/10.1016/j.egyr.2021.06.041.

Questell-Santiago, Y. M., M. V. Galkin, K. Barta, and J. S. Luterbacher. 2020. "Stabilization Strategies in Biomass Depolymerization Using Chemical Functionalization." *Nature Reviews Chemistry* 4 (6): 311–30. https://doi.org/10.1038/s41570-020-0187-y.

Radhika, N., S. Sachdeva, and M. Kumar. 2021. "Lignin Depolymerization and Biotransformation to Industrially Important Chemicals/Biofuels." *Fuel* 312: 122935. https://doi.org/10.1016/j.fuel.2021.122935.

Ragauskas, A. J., G. T. Beckham, M. J. Biddy, R. Chandra, F. Chen, M. F. Davis, B. H. Davison, R. A. Dixon, P. Gilna, M. Keller, P. Langan, A. K. Naskar, J. N. Saddler, T. J. Tschaplinski, G. A. Tuskan, and C. E. Wyman. 2014. "Lignin Valorization: Improving Lignin Processing in the Biorefinery." *Science* 344 (6185). https://doi.org/10.1126/science.1246843.

Rana N., V. Sharma, R. K. Rathour, V. Ahuja, R. K. Bhatia and A. K. Bhatt. 2019. "Utilization of Corn Cob Waste for Xylanase Production by Newly Isolated *Bacillus pumilus* XRL5 and Its Application in Saccharification of Lignocellulosic Biomass." *Trends in Carbohydrate Research* 3: 22–32.

Rath, S., D. Pradhan, H. Du, S. Mohapatra, and H. Thatoi. 2024. "Transforming Lignin into Value-Added Products: Perspectives on Lignin Chemistry, Lignin-Based Biocomposites, and Pathways for Augmenting Ligninolytic Enzyme Production." *Advanced Composites and Hybrid Materials* 7 (1). https://doi.org/10.1007/s42114-024-00836-3.

Rathour, R. K., M. Behl, D. Sakhuja, N. Kumar, N. Chauhan, A. K. Bhatt, and R. K. Bhatia. 2025. "Recent Updates in Biomass Pretreatment Techniques for a Sustainable Biofuel Industry: A Comprehensive Review." *Waste Biomass Valorization.* https://doi.org/10.1007/s12649-025-03016-6.

Rathour, R. K., M. Behl, K. Dhashmana, D. Sakhuja, H. Ghai, N. Sharma, et al. 2023. "Non-Food Crops Derived Lignocellulose Biorefinery for Sustainable Production of Biomaterials, Biochemicals and Bioenergy: A Review on Trends and Techniques." *Industrial Crops and Products* 204: 117220.

Rathour, R. K., M. Devi, P. Dahiya, N. Sharma, N. Kaushik, D. Kumari, P. Kumar, R. R. Baadhe, A. Walia, A. K. Bhatt, and R. K. Bhatia. 2023. "Recent Trends, Opportunities and Challenges in Sustainable Management of Rice Straw Waste Biomass for Green Biorefinery." *Energies* 16(3): 1429. https://doi.org/10.3390/en16031429.

Salaghi, A., L. Zhou, P. Saini, F. Kong, M. Konduri, and P. Fatehi. 2021. "Lignin Production in Plants and Pilot and Commercial Processes." In: *Elsevier eBooks*, pp.551–87. https://doi.org/10.1016/b978-0-12-821888-4.00002-2.

Salvachúa, D., A. Z. Werner, I. Pardo, M. Michalska, B. A. Black, B. S. Donohoe, S. J. Haugen, R. Katahira, S. Notonier, K. J. Ramirez, A. Amore, S. O. Purvine, E. M. Zink, P. E. Abraham, R. J. Giannone, S. Poudel, P. D. Laible, and R. L. Hettich. 2020. "Outer Membrane Vesicles Catabolize Lignin-Derived Aromatic Compounds in *Pseudomonas putida* KT2440." *Proceedings of the National Academy of Sciences* 117 (17): 9302–10. https://doi.org/10.1073/pnas.1921073117.

Schutyser, W., T. Renders, S. Van den Bosch, S.-F. Koelewijn, G. T. Beckham, and B. F. Sels. 2018. "Chemicals from Lignin: An Interplay of Lignocellulose Fractionation, Depolymerisation, and Upgrading." *Chemical Society Reviews* 47 (3): 852–908. https://doi.org/10.1039/c7cs00566k.

Sethupathya, S., G. M. Morales, L. Gao, H. Wang, B. Yang, J. Jiang, J. Sun, and D. Zhu. 2022. "Lignin Valorization: Status, Challenges and Opportunities." *Bioresource Technology* 347: 126696.

Shafiei-Alavijeh, R., M. Aghbashlo, M. Tabatabaei, J. Denayer, and K. Karimi. 2024. "A Critical Review of Sustainable Biorefineries Utilizing High-Solid Processing for Industrial Crop Lignocellulosic Wastes Valorization." *Industrial Crops and Products.* https://doi.org/10.1016/j.indcrop.2024.118236.

Shao, Y., Q. Xia, L. Dong, X. Liu, X. Han, S. F. Parker, et al. 2017. "Selective Production of Arenes via Direct Lignin Upgrading over a Niobium-Based Catalyst." *Nature Communications* 8: 16104.

Sharma, A., P. K. Badola, C. Bhatia, D. Sharma, and P. K. Trivedi. 2020. "Primary Transcript of miR858 Encodes Regulatory Peptide and Controls Flavonoid Biosynthesis and Development in *Arabidopsis*." *Nature Plants* 6 (10): 1262–74. https://doi.org/10.1038/s41477-020-00769-x.

Sharma, S., M. Tsai, V. Sharma, P. Sun, P. Nargotra, B. Bajaj, C. Chen, and C. Dong. 2022. "Environment Friendly Pretreatment Approaches for the Bioconversion of Lignocellulosic Biomass into Biofuels and Value-Added Products." *Environments* 10 (1). https://doi.org/10.3390/environments10010006.

Shu, F., B. Jiang, Y. Yuan, M. Li, W. Wu, Y. Jin, and H. Xiao. 2021. "Biological Activities and Emerging Roles of Lignin and Lignin-Based Products—A Review." *Biomacromolecules* 22: 4905–18.

Silva, J. P., A. R. P. Ticona, P. R. V. Hamann, B. F. Quirino, and E. F. Noronha. 2021. "Deconstruction of Lignin: From Enzymes to Microorganisms." *Molecules* 26 (8): 2299. https://doi.org/10.3390/molecules26082299.

Simões, M. S., G. G. Carvalho, S. S. Ferreira, J. Hernandes-Lopes, N. de Setta, and I. Cesarino. 2020. "Genome-Wide Characterization of the Laccase Gene Family in *Setaria viridis* Reveals Members Potentially Involved in Lignification." *Planta* 251: 1–18. https://doi.org/10.1007/s00425-020-03337-x.

Singh, N., R. R. Singhania, P. S. Nigam, C. Dong, A. K. Patel, and M. Puri. 2021. "Global Status of Lignocellulosic Biorefinery: Challenges and Perspectives." *Bioresource Technology* 344: 126415. https://doi.org/10.1016/j.biortech.2021.126415.

Sudheshwar, A., K. Vogel, G. Nyström, N. Malinverno, M. Arnaudo, C. E. G. Camacho, … C. Som. 2024. "Unraveling the Climate Neutrality of Wood Derivatives and Biopolymers." *RSC Sustainability* 2 (5): 1487–97.

Sulis, D. B., N. Lavoine, H. Sederoff, X. Jiang, B. M. Marques, K. Lan, C. Cofre-Vega, R. Barrangou, and J. P. Wang. 2025. "Advances in Lignocellulosic Feedstocks for Bioenergy and Bioproducts." *Nature Communications* 16 (1). https://doi.org/10.1038/s41467-025-56472-y.

Sundaram, G. A., and G. R. Muniyandi. 2025. "Advancements in Renewable Energy: Innovations and Sustainable Solutions for a Greener Future." *Next Research* 100482. https://doi.org/10.1016/j.nexres.2025.100482.

Tong, Y., H. Fu, F. Xiao, S. Bohm, H. Ma, and Y. Luo. 2023. "Precision Engineering of Precious Metal Catalysts for Enhanced Hydrogen Production Efficiency." *Process Safety and Environmental Protection* 178: 559–79. https://doi.org/10.1016/j.psep.2023.08.066.

Tschulkow, M., M. Pizzol, T. Compernolle, S. Van den Bosch, B. Sels, and S. Van Passel. 2024. "The Environmental Impacts of the Lignin-First Biorefineries: A Consequential Life Cycle Assessment Approach." *Resources, Conservation and Recycling* 204: 107466.

Ullah, M., P. Liu, S. Xie, and S. Sun. 2022. "Recent Advancements and Challenges in Lignin Valorization: Green Routes towards Sustainable Bioproducts." *Molecules* 27 (18): 6055. https://doi.org/10.3390/molecules27186055.

Vasile, C., and M. Baican. 2023. "Lignins as Promising Renewable Biopolymers and Bioactive Compounds for High-Performance Materials." *Polymers* 15 (15): 3177. https://doi.org/10.3390/polym15153177.

Velvizhi, G., K. Balakumar, N. P. Shetti, E. Ahmad, K. K. Pant, and T. M. Aminabhavi. 2021. "Integrated Biorefinery Processes for Conversion of Lignocellulosic Biomass to Value Added Materials: Paving a Path towards Circular Economy." *Bioresource Technology* 343: 126151. https://doi.org/10.1016/j.biortech.2021.126151.

Wei, S., H. Liu, X. Chen, and W. Ke. 2024. "How Does Supplier Integration Influence Supply Chain Robustness and Resilience? The Moderating Roles of Information Technology Agility and Managerial Ties." *Information & Management* 61 (7): 104028. https://doi.org/10.1016/j.im.2024.104028.

Wesseling, J., S. Lechtenböhmer, M. Åhman, L. Nilsson, E. Worrell, and L. Coenen. 2017. "The Transition of Energy Intensive Processing Industries towards Deep Decarbonization: Characteristics and Implications for Future Research." *Renewable and Sustainable Energy Reviews* 79: 1303–13. https://doi.org/10.1016/j.rser.2017.05.156.

Wu, X., J. Jiang, C. Wang, J. Liu, Y. Pu, A. Ragauskas, S. Li, and B. Yang. 2020. "Lignin-Derived Electrochemical Energy Materials and Systems." *Biofuels, Bioproducts and Biorefining* 14 (3): 650–72. https://doi.org/10.1002/bbb.2083.

Xie, S., A. J. Ragauskas, and J. S. Yuan. 2016. "Lignin Conversion: Opportunities and Challenges for the Integrated Biorefinery." *Industrial Biotechnology* 12 (3): 161–7. https://doi.org/10.1089/ind.2016.0007.

Xie, S., X. Qin, Y. Cheng, D. Laskar, W. Qiao, S. Sun, L. H. Reyes, X. Wang, S.-Y. Dai, and S. E. Sattler. 2015. "Simultaneous Conversion of All Cell Wall Components by an Oleaginous Fungus without Chemi-Physical Pretreatment." *Green Chemistry* 17: 1657–67.

Xu, W., S. J. Miller, P. K. Agrawal, and C. W. Jones. 2012. "Depolymerization and Hydrodeoxygenation of Switchgrass Lignin with Formic Acid." *ChemSusChem* 5: 667–75.

Yan, L., A. J. Huertas-Alonso, H. Liu, L. Dai, C. Si, and M. H. Sipponen. 2025. "Lignin Polymerization: Towards High-Performance Materials." *Chemical Society Reviews*. https://doi.org/10.1039/d4cs01044b.

Yang, L., X. Wang, M. Dai, B. Chen, Y. Qiao, H. Deng, D. Zhang, Y. Zhang, C. M. V. B. De Almeida, A. S. Chiu, J. J. Klemeš, and Y. Wang. 2021. "Shifting from Fossil-Based Economy to Bio-Based Economy: Status Quo, Challenges, and Prospects." *Energy* 228: 120533. https://doi.org/10.1016/j.energy.2021.120533.

Zhao, J., M. Zhu, W. Jin, J. Zhang, G. Fan, Y. Feng, Z. Li, S. Wang, J. S. Lee, G. Luan, Z. Dong, and Y. Li. 2025. "A Comprehensive Review of Unlocking the Potential of Lignin-Derived Biomaterials: From Lignin Structure to Biomedical Application." *Journal of Nanobiotechnology* 23 (1). https://doi.org/10.1186/s12951-025-03604-7.

Zhu, D., B. Yang, H. Wang, and M. Shahnawaz. 2023. "Editorial: Lignin Valorization: Recent Trends and Future Perspective." *Frontiers in Bioengineering and Biotechnology* 11: 1190128.

Index

A

B

C

U

V

W

X

Y

Z

For Product Safety Concerns and Information please contact our EU
representative GPSR@taylorandfrancis.com
Taylor & Francis Verlag GmbH, Kaufingerstraße 24, 80331 München, Germany

www.ingramcontent.com/pod-product-compliance
Ingram Content Group UK Ltd.
Pitfield, Milton Keynes, MK11 3LW, UK
UKHW022219190926
475702UK00003B/447
* 9 7 8 1 0 4 1 0 9 9 1 6 1 *